Urban Runoff Control and Sponge City Construction II

Urban Runoff Control and Sponge City Construction II

Editors

Haifeng Jia
Jiangyong Hu
Dafang Fu
Wei-Shan Chen

MDPI

Basel • Beijing • Wuhan • Barcelona • Belgrade • Novi Sad • Cluj • Manchester

Editors
Haifeng Jia
Department of Environmental Planning and Management, School of Environment, Tsinghua University
Beijing
China

Jiangyong Hu
Department of Civil and Environmental Engineering, National University of Singapore
Singapore
Singapore

Dafang Fu
Department of Municipal Engineering, School of Civil Engineering, Southeast University
Nanjing
China

Wei-Shan Chen
Environmental Technology Group, Wageningen University & Research (WUR)
Wageningen
The Netherlands

Editorial Office
MDPI
St. Alban-Anlage 66
4052 Basel, Switzerland

This is a reprint of articles from the Special Issue published online in the open access journal *Water* (ISSN 2073-4441) (available at: https://www.mdpi.com/journal/water/special_issues/UrbanRunoff_Control2).

For citation purposes, cite each article independently as indicated on the article page online and as indicated below:

Lastname, A.A.; Lastname, B.B. Article Title. *Journal Name* **Year**, *Volume Number*, Page Range.

ISBN 978-3-7258-0625-6 (Hbk)
ISBN 978-3-7258-0626-3 (PDF)
doi.org/10.3390/books978-3-7258-0626-3

Contents

About the Editors

Haifeng Jia

Haifeng Jia conducted his research and teaching in watershed–urban water environmental planning and management, water quality and hydrologic modeling, urban runoff control and sponge city, and environmental remote sensing and GIS. He has authored 110 research projects and published over 200 peer-reviewed journal papers, conference papers, and 12 books. He has received 40 different academic and engineering awards and honors. He is active in international academic activities and collaborations and has organized and attended many international conferences.

Jiangyong Hu

Jiangyong Hu specializes in innovative water treatment technology, emerging contaminants detection and removal, water disinfection and biofilm control, and stormwater management. She has published more than 170 papers in various international journals and has been invited to deliver more than 70 invited speeches. She has been the organizing committee chair and co-president for several international conferences and symposiums. She is an editor, guest editor, Editorial Board Member for several journals, and reviewer for funding agencies. She is an International Water Association Fellow at the Institute of Engineers, Singapore, the President of the Environmental Engineering Society of Singapore, a Board member of the International Ultraviolet Association, and Chair of the IWA Specialist Group Management Committee on Assessment and Control of Hazardous Substances in Water.

Dafang Fu

Dafang Fu has researched urban stormwater management, long-term planning of urban drainage infrastructures, water resources, and environmental management. He is the director of the Southeast University-Monash University Joint Research Center for Future Cities and the vice chairman of the Jiangsu Society of Environmental Sciences. He has published more than 210 peer-reviewed journal papers and conference papers and presided over three international cooperation projects of the National Natural Science Foundation of China, one national key research and development project, one major national science and technology project (water pollution control and treatment), and completed several projects on urban pollution control, flood control, and environment restoration. He has won the National Science and Technology Progress Award (8) and the Huaxia Construction Science and Technology Award (3).

Wei-Shan Chen

Wei-Shan Chen is an assistant professor at Wageningen University. His research focuses on designing urban infrastructure transition pathways that facilitate circular resource management in the urban environment. He combines (bio)technology insights and urban digital twins to simulate implementing circular technologies and the required infrastructure transitions in an urban environment. His work addresses three resource streams: urban rainwater reuse, domestic food waste valorization, and commodity plastic recycling in the built environment. His work related to nature-based solutions for improving the resilience and circularity of urban runoff against climate change has been used by several Dutch, Chinese, and African cities to support decision-making on local infrastructure renovation.

Preface

We organized this Special Issue to reflect the latest developments in urban runoff control and sponge city construction. We attempt to discuss and address studies focused on the theories and technologies of sponge city construction; urban hydrology; methods of quantifying the benefits of a sponge city; rainwater utilization; practices that mitigate urban flooding and soil erosion; the performance of GI; the impact of media; preferential flow paths; vegetation; climate; design of hydrological, hydrodynamic and pollutant removal processes; and case studies on sustainable urban design and management using LID–GI principles and practices.

Haifeng Jia, Jiangyong Hu, Dafang Fu, and Wei-Shan Chen
Editors

water

MDPI

Editorial

Urban Runoff Control and Sponge City Construction: Important Topics †

Haifeng Jia [1,*], Jiangyong Hu [2], Dafang Fu [3] and Wei-Shan Chen [4]

1 School of Environment, Tsinghua University, Beijing 100084, China
2 Department of Civil and Environmental Engineering, National University of Singapore, Singapore 119077, Singapore; hujiangyong@nus.edu.sg
3 Department of Municipal Engineering, School of Civil Engineering, Southeast University, Nanjing 211189, China; fdf@seu.edu.cn
4 Environmental Technology Group, Wageningen University & Research (WUR), Bornse Weilanden 9, 6708 WG, Wageningen, The Netherlands; wei-shan.chen@wur.nl
* Correspondence: jhf@tsinghua.edu.cn

† This article belongs to the Special Issue Urban Runoff Control and Sponge City Construction II.

Citation: Jia, H.; Hu, J.; Fu, D.; Chen, W.-S. Urban Runoff Control and Sponge City Construction: Important Topics. *Water* **2024**, *16*, 497. https://doi.org/10.3390/w16030497

Received: 17 January 2024
Accepted: 1 February 2024
Published: 4 February 2024

1. Introduction

Rapid urbanization, which leads to a lack of adequate planning and design, has led to worsening city syndrome situations [1], such as urban flooding, water pollution, heat-island effects, and ecological deterioration. To mitigate these impacts, various technology systems have been proposed in different countries [2,3]. Additionally, a new concept in urban stormwater management strategies was announced by the Chinese government in 2013 called a "sponge city" [4,5]. The Chinese central government selected 30 pilot cities, considering their different natural and social conditions, for sponge city construction exploration in 2015 and 2016. Furthermore, in 2021, based on the experiences of these pilot cities, China began to systematically promote the sponge city concept on a national scale [6,7]. Now, many studies have been conducted and practices have been implemented related to sponge city construction in China [8,9]. More importantly, this new paradigm for a sustainable urban runoff control strategy has become a widespread focus in urban water management research and practices globally.

Along with the demonstration of sponge city construction, many related research achievements were obtained. In this context, in order to present the latest developments, technologies, and case studies related to urban runoff control and sponge city construction, following the success of "Urban Runoff Control and Sponge City Construction I" [10], this Special Issue, "Urban Runoff Control and Sponge City Construction II", is a follow-up. We aimed to discuss and address studies focused on the theories and technologies of sponge city construction; urban hydrology; methods of quantifying the benefits of a sponge city; rainwater utilization; practices that mitigate urban flooding and pollution; the performance of GI; the impact of media; vegetation; climate; the design of hydrological, hydrodynamic, and pollutant removal processes; and case studies on sustainable urban design and management using LID-GI principles and practices. We would like to express our gratitude to all the contributors who made this Special Issue so successful.

2. Summary of This Special Issue

In total, 11 papers were published in this Special Issue. The article types, authors, titles, keywords, and study areas are summarized in Table 1.

Table 1. Summary of the papers published in the Special Issue "Urban Runoff Control and Sponge City Construction II" in *Water* (https://www.mdpi.com/journal/water/special_issues/UrbanRunoff_Control2 (assessed on 16 December 2023)).

Article Type	Authors	Title	Keywords	Study Area
methods and tools	Aldrees, A.; Dan'azumi, S.	Application of Analytical Probabilistic Models in Urban Runoff Control Systems' Planning and Design: A Review.	best management practices; low-impact development; water-sensitive urban design; blue-green infrastructure; sponge cities	/
methods and tools	Zhao, Y.; Price, M.; Trowsdale, S.	Comparison of the Transition to More Sustainable Stormwater Management in China and the USA	stormwater management; transition; multi-level perspective	United States and China
methods and tools	Le, J. T.; Gonzalez, J. P.; Carson, R. T.; Ambrose, R. F.; Levin; L. A.	Integrating Non-Targeted Ecosystem Services into Assessment of Natural Stormwater Treatment Systems.	urban runoff; urban ecology; nature-based solutions; natural treatment systems; biofilters; ecosystem services; monitoring and evaluation; planning and management	Los Angeles County, United States
methods and tools	Yang, Y.; Shao, Z.; Xu, X.; Liu, D.	Impact of Storm Characteristics on Infiltration Dynamics in Sponge Cities Using SWMM.	Chicago storm; Horton; porous pavement; return period; time-to-peak coefficient	Fengxi, China
methods and tools	Xie, M.; Cheng, Y.; Dong, Z.	Study on Multi-Objective Optimization of Sponge Facilities Combination at Urban Block Level: A Residential Complex Case Study in Nanjing, China.	Multi-objective optimization; sponge city planning and design; urban block; sponge facility combination	Nanjing, China
methods and tools	Cao, Q.; Cao, J.; Xu, R.	Optimizing Low Impact Development for Stormwater Runoff Treatment: A Case Study in Yixing, China.	first flush effect; InfoWorks ICM; LID optimization; generalized likelihood uncertainty estimation	Yixing, China
methods and tools	Wang, J.; Zhou, X.; Wang, S.; Chen, L.; Shen, Z.	Simulation and Comprehensive Evaluation of the Multidimensional Environmental Benefits of Sponge Cities.	sponge city; grey and green infrastructure; stormwater management model; integrated environmental benefits; monetary value; stormwater use	Beijing, China
typical source control facility	Lee, Y. T.; Ho, M. C.; Chiou, Y. S.; Huang, L. L.	Assessing the Performance of Permeable Pavement in Mitigating Flooding in Urban Areas.	permeable pavement; monitoring instruments and management systems; flow law formula; low-impact development	Taoyuan City, Taiwan, China
typical source control facility	Chen, X.; Liu, R.; Liu, D.; Xin, X.	Analysis of Preferential Flow in Artificial Substrates with Sedum Roots for Green Roofs: Experiments and Modeling	green roofs; preferential flow; artificial substrate; Sedum roots; solute breakthrough experiments; HYDRUS-1D	/
typical source control facility	Chen, C.; Li, Y.; Le, W.; You, C.; Liu, C.; Liu, W.; Zhang, R.	Field Performance of Rain Garden in Red Soil Area in Southern China.	rainfall runoff; low impact development; runoff control; pollutant removal; Sponge City	Nanchang, China
combination of source control facility	Zhang, C.; Lv, Y.; Chen, J.; Chen, T., Liu, J.; Ding, L.; Zhang, N.; Gao, Q.	Comparisons of Retention and Lag Characteristics of Rainfall–Runoff under Different Rainfall Scenarios in Low-Impact Development Combination: A Case Study in Lingang New City, Shanghai.	Sponge City; low-impact development; stormwater management; retention time; lag time; Lingang New City	Lingang New City, Shanghai, China

Covering the methods and tools aspects, Aldrees et al. (contribution 1) wrote a comprehensive review on the application of Analytical Probabilistic Models (APMs) in urban runoff control systems' planning and design. APMs are closed-form mathematical expressions representing a long-term system's output performance derived from the probability distribution of the system's input variables. Once derived, APMs are easy to handle, allow for sensitive analysis, and can be co-opted into optimization frameworks. The implementation of APMs in the planning and design of runoff control systems will not only help address the runoff quantity and quality problems of urban stormwater but will also go a long way in optimizing the benefits derived from these systems. Zhao et al. (contribution 2) presented a comparative cross-nation study of the transition to more sustainable stormwater management (SSWM) in the United States and China. Multi-level perspective and multiphase models were used to examine the transition dynamics and reflect on how transition theory explains the changes within federal and socialist contexts. The main difference between the transition processes in the United States and China is the extent to which niche level innovations are developed, especially in the type of actors and activities investigated. Le et al. (contribution 3) proposed integrating Non-Targeted Ecosystem Services into Assessment of Natural Stormwater Treatment Systems. Usually, the design of Natural Stormwater Treatment Systems (NTSs) targets water services; however, the biological communities associated with NTSs (i.e., plants, animals, and microbes) can provide non-targeted functions that result in ecosystem services, such as biodiversity, pollination, and climate regulation, or, in some cases, disservices. Additional co-benefits of NTSs include recreation, education and outreach opportunities, and aesthetic value. As NTSs become globally widespread, best practices must include the ability to holistically assess NTS performance in ways that extend beyond water treatment services.

In order to identify the Impact of Storm Characteristics on Infiltration Dynamics in Sponge Cities, Yang et al. (contribution 4) used the Horton method within the stormwater management model to investigate how uniform and Chicago storm parameters affect infiltration rates. Their findings provide the following valuable insights: (1) Increasing the porous pavement area proportionally reduces subarea sizes within subcatchments, and the infiltration rates of porous pavements are supply-controlled. (2) Uniform storms result in consistent initial infiltration rates across pervious areas, subcatchments, and the entire catchment. The duration of this stable state decreases with higher return periods. Catchment infiltration volumes exhibit linear growth with greater storm intensities. (3) Peak infiltration rates and moments for pervious areas, subcatchments, and the overall catchment exhibit correlations with both the return period and the time-to-peak coefficient. This study quantifies the influence of design storm parameters for infiltration, providing valuable insights for stormwater infrastructure design and urban stormwater control.

In sponge city construction, the optimization of sponge facility combinations at the urban block-level is a very important aspect. Xie et al. (contribution 5) utilized a residential complex in Nanjing as a practical example, selected six types of typical sponge facilities to construct a multi-objective optimization combination model for sponge facilities, and employed the Strength Pareto Evolutionary Algorithm (SPEA-2) to determine the optimal combination of sponge facility types and quantities. Cao et al. (contribution 6) used InfoWorks ICM to simulate the properties of runoff and determine the optimal LID design of a residential site in Yixing, China, based on four practical rainfall events. In this study, the software was redeveloped using Ruby object-oriented programming to improve its efficiency in uncertainty analysis using the Generalized Likelihood Uncertainty Estimation method. The simulated runoff was in good agreement with the observed discharge.

The coupling of gray and green infrastructure is another focus in sponge city construction; however, due to the complexity of the process and the diversity of the benefits, there are no measurements of the comprehensive benefits. Adopting a typical university campus in Beijing as an example, Wang et al. (contribution 7) simulated the multidimensional benefits to the water quantity, water quality, and ecology of a gray and green facility renovation by coupling the stormwater management model (SWMM) and InfoWorks Integrated Catchment Management (ICM). Monetization methods and economical means were employed to characterize the comprehensive benefits.

Within the studies of typical source control facilities, Lee et al. (contribution 8) assessed the performance of permeable pavement in mitigating flooding in urban areas. In this study, demonstration roads using a general pavement and a permeable pavement were built on Dahua North Street, Taoyuan City; rainwater was stored in a central irrigation ditch and a permeable pavement through an innovative construction method for reuse in agricultural irrigation. Monitoring instruments and management systems were built to analyze the actual discharge and peak discharge of the permeable pavement and general pavement. The results show that the permeable pavement can effectively reduce the peak discharge by 60~75%, which can not only achieve the benefit of low-impact development but can also reuse rainwater. To understand the occurrence of preferential flow in the vegetated artificial substrates of green roofs, Chen et al. (contribution 9) established an experiment with various plant substrate combinations that involved two Sedum species and two artificial substrates for three depths of 6, 10, and 14 cm. Thereafter, solute breakthrough experiments were conducted, followed by inverse and forward modeling in Hydrus-1D. To assess the performance of a rain garden in a red soil area in southern China, Chen et al. (contribution 10) built a rain garden in Nanchang city, where the local soil is red soil and has low organic carbon, strong acidity, and low permeability rainfall characteristics. Rainfall runoff control and pollutant removal efficiencies were studied based on the on-site conditions. The analysis of almost 2 years of field data showed that the volume capture ratio of annual rainfall and the mean load removal of TSS, NH_3-N, TP, TN, COD, and NO_3-N met the technical guidelines for sponge city construction in Nanchang.

In order to evaluate the actual runoff control effects of LID combination, Zhang et al. (contribution 11) used the real hydrological monitoring data collected from Lingang New City in Shanghai to analyze the retention and lag characteristics of rainfall–runoff in LID combinations under three rainfall-intensity scenarios (light–moderate, heavy, and torrential rainfall). The LID facilities were constructed over three years in the target study area, including rain gardens, retention ponds, green parking, porous pavement, and grass swales. The results confirmed the vital role of the LID combination in stormwater management and the hydrological impact of the LID combination on rainfall-induced runoff retention and lag effects.

Author Contributions: Conceptualization, H.J.; writing—original draft preparation, H.J.; Editing, J.H., D.F. and W.-S.C. All authors have read and agreed to the published version of the manuscript.

Funding: This research received no external funding.

Data Availability Statement: Not applicable.

Acknowledgments: The authors are thankful for all of the contributions to this Special Issue, the time invested by each author, and the anonymous reviewers and editorial managers who contributed to the development of the articles in this Special Issue. All the Guest Editors are very happy with the review process and management of the Special Issue and offer their thanks.

Conflicts of Interest: The authors declare no conflicts of interest.

List of Contributions

1. Aldrees, A.; Dan'azumi, S. Application of Analytical Probabilistic Models in Urban Runoff Control Systems' Planning and Design: A Review. *Water* **2023**, *15*, 1640. https://doi.org/10.3390/w15091640.
2. Zhao, Y.; Price, M.; Trowsdale. S. Comparison of the Transition to More Sustainable Stormwater Management in China and the USA. *Water* **2022**, *14*, 1960. https://doi.org/10.3390/w14121960.
3. Le, J. T.; Gonzalez, J. P.; Carson, R. T.; Ambrose, R. F.; Levin; L. A. Integrating Non-Targeted Ecosystem Services into Assessment of Natural Stormwater Treatment Systems. *Water* **2023**, *15*, 1460. https://doi.org/10.3390/w15081460.
4. Yang, Y.; Shao, Z.; Xu, X.; Liu, D. Impact of Storm Characteristics on Infiltration Dynamics in Sponge Cities Using SWMM. *Water* **2023**, *15*, 3367. https://doi.org/10.3390/w15193367.

5. Xie, M.; Cheng, Y.; Dong, Z. Study on Multi-Objective Optimization of Sponge Facilities Combination at Urban Block Level: A Residential Complex Case Study in Nanjing, China. *Water* **2022**, *14*, 3292. https://doi.org/10.3390/w14203292.
6. Cao, Q.; Cao, J.; Xu, R. Optimizing Low Impact Development for Stormwater Runoff Treatment: A Case Study in Yixing, China. *Water* **2023**, *15*, 989. https://doi.org/10.3390/w15050989.
7. Wang, J.; Zhou, X.; Wang, S.; Chen, L.; Shen, Z. Simulation and Comprehensive Evaluation of the Multidimensional Environmental Benefits of Sponge Cities. *Water* **2023**, *15*, 2590. https://doi.org/10.3390/w15142590.
8. Lee, Y. T.; Ho, M. C.; Chiou, Y. S.; Huang, L. L. Assessing the Performance of Permeable Pavement in Mitigating Flooding in Urban Areas. *Water* **2023**, *15*, 3551. https://doi.org/10.3390/w15203551.
9. Chen, X.; Liu, R.; Liu, D.; Xin, X. Analysis of Preferential Flow in Artificial Substrates with Sedum Roots for Green Roofs: Experiments and Modeling. *Water* **2023**, *15*, 914. https://doi.org/10.3390/w15050914.
10. Chen, C.; Li, Y.; Le, W.; You, C.; Liu, C.; Liu, W.; Zhang, R. Field Performance of Rain Garden in Red Soil Area in Southern China. *Water* **2023**, *15*, 267. https://doi.org/10.3390/w15020267.
11. Zhang, C.; Lv, Y.; Chen, J.; Chen, T., Liu, J.; Ding, L.; Zhang, N.; Gao, Q. Comparisons of Retention and Lag Characteristics of Rainfall–Runoff under Different Rainfall Scenarios in Low-Impact Development Combination: A Case Study in Lingang New City, Shanghai. *Water* **2023**, *15*, 3106. https://doi.org/10.3390/w15173106.

References

1. Chen, M.X.; Lu, D.D.; Zhang, H. Comprehensive evaluation and the driving factors of China's urbanization. *Acta Geogr. Sin.* **2009**, *64*, 387–398. (In Chinese)
2. Hager, J.; Hu, G.J.; Hewage, K.; Sadiq, R. Performance of low-impact development best management practices: A critical review. *Environ. Rev.* **2019**, *27*, 17–42. [CrossRef]
3. Davis, A.P. Green engineering principles promote low-impact development. *Environ. Sci. Technol.* **2005**, *39*, 338A–344A. [CrossRef]
4. Xia, J.; Zhang, Y.; Xiong, L.; He, S.; Wang, L.F.; Yu, Z.B. Opportunities and challenges of the Sponge City construction related to urban water issues in China. *Sci. China Earth Sci.* **2017**, *60*, 652–658. [CrossRef]
5. Chan, F.K.S.; Griffiths, J.A.; Higgitt, D.; Xu, S.Y.; Zhu, F.F.; Tang, Y.T.; Xu, Y.Y.; Thorne, C.R. "Sponge City" in China—A breakthrough of planning and flood risk management in the urban context. *Land Use Policy* **2018**, *76*, 772–778. [CrossRef]
6. Ministry of Finance; Ministry of House and Construction; Ministry of Water Resources. Notice on Systematically Promoting Sponge City Construction Demonstration Work (Finance Office Construction (2021) No. 35), 2021). Available online: https://www.gov.cn/zhengce/zhengceku/2021-04/26/content_5602408.htm (accessed on 16 December 2023).
7. Yin, D.; Xu, C.; Jia, H.; Yang, Y.; Sun, C.; Wang, Q.; Liu, S. Sponge City Practices in China: From Pilot Exploration to Systemic Demonstration. *Water* **2022**, *14*, 1531. [CrossRef]
8. Hamidi, A.; Ramavandi, B.; Sorial, G.A. Sponge City—An emerging concept in sustainable water resource management: A scientometric analysis. *Resour. Environ. Sustain.* **2021**, *5*, 100028. [CrossRef]
9. Yin, D.; Chen, Y.; Jia, H.; Wang, Q.; Chen, Z.; Xu, C.; Li, Q.; Wang, W.; Yang, Y.; Fu, G.; et al. Sponge City Practice in China: A Review of Construction, Assessment, Operational and Maintenance. *J. Clean. Prod.* **2021**, *280*, 124963. [CrossRef]
10. Jia, H.; Hu, J.; Huang, T.; Chen, A.S.; Ma, Y. Urban Runoff Control and Sponge City Construction. *Water* **2022**, *14*, 1910. [CrossRef]

water

Article

Assessing the Performance of Permeable Pavement in Mitigating Flooding in Urban Areas

Yueh-Tan Lee, Min-Che Ho, Yi-Shain Chiou * and Li-Ling Huang

Department of Civil Engineering, National Central University, Taoyuan 320317, Taiwan;
10007950@mail.tycg.gov.tw (Y.-T.L.); b1231520002000@gmail.com (M.-C.H.); irenahf@gmail.com (L.-L.H.)
* Correspondence: lettermail64@gmail.com; Tel.: +886-955075859

Abstract: In the case of rapid urban development, the impact of extreme climates on the world is gradually increasing, resulting in frequent flood events. However, Taiwan is still in the stage of urban development, and it is necessary to develop more roads. Therefore, determining how to reduce the impact of road engineering on the environment is one of the major issues currently faced. Therefore, a demonstration road of a general pavement and a permeable pavement was built in Dahua North Street, Taoyuan City, Taiwan, and rainwater was stored in a central irrigation ditch and a permeable pavement through an innovative construction method for reuse in agricultural irrigation. In addition, monitoring instruments and management systems were built, and the flow law formula was established, with R^2 greater than 0.9. The actual discharge and peak discharge of the permeable pavement and general pavement were analyzed. According to the data analysis results, it can be seen that the permeable pavement can effectively reduce the peak discharge of 60~75%, which not only can achieve the benefit of low-impact development but also can reuse rainwater. The patent application can be used as an example for the application of permeable pavement in Taiwan in the future.

Keywords: permeable pavement; monitoring instruments and management systems; flow law formula; low-impact development

Citation: Lee, Y.-T.; Ho, M.-C.; Chiou, Y.-S.; Huang, L.-L. Assessing the Performance of Permeable Pavement in Mitigating Flooding in Urban Areas. *Water* **2023**, *15*, 3551. https://doi.org/10.3390/w15203551

Academic Editors: Dafang Fu, Haifeng Jia, Jiangyong Hu and Wei-Shan Chen

Received: 8 September 2023
Revised: 6 October 2023
Accepted: 9 October 2023
Published: 11 October 2023

1. Introduction

As Taiwan is still in the stage of urban development, determining how to reduce the impact of urban floods brought by urbanization is a current issue facing Taiwan. Asif Iqbal [1] used the design of permeable pavement to reduce the impact of urban floods. Therefore, this paper hopes to explore the low-impact development benefits of a permeable pavement, which can reduce the burden of the overall drainage system under extreme climate conditions and then reduce the occurrence of disasters. Mariacrocetta Sambito [2] studied the literature on previous pavements in the past and found that there were few actual studies on the site in the past, and there was no long-term rainfall event to confirm the benefits of a pervious pavement. Therefore, this paper built a pervious pavement and a general pavement on the site and established a real-time monitoring system for management to conduct long-term data analysis.

Before the establishment of the demonstration road with permeable pavement in this study, relevant studies at home and abroad tested the permeable pavement and monitored its drainage flow data to understand the benefits of permeable pavement in reducing flood peak flow. Therefore, this study first discusses the results of relevant domestic and foreign test roads, hoping that the analysis of this demonstration road can be used as support for past studies and carry out the extension of the discussion, making up for the past research. In the past, there have been many studies on permeable pavements. Alharbi, F applied porous clay bricks mixed with bran as a permeable pavement, which can reduce pollutants and surface runoff [3]. Vaillancourt used a permeable intermingled concrete pavement (PICP) in urban areas [4] and discussed its benefit of reducing peak flood flow. After 12 months of monitoring, it was found that PICP could effectively reduce surface runoff by 26% to 98%, and it was found that the use of a more permeable soil subgrade would be more effective. Lin Wei Hong established a porous asphalt concrete pervious pavement and a general asphalt concrete pavement on-site [5], and the research results show that the water retention of the pervious pavement can be increased by approximately 35% to 65% compared with a general asphalt concrete pavement. The efficiency of low-impact development in two regions of Indiana, USA [6], was evaluated and analyzed in six sets of scenarios, including rainwater storage tanks and a permeable pavement with different proportions. The results of monitored runoff and water retention showed that under the above scenarios, low-impact development could reduce surface runoff by 2% to 12% in the two study areas. It can reduce urban flooding caused by heavy rainfall and water fog caused by frequent rain. Referring to the research results of permeable pavement in Xindian [7], four different permeable pavements were built in Xindian, and pervious concrete, porous asphalt concrete, grass-planting brick, and pervious brick were used as permeable pavement materials. It was found that the four pavement types could effectively reduce flood peak time, and the results were as follows: pervious concrete > porous asphalt concrete > grass-planting brick > permeable brick. In the past, scholars built permeable pavements on Arizona highways [8] and built monitoring instruments such as rain meters, flow meters, and moisture content meters to collect on-site data and analyze the data during rainstorms. It was found that the permeable pavement could effectively retain water after rainfall, and upon inspection of the pavement on site after rainfall, it was found that permeable pavement could dissipate surface runoff more quickly. The general pavement will have a water phenomenon. Korean scholars built permeable pavement in Seoul [9] and built on-site monitoring instruments and found that permeable pavement can effectively reduce surface runoff by 30–65%. Another study established a porous asphalt concrete bicycle path and pervious concrete brick sidewalk in Taipei City [10] and imported the field test data into SWMM for simulation analysis. The results showed that when heavy rain fell, pervious pavement could effectively reduce surface runoff by 35% to 41%.

To sum up, previous studies can effectively reduce surface runoff by analyzing the flow of permeable pavement. However, most studies still focus on the analysis of surface runoff and do not analyze the actual discharge after rainfall of permeable pavement and general pavement. However, Gauss found in his study [11] that the setting of low-impact development can store part of the rainwater of a rainfall event. It will delay its outflow time, but for long-delay heavy-rainfall events, the initial storage of rainwater in the low-impact development will be drained along with the subsequent rainfall of the long-delay rainfall event, resulting in increased drainage. Therefore, this study built permeable pavement and general pavement, used a central irrigation ditch and permeable pavement to reduce the actual discharge through innovative construction methods, buried rain meters and flowmeters under general pavement and permeable pavement, and established a flow law formula, which can calculate the actual discharge and flood peak discharge of permeable pavement and general pavement for a long time. The benefits of pervious pavement in reducing peak flood flow and reducing the burden on the overall drainage system during heavy rainfall are confirmed. The flow chart of this paper is shown in Figure 1.

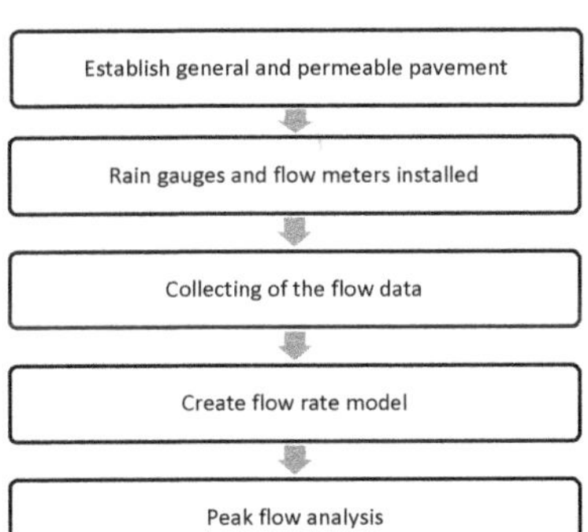

Figure 1. Research Flow Chart.

2. Materials and Methods

2.1. Difference between Permeable Pavement and Other Pavement

The so-called permeable pavement system applies materials with good permeability and high porosity to the surface layer, foundation, and bottom layer of the pavement, so that rainwater flows through the pavement with excessive porosity and directly penetrates the subgrade soil, and then the water returns to the ground through the action of atmospheric water circulation. The permeable material of the drainage pavement is used as the surface layer, and the rainwater penetrates through the surface layer and then enters the side ditch. The conventional pavement drains rainwater into the side ditch by running on the pavement surface.

As can be seen from Figure 2, three types of pavement, general pavement, drainage pavement, and permeable pavement, are used to discharge rainwater from the road surface. The traditional dense-graded pavement emphasizes the concept of rapid drainage, so the pavement design adopts an inclined angle to direct rainwater into the roadside water collection facilities. The surface layer of the drainage pavement has high porosity, so most of the rainwater can first infiltrate into the surface layer and then be discharged by drainage facilities, such as the permeable pavement express lane in this study. The design principle of permeable pavement is roughly the same as that of drainage pavement, the only difference is that there is no permeable layer in the structure of permeable pavement, so rainwater permeates into the soil through the pores on the pavement, such as the permeable pavement slow lane in this study [12].

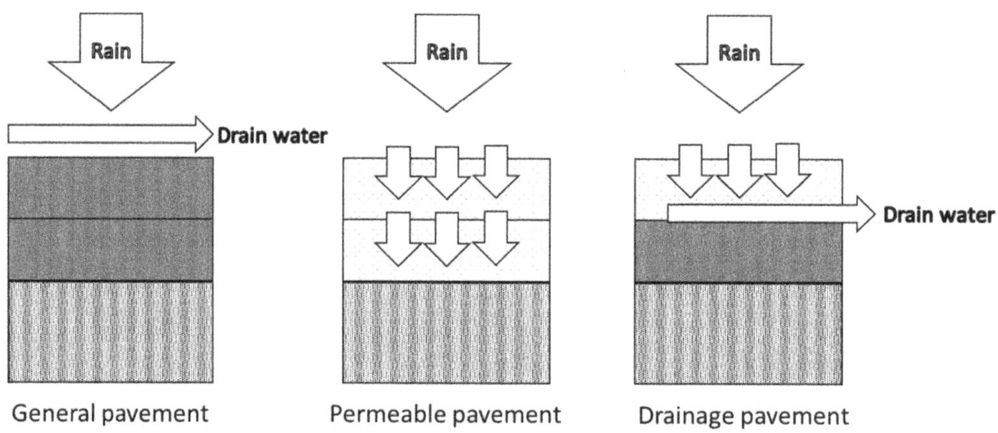

Figure 2. Comparison of permeable pavement with other pavements.

2.2. Test Road Setup

2.2.1. Introduction of Local Environment

1. Location: Dahua North Street, Luzhu District, Taoyuan City, Taiwan.
2. Water catchment area: It is divided into permeable pavement and ordinary pavement test sites. The dividing point is shown by the red dotted line in Figure 3. The ordinary pavement to the west of the red line and the permeable pavement to the east of the red line are approximately 200 m in length and 15 to 20 m in width.
3. The permeable pavement is further divided into the fast lane and the slow lane, aiming to divert traffic volume and enhance traffic safety. Each lane serves different types of vehicles, with the fast lane designed for regular cars and heavy vehicles, while the slow lane is designated for motorcycles, as shown in Figure 4.
4. Climatic conditions: Precipitation collected in the past three years is shown in Figure 5.

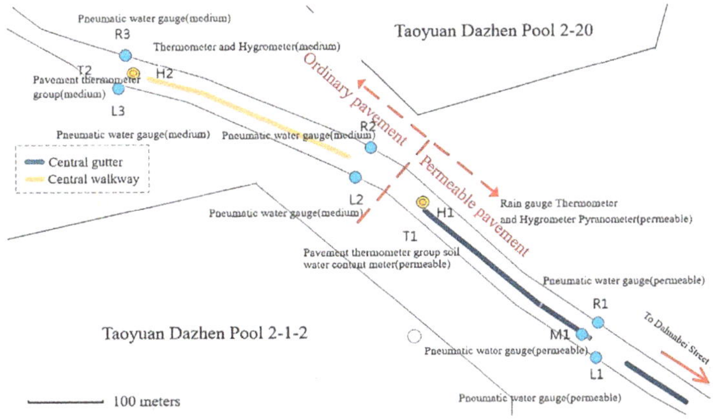

Figure 3. Test road scene situation.

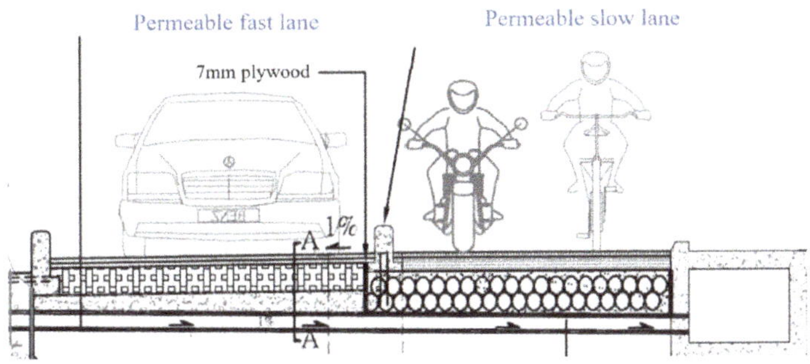

Figure 4. Design of permeable pavement.

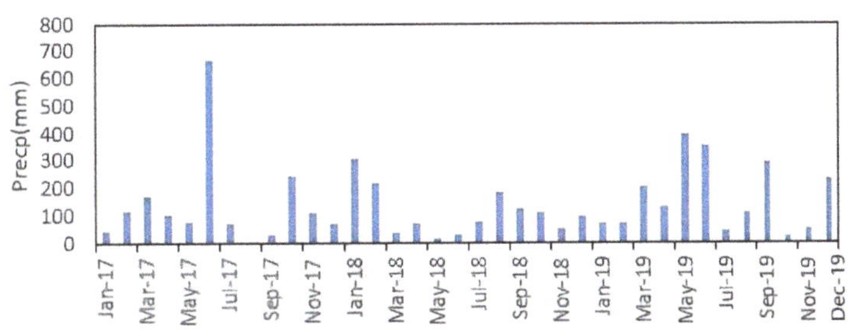

Figure 5. Precipitation in recent 3 years.

2.2.2. Sectional Configuration

The selection and construction of permeable pavement materials adhere to the guidelines outlined in the Urban Road Permeable Pavement User Manual [13] provided by the Urban Construction Department.

This study aims to evaluate the effectiveness of three types of pavement: A general graded asphalt concrete pavement, a permeable pavement for the fast lane, and a permeable pavement for the slow lane.

Figure 6 illustrates the cross-section configuration of the general graded asphalt concrete pavement and the permeable pavement for the fast lane and slow lane.

To analyze temperature variations, earth thermometers were buried within all three types of pavements.

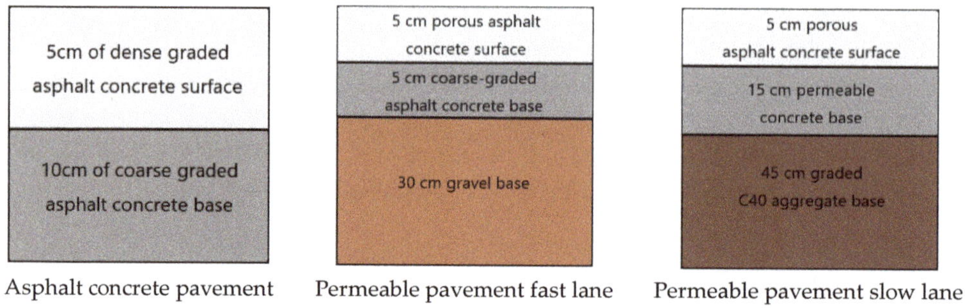

Figure 6. Section design of pavement.

2.2.3. Selection of Permeable Pavement Materials

1. Permeable pavement fast lane

The surface material used for the permeable pavement is porous asphalt concrete. Porous asphalt concrete generally refers to an asphalt mixture that retains over 15% porosity even after compaction. Its internal composition consists of interconnected continuous voids, allowing water to flow freely between them. According to the specifications, the water permeability coefficient should reach 10^{-2} cm/s [14].

The potential harm caused by water to asphalt concrete pavement has always been a topic of concern for pavement designers, as water retention on the asphalt concrete surface can jeopardize driving safety and pavement durability.

Due to its specific gradation, the porous asphalt concrete surface layer lacks fine materials to fill the gaps between particles, resulting in a high porosity. This allows water to flow freely between the gaps, swiftly removing precipitation and reducing the occurrence of surface water on the road.

Moreover, the rough surface of porous asphalt concrete pavement enhances its anti-skid ability, leading to reduced braking distances, decreased vehicle slippage, and improved driving safety.

Furthermore, research on the durability of porous pavement suggests that mixtures with smaller nominal particle sizes are more prone to clogging, potentially impacting the long-term durability of permeable pavement.

The pervious capacity of porous asphalt concrete can be evaluated through in situ pervious tests. According to the "Technical pointer for drainage paving" issued by the Japan Road Association [15], the individual measured value in an in situ pervious test should exceed 900 mL/15 s.

Porous asphalt concrete is used as the surface material for the permeable pavement express lane used in this paper. The porous asphalt concrete ratio is shown in Table 1, and relevant tests are carried out for the asphalt concrete mixture, all of which comply with the specifications in Chapter 02798 of the construction outline specification [14], as detailed in Table 2.

Table 1. Porous asphalt concrete ratio table.

Mesh No.	Passing Weight Percentage of Test Sieve (%)			
	Lower Limit Value	Experimental Value	Upper Limit Value	Conform to Specifications
1″	100	100	100	OK
3/4″	95	97	100	OK
1/2″	64	64	84	OK
3/8″	-	49	-	
No. 4	10	20	31	OK
No. 8	10	15	20	OK
No. 16	-	10	-	

Table 1. Cont.

Mesh No.	Passing Weight Percentage of Test Sieve (%)			
	Lower Limit Value	Experimental Value	Upper Limit Value	Conform to Specifications
No. 30	-	8	-	
No. 50	-	7	-	
No. 100	-	6	-	
No. 200	3	5.1	7	OK
Bitumen content (%)	4	5.1	6	OK

Table 2. Test results of porous asphalt concrete mixture.

Test Project	The Test Results	Specification Values
Stable value (kgf)	650	350
Mobility value (0.1 mm)	24	20~40
Porosity rate (%)	17	15~25
Dynamic stability value (times/mm):	2487	Over 1500
Retention strength index (%)	90	Over 75
Cantabria test (%)	16.8	Below 20
Vertical flow test (%):	0.12	Below 0.3
Permeability coefficient (cm/s)	0.11	Over 0.01

2. Permeable pavement slow lane

The porous asphalt concrete surface is used for the slow lane of the permeable pavement. The material properties are shown in Tables 1 and 2. In addition, pervious concrete and C40 are used as the bottom material.

- Permeable Concrete

Permeable concrete exhibits a porosity ranging from 20% to 35%. The high porosity enhances the water permeability and water retention capabilities of the permeable pavement. According to the specifications, the water permeability coefficient should reach 10^{-3} cm/s. The aggregates used in permeable concrete have a passing percentage of approximately 2% to 3% or less for the No. 4 sieve. Cylindrical specimens are cured at different ages: 1 day, 3 days, 7 days, and 28 days. Additionally, bending specimens are prepared for the 28-day curing period [13].

Various design methods are employed for permeable concrete, including the weight ratio method, volume method, and specific surface area method. For this study, the weight ratio method is used as the design method for permeable concrete. The proportions of aggregates, cement, mixing water, and admixtures are determined after confirming the effectiveness of the trial mix.

- Grade C40 Aggregate

Pao-Ching Chang carried out test paving of three permeable sections and general pavement on Longci Road, Zhongli City [16]. The surface layers of the three permeable pavement sections are all permeable asphalt concrete, and only the bottom material is different. Pervious concrete is used in Section 1, low-density pervious concrete in Section 2, and C40 permeable grade in Section 3. The flow analysis of each pavement after rainfall showed that three permeable pavements had the benefit of flood peak delay, namely, Section 1 (30 min), Section 2 (20 min), and Section 3 (40 min), in which C40 permeable grade was used as the bottom pavement with the best effect. Therefore, C40 was selected as the bottom material of permeable pavement in this paper.

The grade C40 aggregate used for the bottom layer exhibits higher porosity compared to the general graded layer, enabling effective water permeability. The aggregate standard grading for grade C40 aggregate is provided in Table 3 [17].

Table 3. The aggregate standard grading.

	Passing Weight Percentage of Test Sieve (%)						
	37.5 mm	**31.5 mm**	**25.0 mm**	**19.0 mm**	**12.5 mm**	**4.75 mm**	**2.36 mm**
C40	100~95	-	-	80~50	-	40~15	25~5

According to the recommended grading for C40 materials in Japan [18], the suggested particle content passing through the #8 sieve falls between 5% and 25%. For particle sizes that do not fall within the recommended range, a smooth and continuous curve is employed as the design gradation. Standard compaction tests are conducted using crushed stone-grade ingredients with three different fine particle contents of 5%, 15%, and 25% to determine their optimum moisture content (OMC).

Subsequently, permeability tests are performed to assess the permeability of various gradation combinations and evaluate their permeability.

- The material properties and test specifications of permeable pavement slow lanes are shown in Table 4.

Table 4. Properties of the material.

Project	Test Project	The Test Results	Specification Values
Porous asphalt concrete	Porosity rate (%)	17	15~25
	Permeability coefficient (cm/s)	0.11	Over 0.01
Grade C40 Aggregate	Porosity rate (%)	10	6~18
	Permeability coefficient (cm/s)	0.0103	3×10^{-3}~4×10^{-2}
Permeable Concrete	Porosity rate (%)	30	-
	Permeability coefficient (cm/s)	0.378	Over 10^{-3}

2.2.4. On-Site Monitoring Instrument

The monitoring instruments installed on permeable pavements and general dense-graded pavements in this study include the following monitoring instruments. The data collected by earth thermometers and thermometers are used for the analysis of this study.

In order to facilitate subsequent comprehensive analysis, this study set the monitoring frequency of various monitoring instruments to obtain 1 piece of data every 10 min, so 144 data will be collected in one day:

- Earth thermometers (soft adhesive type Thermalpas NR-40-MS).
- Flowmeter (German NIVUS PCM 4 portable ultrasonic flowmeter).
- Thermometer.
- Rain Gauge,

In this study, it is estimated that the instruments to be set on Dahuabei Street in Luzhu District are the pressure water level gauge (eYc L051 water level sensor), earth thermometer, and atmospheric thermometer.

Table 5 shows the buried settings of the equipment after the on-site survey. The research site is Dahua North Street, Luzhu District, the annual temperature is around 21 degrees, the annual average rainfall is approximately 10.8 mm/day, which is lower than the average rainfall in Taiwan, and the number of rainy days per year is approximately 140 days with more rain in summer but more rainy days in winter than in summer because of the strong northeast monsoon in winter.

Table 5. Equipment setting location.

Code	Equipment	Pavement
M1	Pneumatic water gauge	permeable pavement
R1	Pneumatic water gauge	permeable pavement
R2	Pneumatic water gauge	general pavement
R3	Pneumatic water gauge	general pavement
L1	Pneumatic water gauge	permeable pavement
L2	Pneumatic water gauge	general pavement
L3	Pneumatic water gauge	general pavement
T1	Pavement thermometer group	permeable pavement
T2	Pavement thermometer group	general pavement

We set up 4 sets of water level gauges in the side ditch of the permeable pavement and 1 set of water level gauges on the central safety island to measure the surface runoff and outflow runoff, and then calculated the net flow reduction and water retention capacity.

In addition, two sets of water level gauges are installed on the general asphalt concrete pavement to calculate the surface runoff, compare the rainwater effect between the permeable pavement and the general asphalt concrete pavement, and then evaluate the low-impact development effect of the permeable pavement.

The site construction diagram is shown in Figure 7.

Figure 7. Field configuration diagram of the flowmeter.

2.3. Water Flow Analysis Method

1. Flood peak reduction

The permeable pavement can use its full water retention, semi water retention, and institutional drainage mechanisms to reduce the runoff of the pavement; to achieve the benefits of low-impact development, its peak flow effect should be smaller than that of general asphalt concrete pavement.

Follow-up analysis is carried out using the aforementioned monitoring equipment conditions and calculating the runoff of permeable pavement and general asphalt concrete pavement on Dahuabei Street.

In addition, the on-site rain gauge is used to measure and analyze the rainfall, and the rainfall is measured through the rain gauge. Finally, the surface runoff and rainfall are used to calculate the flood peak reduction to evaluate the effect of low-impact development.

In this study, the flood peak flow is calculated as the reduction of the flood peak flow of the permeable pavement relative to the general asphalt concrete pavement, as shown in Equation (1).

$$\text{Peak Flow Reduction} = \text{General Asphalt Concrete Peak Flow} - \text{Permeable Pavement Peak Flow} \quad (1)$$

2. Flow rate determination method

The basic definition of flow rate is the volume of water (Wv) passing through each unit of time (t), which is called the flow rate (Q).

Based on the continuous equation, the flow rate through a certain cross-section can be obtained according to Equation (2) or Equation (3).

Since the channel cross-section is irregular and has no specific shape, the water-passing cross-section is divided into n small sections, and the flow qi of each small section is estimated according to the concept of Equation (4). The above concept can be expressed by Equations (4) and (5) and can be further divided into the mid-section method and the mean-section method.

The interrupted surface method assumes that the water passage section of the channel is composed of several different rectangular sub-sections, and the width of each sub-section is half the distance between the adjacent water depths. The corresponding sub-sectional area is shown in Equation (7), and the multiplication and accumulation are the total flow.

The average cross-section method treats the cross-section as composed of multiple trapezoidal sub-sections (as shown in Figure 8), and the average velocity of each sub-section is the average of the average velocity of the two adjacent vertical lines in Equation (8). The sub-sectional area is shown in Equation (9), and according to Equation (4), it can be accumulated into the total flow.

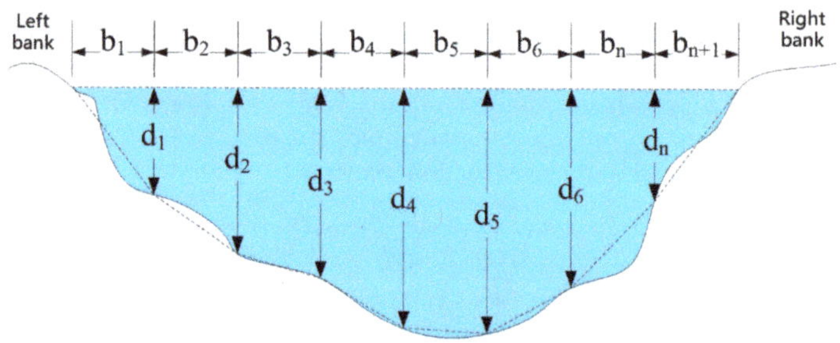

Figure 8. Schematic diagram of the mean section method.

Although the interrupted surface method has the advantage of simplicity, the accuracy is relatively low, while the average cross-section method is complicated but has the advantages of high accuracy (Chen Fengwen et al., 2012) [19].

$$Q = \int_A v dA \tag{2}$$

$$Q = Vav \bullet A \tag{3}$$

$$Q = a_1 v_1 + a_2 v_2 + a_3 v v_3 \ldots + a_n v_n \tag{4}$$

$$Q = q_1 + q_2 + q_3 + \ldots + q_n \tag{5}$$

In the formula: A is the cross-sectional area of the channel measurement surface, v is the flow velocity at any point in the cross-section of the flow zone, V_{av} is the average velocity of the water-passing cross-section, n is the number of divisions of the water sectional area, a_i is the area of the i-th smallest section, and v_i is the flow velocity of the i-th smallest section.

$$v_{\text{mid}} = v_{\text{av,i}} = v_{\text{i}} \tag{6}$$

$$a_{\text{mid}} = a_{\text{i}} = bi \cdot di \tag{7}$$

$$v_{\text{mean}} = v_{\text{av,i}} = \frac{v_{\text{i}} + v_{\text{i+1}}}{2} \tag{8}$$

$$a_{\text{mean}} = a_{\text{i}} = \frac{(b_{\text{i}} + b_{\text{i+1}}) \cdot d_{\text{i}}}{2} \tag{9}$$

where v_{mid} is the average velocity in sub-sections of the medium section method, v_{mean} is the average velocity in sub-sections of the average section method, a_{mid} is the sub-section area of the medium section method, and a_{mean} is the sub-section area of the average section method, as shown in Figure 8.

From the flow measurement and monitoring method, it can be seen that the velocity observation mainly functions to obtain the vertical average velocity of each sub-section.

This study uses the Acoustic Doppler Profiler (ADP), which is a device for directly observing the vertical profile flow velocity of the survey line.

The operation method and related principles of the Doppler flow meter are as follows.

The direct observation of the average vertical flow velocity can be obtained directly through the Doppler flowmeter without further correction or calculation (Chen Fengwen et al., 2012) [19].

Among various types of sonic flowmeters, ultrasonic is the most widely used, and its operating principle is the use of the Doppler effect.

The measurement data can be divided into the Acoustic Doppler Velocimeter (ADV) and the Acoustic Doppler Profiler (ADP) according to the range of sonar scanning.

The difference between the above two is that ADV can only measure the flow velocity of a single point, while ADP can measure the flow velocity of multiple points on a vertical line and can directly measure the flow velocity profile.

ADP is also widely used in stereotyped channel flow observation and has obtained excellent results (Chen Fengwen et al., 2008a; Chen Fengwen et al., 2008b) [20,21].

Hydrological information can obtain stable long-term data through automatic monitoring. In terms of flow, the flow (Q) and the water level (H) at a fixed position of a fixed water cross-section have a certain mathematical relationship.

Therefore, based on the establishment of the H–Q relationship, the water level (H) measured by the water level sensing element can be set up, and the flow information can be obtained through the conversion of the relationship between the water level and the flow rate (Q).

The above water level (H) parameter contains elevation information, so it can be applied to natural rivers where the bottom of the canal is easy to scour or silt, and the elevation changes or the shape in the satin surface is easy to change.

Therefore, the Water Resources Department applies this method to water level or flow measurement stations to obtain the flow information of different regions of each river basin at different times, as in Equation (10).

Since the local channel in this study is a fixed channel and the bottom elevation and section factors are fixed, the water level can be adjusted to the water depth, and the symbol is still represented as H.

The regression analysis of the quadratic equation of one variable is directly applied to establish the water depth (H)-discharge (Q) calibration formula, which is shown in Equation (11).

In order to avoid the occurrence of extreme values or negative values in the formula or the phenomenon of standing water level on-site, the regression analysis process sets the intercept to 0, and Equation (11) can be modified to Equation (12).

In order to avoid the occurrence of extreme values or negative values in the formula or the phenomenon of a standing water level on-site, the regression analysis process sets the intercept to 0, and Equation (11) can be modified to Equation (12).

Through the H–Q curve, the flow value corresponding to different water depths can be estimated.

Considering that there is no reliable water source at the test site of this study, the flow measurement can be adjusted according to different water depth conditions, so the Manning formula and the field-measured data are combined to estimate the H–Q equation.

The main consideration is that the theoretical value estimated by the Manning formula is mainly used initially, and the subsequent flow measurement with different water level conditions of the rainfall event is mainly used.

If there is a lack of flow measurement data at medium and high water depths, the theoretical flow data estimated by Manning's formula should be supplemented, and regression analysis will be carried out after integration.

It is important to ensure that the formula is based on the interpolation results obtained on a scientific basis or from field measurement data within the applicable range, rather than the uncertain extrapolation results.

$$Q = A(H_{EL} - H_B)^B\text{; Form of water level} \tag{10}$$

$$Q = CH^2 + DH \pm E\text{; Form of water depth} \tag{11}$$

$$Q = CH^2 + DH\text{; Form of water depth} \tag{12}$$

In the formula: A, B, C, D, and E are coefficients; Q is the flow rate (cms).

H is the water level in Equation (10) and water depth (m) in Equations (11) and (12).

H_{EL} is the water level, that is, the water surface elevation (m).

HB is the canal bottom elevation (m).

2.4. Construction of Permeable Pavement System

This system can be used to query the temperature monitoring data of Dahuabei Street, including real-time data and historical data.

The management of water level monitoring data can obtain water level information of the monitoring road at any time, and the information includes the monitoring station, equipment code, water level, and data measurement time, as shown in Figure 9.

The measured data use the change in water level and time to calculate the flow of its road and store the data for future queries.

Its main architecture includes a real-time information display, historical query, and flow calculation, while its functional architecture planning is shown in Figure 10 and is described as follows:

1. Real-time information query: The water level will be measured at each fixed time and uploaded to this system immediately,
2. Historical query: After the data are uploaded, they will be stored in the system. The data can be exported by entering the date to be inquired into the system.
3. Flow calculation: The flow is calculated by using the time and water level difference, and the flow between the permeable pavement and the general pavement is compared.

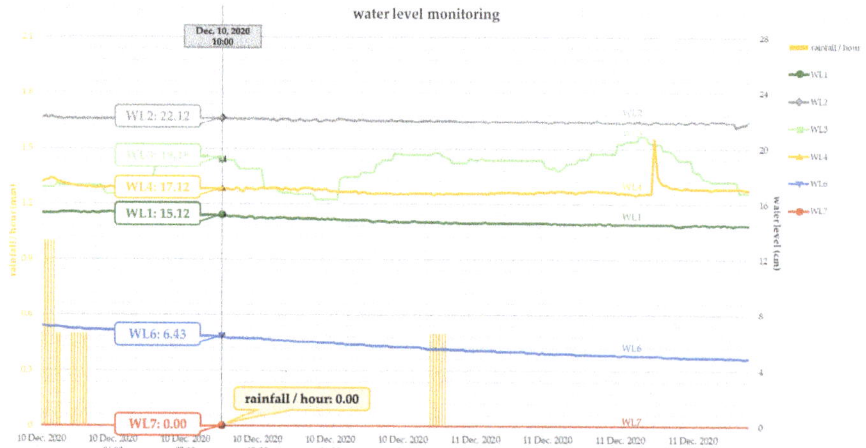

Figure 9. Line chart of historical data of water level monitoring.

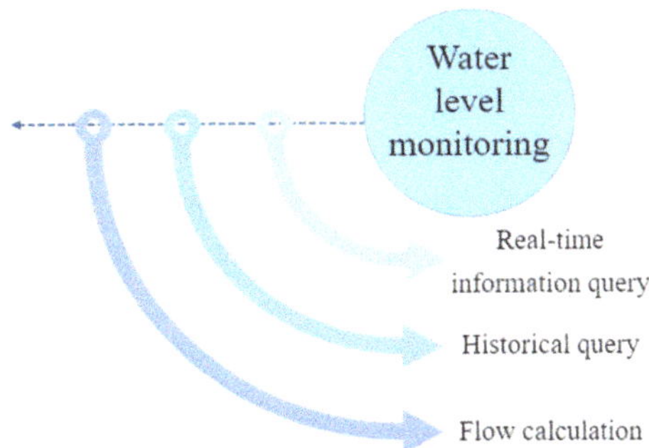

Figure 10. Functional Architecture of Water Level Monitoring.

3. Results

3.1. Analysis of Hydraulic Design Benefits for Permeable Pavement

Taoyuan City is known for its abundance of ponds, with a total of over 2800 ponds. The region also boasts a well-developed water irrigation system with extensive waterways. This project aims to leverage the existing water irrigation channels and transform them into dual-purpose channels that serve as irrigation channels during normal times and provide floodwater detention capacity during heavy-rainfall events. By implementing permeable or water retention measures, the decline in floodwater drainage function caused by development can be reduced and the ability to handle rainfall can be enhanced. Therefore, incorporating the concept of low-impact development into pavement engineering should be a top priority in creating a sustainable environment.

This project aims to achieve the benefits of low-impact development through the use of permeable pavement, focusing on enhancing the effectiveness of floodwater drainage. The design concept includes two key elements: The innovative design of central irrigation channels and the design of the permeable pavement itself to retain water. These two designs enable the permeable pavement to withstand rainfall intensities much higher than conventional pavements can tolerate.

3.1.1. Central Irrigation Channel Design

1. The innovative concept of permeable pavement in this project incorporates the design of central irrigation channels. The design involves directing runoff from rainfall into the central irrigation channels. The material is constructed using cobblestones, so the designer opted for a trapezoidal cross-section in the design, as shown in Figure 11. The collected rainwater is then diverted to adjacent farmland for irrigation purposes. This approach not only allows for the reuse of rainwater but also helps reduce the peak flow in the side ditches.
2. When the water level in the central irrigation channel exceeds 0.6 m, it will be discharged into the side ditches. Therefore, the calculated volume of rainwater that can be accommodated by the side ditches before being discharged is as follows: $Q_1 = 100 \times (1.3 + 1.63) \times 0.6/2 = 90\ m^3$.
3. In this paper, the innovative construction method is used to introduce rainwater into the central irrigation ditch and then introduce it into the surrounding farmland for reuse. Considering that the rainwater may contain harmful elements, the harmful elements are removed by referring to foreign methods. Among them, Azithromycin (AZM) is a harmful substance often found in water, which hurts both aquatic and terrestrial ecology. Muhammad Wahab [22] used activated carbon (AC) and magnetic activated carbon (MAC) to remove Azithromycin (AZM) from water. It can be seen from the research results that magnetic activated carbon (MAC) has the best effect on removing Azithromycin (AZM), so it can be introduced in the future to remove related harmful elements. Another common harmful element is arsenic (As), which

can cause serious harm to both human beings and the environment. Yongchang Sun [23] uses biochar and modified biochar to remove arsenic (As) from water, and biochar has achieved good results. Modified biochar can improve the removal rate more significantly. However, modified biochar still has production technical problems, so biochar can be applied first in the future, and then applied after modified biochar overcomes the technical problems.

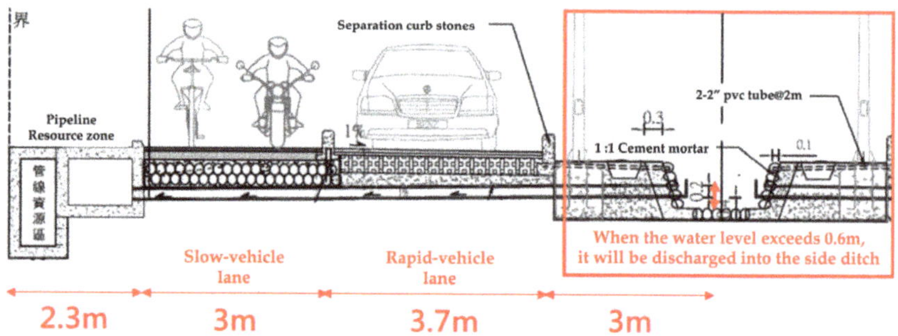

Figure 11. Central Irrigation Channel Design.

3.1.2. Design of Water Retention Capacity in Permeable Pavement Structure

1. Additionally, the permeable pavement itself has the ability to store rainwater within the void spaces between the aggregate particles. This helps to reduce surface runoff and decrease peak flow in the side ditches. Furthermore, the water retention benefits of permeable pavement contribute to mitigating the heat island effect. Among the different pavement sections, the permeable pavement used in slow-traffic lanes serves as the primary water retention area.
2. The structure of the permeable pavement used in slow traffic lanes serves as the primary rainwater storage space, and the section is shown in Figure 6 while the porosity is shown in Table 4.
3. The calculated volume of rainwater that can be accommodated by the road before being discharged into the side ditches (assuming a road width of 3 m and calculating based on a unit length of 100 m) is as follows: $Q_2 = 100 \times 3 \times 0.05 \times 17\% + 100 \times 3 \times 0.15 \times 30\% + 100 \times 3 \times 0.45 \times 10\% = 30\ m^3$.

3.1.3. Calculation of Overall Rainwater Storage Capacity

By utilizing the central irrigation channels and permeable pavement for rainwater storage, the permeable pavement can accommodate higher rainfall intensities. The overall rainwater storage capacity (Q), calculated per 100 m, is the sum of the rainwater storage from the central irrigation channels and the porosity of the permeable pavement: $Q = Q_1 + Q_2 = 120\ m^3$.

3.1.4. Rainfall Intensity Calculation

According to the definition of rainfall intensity by the Central Weather Bureau, heavy rainfall is defined as 100 mm or more within a 3 h period. To calculate the rainfall amount for the heavy rainfall level per 100 m, the following formula can be used:

Calculating the rainwater collection area based on the full width of one side of the road, including the side ditch, permeable pavement in slow traffic lanes, permeable pavement in fast traffic lanes, and central irrigation channels. The full width of one side is 12 m.

Calculating the rainfall amount for the heavy rainfall level per 100 m (Q′) based on the full width of 12 m:

$$Q' = \text{Full width} \times 100\ \text{m} \times \text{Rainfall intensity} = 12 \times 100 \times 0.1 = 120\ m^3.$$

Based on the above, it can be concluded that the rainwater storage capacity of the permeable pavement alone is sufficient to accommodate the rainfall amount at a heavy-rainfall level. Therefore, it is evident that the permeable pavement in this project is indeed capable of effectively handling higher rainfall intensities.

3.2. Flow Analysis with Water Level Gauge

3.2.1. Field Test and Results

This plan selects 7 flow measuring stations: 3 are located on the south side of Dahua North Street (the left bank of the canal), 3 are located on the north side of Dahua North Street (the right bank of the canal), and 1 is located on the central waterway of Dahua North Street.

We use Manning's formula to estimate the theoretical flow rate of each station under different water depth conditions and the maximum water flow rate at full water level. The calculation results are detailed in Table 6. The cross-sectional view of the side ditch is detailed in Figure 12.

Table 6. List of water parameters of the station channel.

Station Number	Canal Elevation (m)				Distance (m)	Slope (m/m)	Channel Status	Manning's *n*-Value
	Up-Stream (A)	Mea-Suring (B)	Down-Stream (C)	Elevation Difference (A-C)				
L1	103.040	102.845	101.507	1.533	264.59	0.0058	Smooth	0.025
L2	102.845	101.507	100.193	2.652	391.89	0.0068	Smooth	0.025
L3	101.507	100.803	100.041	1.466	218.40	0.0067	Smooth	0.025
R1	103.040	102.748	101.842	1.198	255.26	0.0047	Smooth	0.025
R2	102.748	101.842	100.136	2.612	391.79	0.0067	Smooth	0.025
R3	101.842	100.136	99.960	1.882	240.80	0.0078	Smooth	0.025
M1	102.834	102.800	102.267	0.567	206.47	0.0027	Gravel surface	0.030

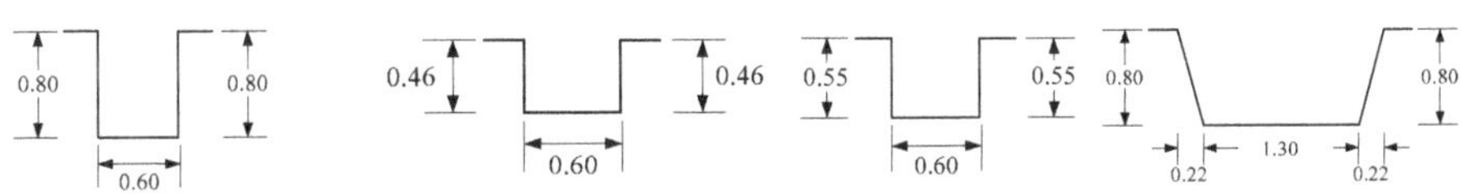

(**A**). L1, L2, R1, R2 Section (**B**). L3 Section (**C**). R3 Section (**D**). M1 Section

Figure 12. Sectional view of each station.

3.2.2. Establishment of Water Depth-Flow Formula

For the estimation of the bathymetric discharge rate formula, the corresponding curves of theoretical bathymetric discharge at 7 places were established by the Manning formula and in situ measurement results. According to the results of meteorological prediction, the days when heavy rain was predicted were selected for on-site measurement. However, the rainfall intensity at that time could not cause the lateral ditch to reach a high water level, so the measured data were used to estimate the discharge at the low water level. The corresponding flow of the high water level and full water level are estimated by the Manning formula, and the data are combined to estimate the water H–Q equation.

It Is hereby estimated that the water depth-discharge rate formula and its applicable restrictions are obtained by combining the theoretical method (Manning formula) with the on-site flow measurement method for each station, as shown in Table 7 and Figure 13.

Table 7. List of water depth-flow and formula at each station.

Station	Water Depth (H)—Flow Rate (Q) Formula	Coefficient of Determination	Suitable Range (m)
L1	$Q = 1.3184H^2 - 0.3768H$	$R^2 = 0.9979$	0~0.80
L2	$Q = 1.0192H^2 - 0.1997H$	$R^2 = 0.9983$	0.20~0.80
L3	$Q = 2.9422H^2 - 0.5290H$	$R^2 = 0.9983$	0.20~0.46
R1	$Q = 1.1593H^2 - 0.3092H$	$R^2 = 0.9974$	0~0.80
R2	$Q = 0.8687H^2 - 0.1516H$	$R^2 = 0.9975$	0.183~0.80
R3	$Q = 3.4755H^2 - 0.4800H$	$R^2 = 0.9994$	0.16~0.55
M1	$Q = 1.2248H^2 + 0.5055H$	$R^2 = 0.9987$	0~0.80

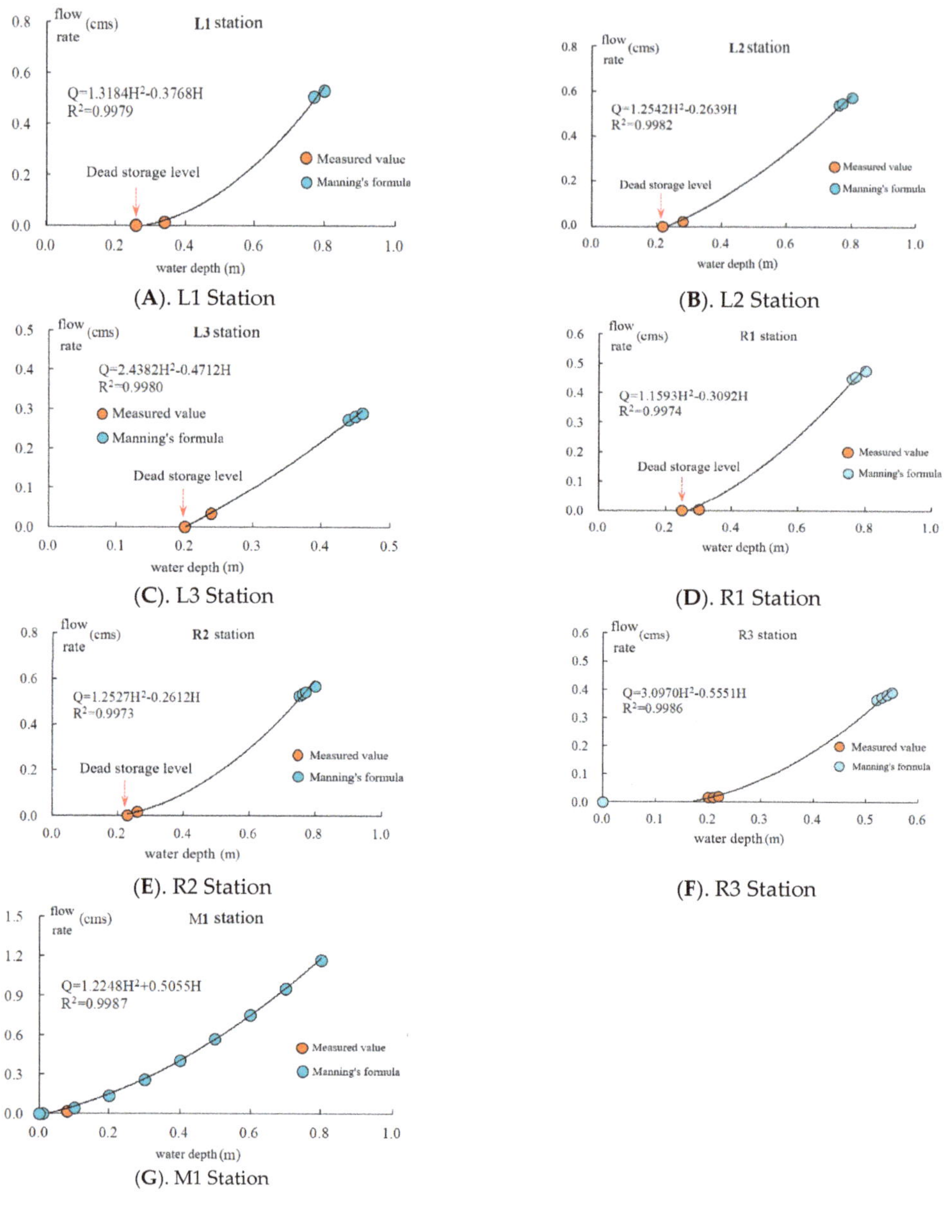

Figure 13. Comparison of estimated discharge and water depth using Manning's formula.

3.3. *Outflow Analysis*

In this study, the flow law formula in Section 5 is used for analysis, and the water level gauge data of the general pavement and the permeable pavement are analyzed.

We use the on-site investigation data to conduct flow analysis, derive the relationship between water level and flow, and import this calculation model into the water circulation environment system so that the system can calculate the flow of each measuring ditch,

The configuration of the water level gauge in this study shows that the upstream and downstream of the left-side groove are R1, R2, and R3, respectively.

The range from R1 to R2 collects the water flow from the permeable pavement, so the flow through R2 minus the flow through R1 is the outflow of the permeable pavement.

The range from R2 to R3 collects the water flow out of the general pavement, so the flow through R3 minus the flow through R2 is the outflow of the general pavement.

The outflow of the permeable pavement and the general pavement can be calculated from the following relationship:

1. Outflow of permeable pavement = R2 flow − R1 flow
2. Outflow of general pavement = R3 flow − R2 flow

In this study, the outflow of the pavement was analyzed on the days with daily rainfall higher than 40 mm/day in 2020, and the outflow of the permeable pavement, the outflow of the general pavement, and the rainfall were calculated by the system on the left- and right-side ditches. The results of one of the days are shown in Figure 14, and the other results are detailed in the attachment.

From the analysis results of each day, it can be seen that the outflow of the left-side ditch or the right-side ditch is greater than the outflow of the permeable pavement, and the permeable pavement can effectively reduce the flood peak.

In addition, the daily flood peak flow of the permeable pavement and the general pavement is sorted out, and the flood peak reduction amount of the permeable pavement is calculated, as shown in Table 8.

It can be seen that the permeable pavement can reduce the flood peak by 60~75% for the general pavement. In terms of the overall effect, the permeable pavement has a better water control effect than the general pavement, which is related to its water permeability and water retention capacity and is effective in low-impact development.

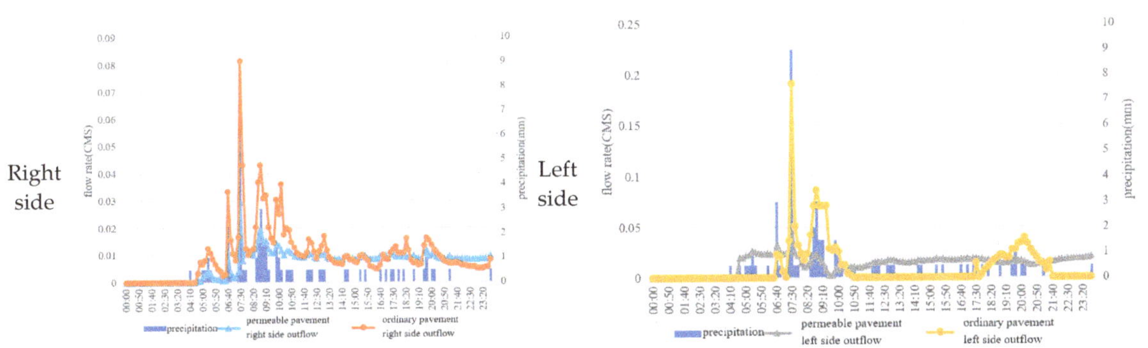

Figure 14. Flow analysis on January 26.

Table 8. Peak reduction amount of permeable pavement.

Date	Peak Flow of Permeable Pavement Right Side	Peak Flow of General Pavement Right Side	Peak Flow of Permeable Pavement Left Side	Peak Flow of General Pavement Left Side	Average Peak Reduction
26 January 2020	0.03	0.08	0.05	0.19	70%
13 March 2020	0.14	0.53	0.07	0.26	73%
18 May 2020	0.07	0.28	0.08	0.23	70%
28 May 2020	0.12	0.29	0.11	0.29	61%
2 July 2020	0.2	0.81	0.2	0.61	71%
3 August 2020	0.33	1.50	0.50	1.62	74%

4. Discussion

1. Referring to past studies on pervious pavements by domestic and foreign scholars, most of them only calculated the benefits of pervious pavement in reducing surface runoff [3–10]. However, previous scholars also mentioned the benefits of low-impact development in the past, and it is necessary to analyze the impact of rainfall on the

overall drainage system [11]. Therefore, this study uses the innovative construction method of the central ditch to combine it with a pervious pavement. By reducing the burden of the whole drainage system and calculating the actual discharge of permeable pavement and general pavement, the low-impact development benefit of permeable pavement can be analyzed more directly.

2. In this study, innovative engineering methods were used to introduce rainwater into central irrigation ditches and then into surrounding farmland for reuse. Considering that rainwater may contain harmful elements, magnetic activated carbon (MAC) and biochar will be used to remove harmful elements in reference to foreign methods [22,23] in the future. Moreover, regular water quality monitoring operations will ensure the quality of rainwater reuse.
3. According to the analysis of the data of this project, the permeable pavement can reduce the flood peak flow, but the scope of the permeable pavement in this project is not large enough, so the effect of delaying the flood peak cannot be achieved. It is suggested that permeable pavements can be used in large areas such as urban renewal or in rezoning in the future, and the effects should be monitored.

5. Conclusions

1. This study used rainy days to estimate the flow of the water level and velocity on site and established the flow law formula with Manning's formula and the regression model of subsequent water level converted flow by two methods, with R^2 values greater than 0.9. Then the actual discharge and peak discharge of the permeable pavement and the general pavement after rainfall can be calculated.
2. In this paper, an additional monitoring system is set up for the analysis of real-time return data, which can not only analyze the flood peak reduction of permeable pavement but also carry out pavement management. Monitoring data with large rainfall in 2020 are screened for the analysis of the discharge of permeable pavement and general pavement. For the general pavement, the flood peak reduction can reach 60~75%, so the permeable pavement can effectively achieve the benefit of low-impact development.
3. The biggest difference between the permeable pavement built in this paper and those studied in the past is that the permeable pavement built in this paper can store rainwater through the innovative construction method of the central irrigation ditch and permeable pavement and can temporarily store approximately 120 m^3 of rainwater based on 100 m as a unit and can accommodate heavy-rainfall intensity by using only the functions of water storage and water retention. The temporary water can be imported into farmland irrigation for reuse.
4. To summarize, by constructing a monitoring system, this paper can effectively calculate the actual discharge of permeable pavement and general pavement and analyze the length of time. In addition, through innovative construction methods, the permeable pavement in this paper can not only reduce the burden of the overall drainage system in extreme climates but also reuse rainwater. This can be used as a model for permeable pavement in the future.

Author Contributions: Validation, L.-L.H.; writing—review and editing, M.-C.H.; Data curation, Y.-S.C.; Analysis, Y.-S.C.; Writing—original draft, Y.-T.L. All authors have read and agreed to the published version of the manuscript.

Funding: This research received no external funding.

Data Availability Statement: The data are not publicly available due to privacy reasons.

Conflicts of Interest: The authors declare no conflict of interest.

References

1. Iqbal, A.; Rahman, M.M.; Beecham, S. Permeable Pavements for Flood Control in Australia: Spatial Analysis of Pavement Design Considering Rainfall and Soil Data. *Sustainability* **2022**, *14*, 4970. [CrossRef]
2. Sambito, M.; Severino, A.; Freni, G.; Neduzha, L. A Systematic Review of the Hydrological, Environmental and Durability Performance of Permeable Pavement Systems. *Sustainability* **2021**, *13*, 4509. [CrossRef]
3. Alharbi, F.; Almoshaogeh, M.; Shafiquzzaman, M.; Haider, H.; Rafiquzzaman, M.; Alragi, A.; ElKholy, S.; Bayoumi, E.A.; EL-Ghoul, Y. Development of Rice Bran Mixed Porous Clay Bricks for Permeable Pavements: A Sustainable LID Technique for Arid Regions. *Sustainability* **2021**, *13*, 1443. [CrossRef]
4. Vaillancourt, C.; Duchesne, S.; Pelletier, G. Hydrologic Performance of Permeable Pavement as an Adaptive Measure in Urban Areas: Case Studies near Montreal, Canada. *J. Hydrol. Eng.* **2019**, *24*, 05019020. [CrossRef]
5. Lin, W.-H. Performance Analysis of Permeable Pavement Monitoring and Testing on Urban Roads-Case Study by Taoyuan Luzhu District Da-hua North Street. Master's Thesis, Department of Civil Engineering, National Central University, Taoyuan City, Taiwan, 2018; pp. 66–139.
6. Ahiablame, L.M.; Engel, B.A.; Chaubey, I. Effectiveness of low impact development practices in two urbanized watersheds: Retrofitting with rain barrel/cistern and porous pavement. *J. Environ. Manag.* **2013**, *119*, 151–161. [CrossRef]
7. Wu, C.-S. Benefit Analysis of the Impact of Permeable Pavement on the Engineering Environment. Master's Thesis, Department of Civil Engineering, National Central University, Taoyuan City, Taiwan, 2005; pp. 43–110.
8. Hossain, M.; Scofield, L.A.; Meier, W.R. Porous Pavement for Control of Highway Runoff in Arizona: Performance to Date. *Transp. Res. Rec.* **1992**, *1354*, 45–54.
9. Shafique, M.; Kim, R.; Kyung-Ho, K. Rainfall Runoff Mitigation by Retrofitted Permeable Pavement in an Urban Area. *Sustainability* **2018**, *10*, 1231. [CrossRef]
10. Cheng, Y.Y.; Lo, S.L.; Ho, C.C.; Lin, J.Y.; Yu, S.L. Field Testing of Porous Pavement Performance on Runoff and Temperature Control in Taipei City. *Water* **2019**, *11*, 2635. [CrossRef]
11. Gao, S. Efficiency Analysis of LID on Reducing Urban Flood. Master's Thesis, Department of Civil Engineering, National Chiao Tung University, Taoyuan City, Taiwan, 2014; pp. 51–115.
12. Wu, Z.-X. The Study of Porous Asphalt's Temperature Behavior. Master's Thesis, Feng Chia University, Taichung, Taiwan, 2007; pp. 4–33.
13. Chinese Society of Pavement Engineering. *Urban Road Permeable Pavement User Manual*; Construction and Planning Agency, R.O.C.: Taipei City, Taiwan, 2015.
14. Public Construction Commission, Executive Yuan. *Chapter 02798 v6.0 Porous Asphalt Concrete Pavement*; Public Construction Commission, Executive Yuan: Taipei City, Taiwan, 2013.
15. Japan Road Association. *Technical Pointer for Drainage Paving*; Japan Road Association: Tokyo, Japan, 1999.
16. Chang, P.-C. Assessment of Water Holding and Differential Temperature Response for Permeable Pavement—In Chungli City, Long Chee Road as an Example. Master's Thesis, Department of Civil Engineering, National Central University, Taoyuan City, Taiwan, 2014; pp. 55–76.
17. Japanese Industrial Standards. *JIS A5001 Road Crushed Stone*; Japanese Industrial Standards: Tokyo, Japan, 1995.
18. Japan Road Association. *Guide Book for Permeable Surface Construction*; Japan Road Association: Tokyo, Japan, 2007.
19. Chen, F.-W.; Lin, H.-T.; Jiang, C.-S.; Tsai, S.-M. *Comparison Study on Mid-Section and Mean-Section Methods for Measuring Flow in Canals and Rivers*; Seminar Paper; Taiwan Agricultural Engineers Society: Taipei City, Taiwan, 2012.
20. Chen, F.-W. *Flow Rate Determination of Jingshan Creek Downstream of Liyutan Reservoir*; AERC-08-RR-14; Taiwan Agricultural Engineers Society: Taipei City, Taiwan, 2008.
21. Chen, F.-W. *Flow Rate Determination in Yuanli Town of Workstation at the Foot of the Mountain*; AERC-08-RR-15; Taiwan Agricultural Engineers Society: Taipei City, Taiwan, 2008.
22. Wahab, M.; Zahoor, M.; Muhammad Salman, S.; Kamran, A.W.; Naz, S.; Burlakovs, J.; Kallistova, A.; Pimenov, N.; Zekker, I. Adsorption-Membrane Hybrid Approach for the Removal of Azithromycin from Water: An Attempt to Minimize Drug Resistance Problem. *Water* **2021**, *13*, 1969. [CrossRef]
23. Sun, Y.; Yu, F.; Han, C.; Houda, C.; Hao, M.; Wang, Q. Research Progress on Adsorption of Arsenic from Water by Modified Biochar and Its Mechanism: A Review. *Water* **2022**, *14*, 1691. [CrossRef]

water

MDPI

Article

Impact of Storm Characteristics on Infiltration Dynamics in Sponge Cities Using SWMM

Yuanyuan Yang *, Zijian Shao, Xiaoyan Xu and Dengfeng Liu

State Key Laboratory of Eco-Hydraulics in Northwest Arid Region of China, Xi'an University of Technology, Xi'an 710048, China; 2220421325@stu.xaut.edu.cn (Z.S.); liudf@xaut.edu.cn (D.L.)
* Correspondence: yuanyuanyang@xaut.edu.cn

Abstract: Effective stormwater management in urban areas requires enhancing the permeability of underlying surfaces. However, the impact of storm characteristics on infiltration processes in sponge cities remains insufficiently explored. This study uses the Horton method within the storm water management model to investigate how uniform and Chicago storm parameters affect infiltration rates. Our findings provide valuable insights: (1) Increasing porous pavement area proportionally reduces subarea sizes within subcatchments, and infiltration rates of porous pavements are supply-controlled. (2) Uniform storms result in consistent initial infiltration rates across pervious areas, subcatchments, and the entire catchment. The duration of this stable state decreases with higher return periods. Catchment infiltration volumes exhibit linear growth with greater storm intensities (R-squared = 0.999). (3) Peak infiltration rates and moments for pervious areas, subcatchments, and the overall catchment exhibit correlations with both the return period and the time-to-peak coefficient, with correlation coefficients ranging from −0.9914 to 0.9986 and p-values ranging from 0.0334 to 0.6923. This study quantifies the influence of design storm parameters on infiltration, providing valuable insights for stormwater infrastructure design and urban stormwater control.

Keywords: Chicago storm; Horton; porous pavement; return period; time-to-peak coefficient

1. Introduction

Infiltration, the process of water movement from the surface into the soil and subsurface driven by gravity and soil capillarity, plays a vital role in the redistribution of water resources and significantly impacts various hydrologic processes (e.g., runoff generation [1], groundwater recharge [2]) in urban catchments. Accurate simulation of infiltration is a subject of interest in hydrological modeling, particularly in the context of rainfall–runoff models. Understanding infiltration dynamics and mechanisms in sponge cities, where low impact development facilities (LIDs) are employed, holds substantial potential to enhance urban stormwater modeling and management [3,4]. Sponge cities typically employ infiltration-based and retention-based strategies. Therefore, gaining insight into infiltration characteristics is pivotal for comprehending their hydrological responses and achieving effective stormwater control.

Directly measuring infiltration at a large-scale field is time-consuming, costly, and subject to significant spatial and temporal variability. Consequently, numerous theoretical and empirical infiltration models have been developed for indirect estimation [5,6]. Infiltration models can be categorized into two types [7]: physically-based equations such as Horton [8–10], Green–Ampt [11], Soil Conservation Service [12], Swartzendruber [13], Kostiakov, Kostiakov–Lewis, and Philip; and empirical and data-driven methods including artificial neural networks [14], support vector machines [15], random-forest models [16], and Gene Expression Programming [17].

Theoretically, the process of soil infiltration is governed by the Richards equation. The equation is a highly nonlinear partial differential equation and challenging to solve. So, the

Citation: Yang, Y.; Shao, Z.; Xu, X.; Liu, D. Impact of Storm Characteristics on Infiltration Dynamics in Sponge Cities Using SWMM. *Water* **2023**, *15*, 3367. https://doi.org/10.3390/w15193367

Academic Editor: Enedir Ghisi

Received: 6 August 2023
Revised: 18 September 2023
Accepted: 22 September 2023
Published: 25 September 2023

Storm Water Management Model (SWMM) [18] is used in this study, which employs various basic algebraic infiltration models that represent the general dependency of infiltration capacity on soil properties and the volume of water previously infiltrated during a storm event. There is no consensus over the optimal algebraic infiltration model; that is, the physically-based infiltration models show varying levels of effectiveness and applicability. For example, the Horton and Green–Ampt methods underperform the modified Philip's model [19]. Therefore, SWMM allows the user to select from five of the most popular models: the Horton method, the modified Horton method, the Green–Ampt method, the modified Green–Ampt method, and the Curve Number method.

The Horton method in SWMM is chosen in this paper to synthetically produce infiltration data on urban permeable surfaces for three reasons. First, the Horton method, as the default infiltration model in SWMM, is widely used and offers reliable predictability for estimating rainwater infiltration into the upper soil zone [20]. For example, the Horton model outperforms Kostiakov and Philip models in built-up surfaces [21] and in semiarid regions; the Horton model outperforms the Curve Number method for grass soils [22]. Second, the Horton model often fits experiment data well [23] and has a few parameters that can be obtained with easy monitoring [24]; in contrast, the fitting accuracy of other models requires advanced field investigations [25]; for example, the performance of Green–Ampt model is considerably affected by the monitoring area and hydraulic gradients [26]. Third, our study site is located in a semiarid region [27] where storms predominantly result in infiltration-excess (or Hortonian) overland flow rather than saturation overland flow [28].

Infiltration capacity and rate on urban permeable surfaces are influenced by soil conditions and properties, such as moisture content [29–31] and structure [32]. Additionally, storm characteristics play a significant role [33]. The Horton method is susceptible to rainfall intensity [34] and temporal distribution [35,36] in semiarid regions. Our previous studies have shown that the performance of LIDs generally declines with less frequent and more intense storms [37], and the time-to-peak coefficient of rainfalls impacts runoffs in sponge cities [38].

This study's core focus and novelty reside in investigating how storm parameters influence infiltration rates. We employ both the Horton and Green–Ampt methods within SWMM [39]. Notably, the Horton method is applied for permeable surfaces, while the Green–Ampt method is utilized for modeling LIDs (i.e., porous pavements) in sponge cities. Our findings unveil the profound influence of storm characteristics on infiltration processes. These results underscore the potential benefits of augmenting porous pavements and gaining comprehensive insights into infiltration behavior under various storm scenarios, ultimately enhancing urban stormwater management practices.

2. Study Area and Data

Our research focuses on the WR8 site (8.5×10^5 m^2, Figure 1), an urban drainage basin in the experimental sponge city of Fengxi, China, designated as a UNESCO Ecohydrology demonstration site [40]. The climate of WR8 falls under the warm temperate semiarid continental monsoon classification, characterized by pronounced seasonal variations in temperature and humidity. Over a year, the region receives a total of 1983.4 sunshine hours, with an average annual temperature of 13.6 °C. Notably, July exhibits the highest temperatures, averaging 26.8 °C, while January is the coldest month, with an average temperature of −0.5 °C. Precipitation displays substantial interannual fluctuations, with values notably surpassing evaporation. This study area experiences an average annual precipitation of 552.0 mm (averaged from 1981 to 2016, excluding 1986 and 2011), with a notable concentration of 50–60% falling between July and September [41]. Additionally, the average wind speed registers at 1.5 m/s.

WR8 features a prominent loess layer spanning elevations from 380.5 to 384.3 m above sea level. The soil composition predominantly comprises loamy clay, characterized by a compact structure with a yellowish-brown appearance, sparsely inhabited by plant roots, and punctuated by needle-shaped holes and insect burrows. The groundwater table depth typically ranges from 10 to 20 m. The land use in the WR8 site encompasses diverse categories, encompassing parks and green spaces, residential lands, transportation lands, educational lands, industrial lands, and undeveloped areas. Stormwater finds its way to the Fenghe River via a designated outfall.

The drainage system in WR8 was mainly designed to accommodate storms with 1- or 2-year return intervals before 2014, resulting in frequent waterlogging events due to inadequate drainage capacity. Since then, the region has implemented LID-based stormwater management technology to mitigate storm-related problems. Numerous porous pavements (PP or permeable pavements, Figure 2) [42,43] have been implemented, covering 134,522 m^2, accounting for 15.8% of the total catchment area. PP has a stratified system including surface, pavement, storage, and underdrain components. Stormwater permeates each layer vertically. If the drainage rate exceeds the capacity of the underdrain, the water level will rise until it reaches the ground's surface, resulting in runoff. The water in PP can leave the bottom via percolation and evapotranspiration and be routed to a sewer junction or pervious area via the drain.

Crucial data, including precipitation data, land use, elevation information, details about storm-related facilities, and surface and pipe flow data, were provided by the Fengxi New City Management Committee [44]. For analytical purposes, the WR8 site was divided into nine subcatchments, 21 nodes, and one outfall in SWMM.

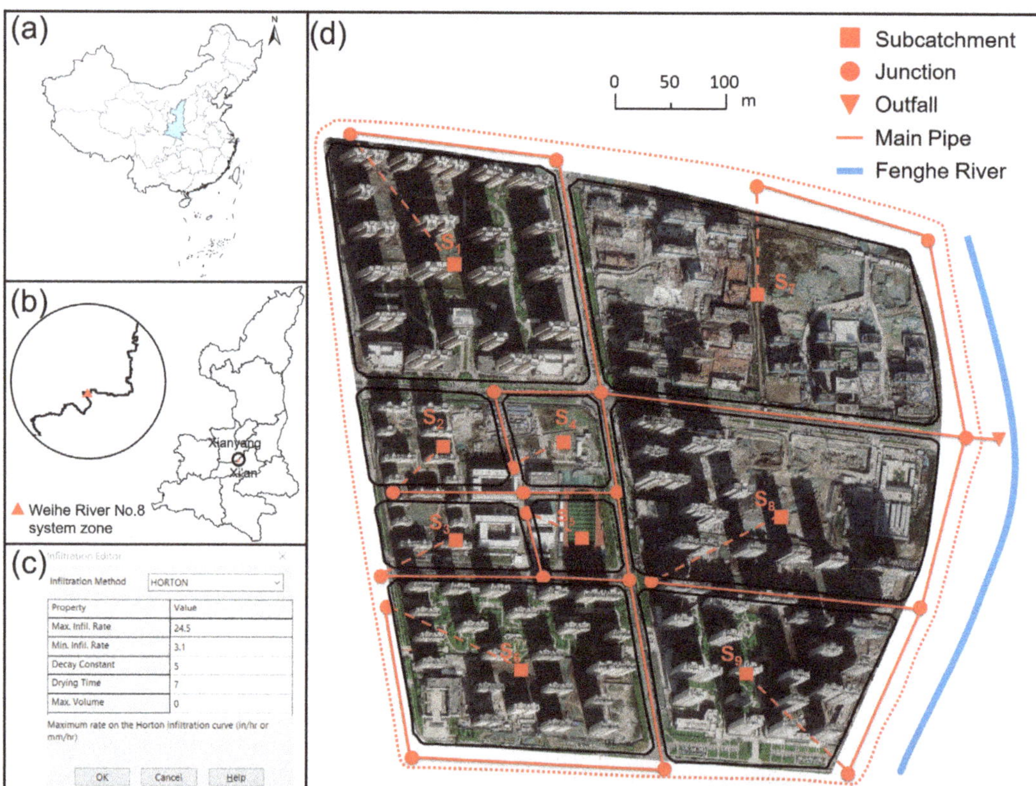

Figure 1. Study area (adapted with permission from Yang et al., 2023 [45]. 2023, Elsevier). (**a**) Shannxi Province, China; (**b**) Weihe River No. 8 system zone (WR8), Fengxi New City; (**c**) Infiltration editor in storm water management model (SWMM); (**d**) Aerial photograph of WR8 overlapped with SWMM generations.

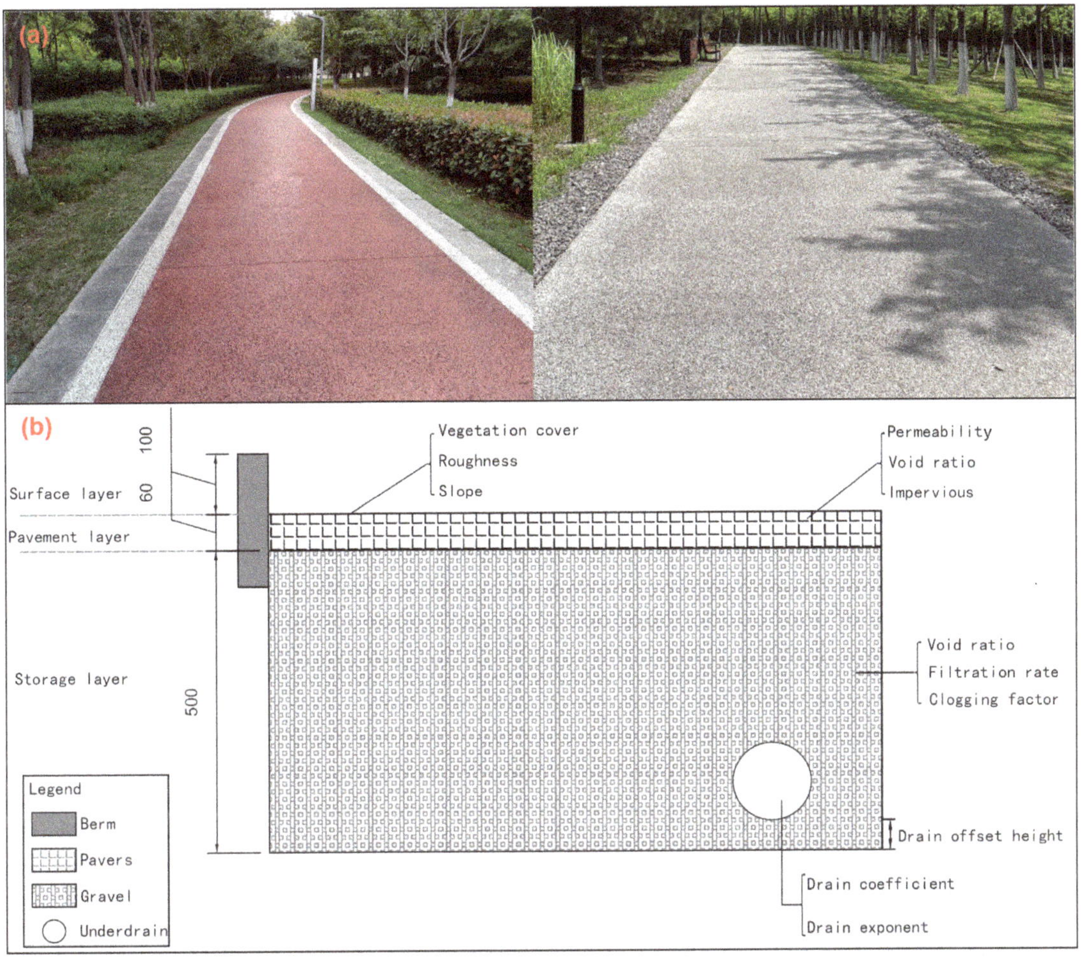

Figure 2. Porous pavements in study area. (**a**) Photograph. (**b**) Profile.

3. Methods

To analyze the influence of storm parameters on infiltration dynamics, we have established a framework utilizing the SWMM engine in Visual Studio 2022 for conducting stormwater simulations (Figure 3). In this framework, MATLAB is employed for storm design. The framework comprises four main components: (1) Designing uniform and Chicago storms with various parameter values. (2) Executing SWMM simulations to compute the time series of infiltration rates in each subcatchment and their corresponding subareas. (3) Calculating infiltration statistics, including peak rate, peak time, and volume. (4) Assessing the impact on the infiltration statistics.

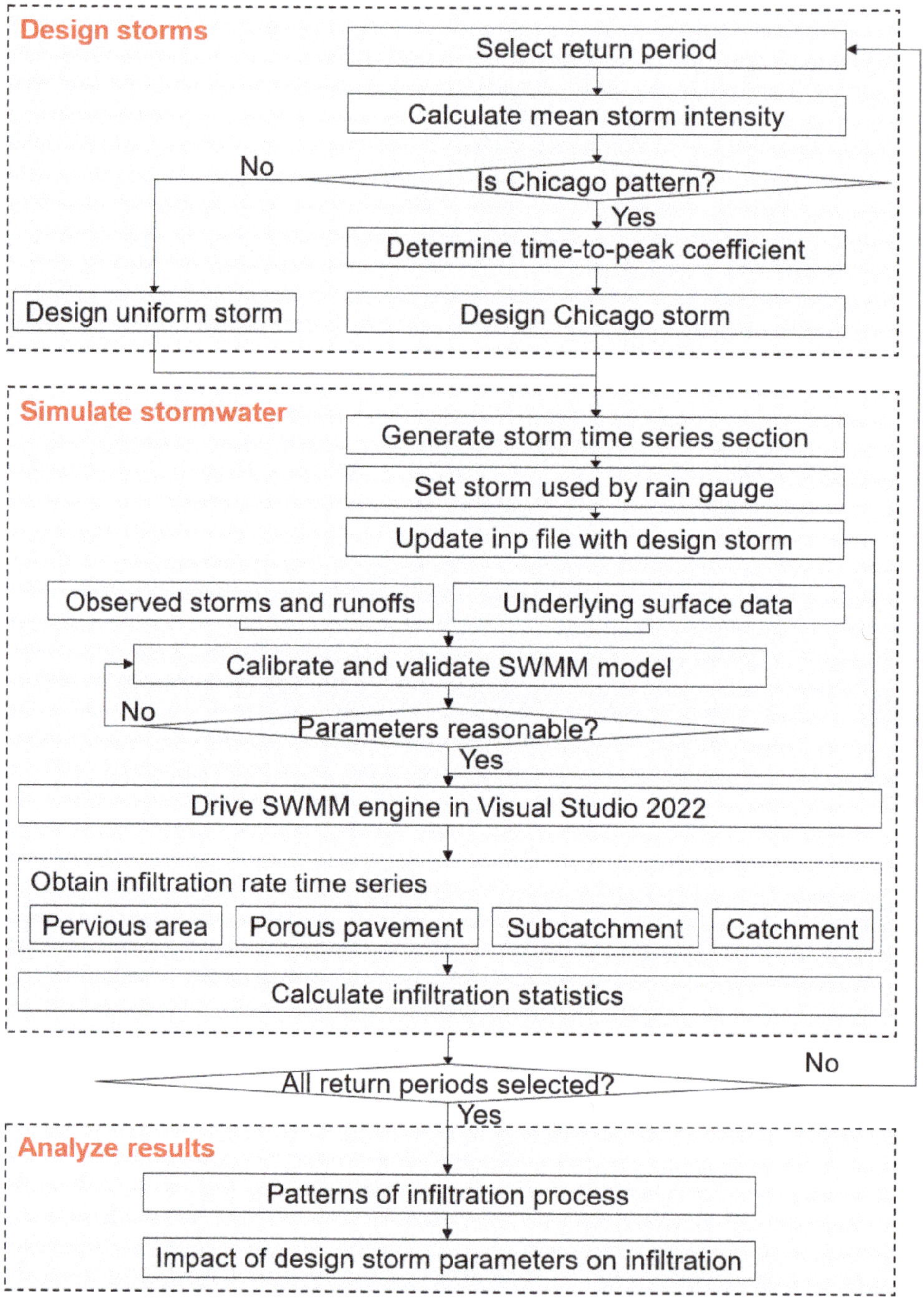

Figure 3. Analytical framework.

3.1. Design Storm

The uniform storms were designed according to the duration–intensity–frequency approach:

$$q = \frac{a(1 + c\lg T)}{(d + b)^n} \tag{1}$$

where q is the average intensity, mm/min; a is the storm coefficient; c is the coefficient of variation; b is the duration correction factor; n is the attenuation index; T is the return period, years; d is the duration, min.

The Chicago storms were designed as follows:

$$i(t) = \begin{cases} \dfrac{\frac{(1-n)(t_p-t)}{r}+b}{\left(\frac{t_p-t}{r}\right)^{n+1}} a(1 + c\lg T), \text{ if } t \leq t_p \\ \dfrac{\frac{(1-n)(t-t_p)}{1-r}+b}{\left(\frac{t-t_p}{1-r}\right)^{n+1}} a(1 + c\lg T), \text{ if } t > t_p \end{cases} \tag{2}$$

where $i(t)$ is the average intensity at the t-th time-step, mm/min; r is the time-to-peak coefficient, which is the ratio of the peak time (t_p, min) to the duration (d).

For this study, the values chosen are T = 1, 2, 5; r = 0.2, 0.4, or 0.6; d = 120; a = 16.715; c = 1.1658; b = 16.813; n = 0.9302 [37]. Here are the reasons for each parameter setting: (1) T = 1, 2, 5: Sponge cities typically design their LIDs to handle storms with short return periods (usually less than 5 years), as more intense storms exceeding this threshold can lead to overspending. (2) r = 0.2, 0.4, 0.6: These values are based on local rainfall observations, where time-to-peak coefficients typically fall within the ranges of 0.1~0.2, 0.3~0.4, and 0.5~0.6. (3) d = 120 min: This duration aligns with urban drainage system standards, which often focus on short-duration storms. Although a 180-min duration could be considered, we chose 120 min to emphasize infiltration characteristics. (4) a = 16.715, c = 1.1658, b = 16.813, n = 0.9302: These parameters are provided by the local weather bureau based on extensive, long-term rainfall observations.

Storm intensity is the primary factor influencing available water for infiltration, namely ponded surface water depth. Meanwhile, the temporal distribution of the Chicago storm is determined by Equation (2). Therefore, our analysis centers on the return period and time-to-peak coefficient.

3.2. Storm Water Management Model

SWMM stands out among urban stormwater models, rendering it the primary choice for this study. It offers an extensive suite of capabilities. These encompass infiltration simulation, surface runoff modeling, hydrological response assessment of LIDs, drainage network flow calculations, pollutant tracking, treatment evaluation, and overflow prediction [46]. SWMM is versatile, accommodating both single-event and long-term simulations, and excels in accurately modeling water dynamics within stormwater management [47]. Furthermore, its open-source nature allows for code redevelopment. Within SWMM, diverse methods are integrated to facilitate infiltration simulation, including the default Horton formula and the Green–Ampt method [48].

Table 1 provides an overview of the critical parameter values specifically adopted for porous pavements in SWMM. These values were derived from experiments and on-site observations and were provided by the Fengxi Management Committee [49]. These parameters are paramount in effectuating precise modeling and simulation of the infiltration and runoff processes within the catchment [50].

Table 1. Main parameter values of porous pavement in storm water management model.

Layer	Parameter	Value	Layer	Parameter	Value
Surface	Berm height (mm)	100	**Storage**	Thickness (mm)	500
	Vegetation volume fraction	0		Void ratio (voids/solids)	0.75
	Surface roughness	0.01		Seepage rate (mm/h)	1000
	Surface slope (percent)	0.5		Clogging factor	0
Pavement	Thickness (mm)	60	**Drain**	Flow coefficient	0.5
	Void ratio (voids/solids)	0.15		Flow exponent	0.5
	Impervious surface fraction	0		Offset (mm)	0
	Permeability (mm/h)	1000		Open level (mm)	6
	Clogging factor	0		Closed level (mm)	0
	Regeneration interval (days)	0		Control curve	0
	Regeneration fraction	0			

In pursuit of the most theoretically accurate outcomes, the dynamic wave model was deliberately chosen from among the routing models in SWMM. This model achieves heightened precision by solving the one-dimensional Saint Venant equations and adeptly replicates backwater flow effects by incorporating pipe storage, water return, import and export losses, and due consideration of countercurrent and pressure flow [51].

In each subcatchment, we derived most parameters through measurements or estimations based on underlying surface data and field investigations. These parameters encompass subcatchment area, imperviousness, slope, roughness, and facility sizes. Calibration of other SWMM parameters followed two criteria: first, minimizing errors in simulated outflow time series using Nash–Sutcliffe efficiency (NSE) [52], and second, minimizing errors in simulated peak flow rate using relative error. The parameters subjected to calibration [53] included subcatchment width, infiltration parameters, depression storage, and the percentage of runoff routed from impervious to pervious areas. It is important to note that the values of these parameters were constrained within limits recommended in the SWMM manual [39] and corroborated by relevant literature.

3.3. Horton Infiltration Method

3.3.1. Governing Equations

The Horton formula has held a pivotal position within SWMM since its first release. Its classical form utilizes an exponential equation to calculate the reduction in infiltration capacity over time during rainfall events [54]:

$$f_p = f_\infty + (f_0 - f_\infty)e^{-k_d t} \quad (3)$$

where t is the elapsed time (from the storm onset), h; f_p is the infiltration capacity into the soil, mm/h; f_∞ is the minimum (or equilibrium) value of f_p at infinite time, mm/min; f_0 is the maximum (or initial) value of f_p at the start of the storm, mm/h; k_d is the decay coefficient, a constant reflecting how fast the infiltration rate decreases over time, 1/h. Soil conditions primarily influence the values of these parameters. Consequently, the actual infiltration rate (f) is determined as the lesser value between the infiltration capacity and actual storm intensity:

$$f(t) = \min\left[f_p(t), i(t)\right] \quad (4)$$

SWMM uses the integrated form to determine the cumulative infiltration capacity:

$$F(t_p) = \int_0^{t_p} f_p \mathrm{d}t = f_\infty t_p + \frac{(f_0 - f_\infty)}{k_d}(1 - e^{-k_d t_p}) \quad (5)$$

The actual cumulative infiltration (F) is calculated as follows:

$$F = f_\infty t_p + \frac{(f_0 - f_\infty)}{k_d}(1 - e^{-k_d t_p}) \quad (6)$$

Estimating the values of f_0, f_∞, and k_d for each subcatchment requires considering the physical properties of the soil and fitting the equation to multiple field or laboratory datasets from different sites. The value of f_0 is influenced by soil type, initial soil moisture content, and vegetation conditions, while f_∞, the most sensitive parameter in the Horton method, corresponds to saturated hydraulic conductivity. The k_d value depends on the soil's initial moisture content. Additionally, the recovery rate is not considered here due to the use of design storms with a duration of 120 min for the SWMM simulation.

3.3.2. Computational Scheme in Storm Water Management Model

The SWMM engine employs a computational scheme to calculate infiltration for the Horton method, as depicted in Figure 3 of Parnas et al. (2021) [55]. The process for determining the infiltration rate (f) in a subcatchment during a time step (Δt) under a storm is outlined as follows:

(1) Input the necessary variables, including rainfall rate ($i(t)$), ponded surface water depth (d), equivalent time (t_p) on the Horton curve, and constants f_0, f_∞, and k_d.
(2) Calculate the available storm rate (i_a).
(3) If i_a equals 0, update the current time (t_p) on the infiltration curve and set f to 0. Otherwise, compute the cumulative infiltration volume using Equations (5) and (6) at times t_p and $t_p + \Delta t$.
(4) Calculate the average infiltration rate over the time step.
(5) Update t_p and update f using Equation (4).

Subsequently, the following steps are performed for the catchment infiltration calculation for each time step within the SWMM engine:

(1) Determine if the area is pervious. If it is, apply the Horton formula to calculate the infiltration rate and volume for the time step. If it is not pervious, set the infiltration rate and volume to 0.
(2) Check for the presence of LIDs. If one exists, use the Green–Ampt model (allowing the consideration of surface ponding) to calculate the infiltration rate and volume for each LID facility. The infiltration volume of the subcatchment is obtained by summing the infiltration volumes of the pervious area and each LID facility.
(3) Compute the infiltration volume for the entire study area by summing up the infiltration volumes of each subcatchment.
(4) Determine the infiltration rate for the study area by dividing the infiltration volume of the study area by the area and time steps.

3.4. Field Investigation

The performance of the Horton model exhibits site-dependent behavior, closely linked to the soil textures prevalent in the monitoring sites. Soil infiltration monitoring was conducted at three distinct sites within WR8 in 2017, utilizing a portable double-ring infiltrometer to generate site infiltration curves [50]. The selected sites for monitoring included a lawn near Qinhuang Avenue, a wooded area near Xingxian Road, and a barren area near Tongxin Road, each representing distinct soil textures. Among the five infiltration models available in SWMM, the Horton model demonstrated superior fitting performance, as evidenced by its favorable performance across various evaluation metrics. This outcome underscores the suitability of the Horton model for characterizing infiltration dynamics at the WR8 site.

4. Results and Discussion

4.1. Area Changes after Adding Porous Pavements

Table 2 provides an overview of the areas allocated for schemes without porous pavements (no-PP scheme) and schemes with porous pavements (PP scheme). Our observations revealed that for each subcatchment, an increase of n percent in the PP area resulted in proportional decreases of np_1 percent, np_2 percent, and np_3 percent in the impervious area without depression storage, impervious area with depression storage, and pervious area, respectively. p_1, p_2, and p_3 denote the percentages of the three underlying surfaces in the no-PP scheme.

Table 2. Areas (m^2) of subareas in subcatchments for no porous pavements (no-PP) and porous pavements (PP) schemes.

Subcatchment	Area	No-PP Scheme			PP Scheme			
		IA-NO [1]	IA	Pervious Area	IA-NO	IA	Pervious Area	Porous Pavements (% [2])
s1	162,329	30,843	92,528	38,958	26,117	78,352	32,990	24,870 (15.3%)
s2	40,384	7673	23,019	9692	6373	19,120	8050	6841 (16.9%)
s3	21,504	4086	12,257	5161	3384	10,153	4275	3692 (17.2%)
s4	27,903	5302	15,905	6696	4861	14,583	6140	2319 (8.3%)
s5	14,799	2812	8435	3552	2233	6700	2821	3045 (20.6%)
s6	118,273	22,472	67,416	28,385	17,157	51,472	21,672	27,972 (23.7%)
s7	202,587	38,492	115,475	48,620	32,579	97,738	41,153	31,117 (15.4%)
s8	153,206	29,109	87,327	36,770	25,750	77,250	32,526	17,680 (11.5%)
s9	104,350	19,827	59,480	25,043	16,599	49,796	20,967	16,988 (16.3%)

Notes: [1] IA-NO represents the impervious area with no depression storage; IA represents the impervious area with depression storage. [2] Percentage share, namely, the area ratio of porous pavements to the subcatchment.

4.2. Calibration Results of Storm Water Management

The SWMM model underwent calibration using data from three recorded storm events [38]. Table 3 lists the values of critical parameters for different subcatchments in SWMM. For the outflow series, the NSE values were 0.63, 0.84, and 0.76, while the relative errors for peak flow rates were 0.0038, 0.1552, and 0.0153 m^3/s, respectively. These results affirm the effectiveness of SWMM in accurately representing the hydrological processes within the study area. For additional information concerning the calibration and validation of SWMM, please refer to Section 3.3 of the study [38].

Table 3. Key parameters for different subcatchments in storm water management model.

Parameter	Value
Width (m)	121.7~450.1
Slope (%)	0.5
Imperviousness (%)	0.76
Manning's n for overland flow in impervious area	0.013
Manning's n for overland flow in pervious area	0.15
Depression storage in impervious areas (mm)	1
Depression storage in pervious areas (mm)	3.2
Conduit roughness	0.013
Conduit diameter (m)	0.6~2.2
Conduit length (m)	105.0~641.0
Junction elevation (m)	380.6~384.3
Outfall elevation (m)	380.5

The measured minimum infiltration capabilities at the three monitoring sites were 38.5, 94.4, and 118.6 mm/h at 10 degrees Celsius and 45.1, 110.4, and 138.8 mm/h at 20 degrees Celsius, respectively. These site-specific data were utilized for calibrating the SWMM model in conjunction with other observed storm-related data, such as storm and outflow time series. Consequently, the catchment's initial infiltration capacity (f_0), minimum infiltration capacity (f_∞), and decay constant (k_d) were estimated at 24.5 mm/h, 3.1 mm/h, and 5 h^{-1}, respectively.

The soils in WR8 predominantly consist of loamy clays, as mentioned in Section 2. Following the SWMM manual [40], clay and loam soil exhibit initial capacities of 25.4 and 76.2 mm/h (1 in/h and 3 in/h), respectively. The derived values for the Horton model align reasonably with those specified in the SWMM manual, thus providing further validation.

While this approach provides valuable information on local infiltration characteristics, it may capture a fraction of the spatial heterogeneity within the catchment under actual conditions. Therefore, future research should involve extensive investigations and monitoring at various locations, considering the diverse soil textures. Moreover, for improved calibration of the Horton model in SWMM, minimizing the bias in simulating runoff responses at point, subcatchment, and catchment scales using measured storm–runoff data at multiple sites requires further research to reduce uncertainty [36]. By adopting such a comprehensive approach, we can better elucidate the intricate dynamics of infiltration and enhance stormwater management strategies.

4.3. Uniform Storm Parameters Impact on Infiltration

4.3.1. Catchment Scale

Figure 4 presents the catchment's infiltration capacities and intensities under uniform storms with 120 min and 1-, 2-, or 5-year return periods for porous pavements scheme. The three uniform storms (yellow bars) featured total depths of 20.7, 27.9, and 37.5 mm, respectively, accompanied by corresponding intensities of 0.1722, 0.2327, and 0.3126 mm/min. Notably, the depth and intensity of the 2-year (or 5-year) uniform storm were approximately 1.35 times (or 1.81 times) those of the 1-year uniform storm. Importantly, all three storm intensities remained below the maximum infiltration capacity of 0.4083 mm/min (equivalent to 24.5 mm/h).

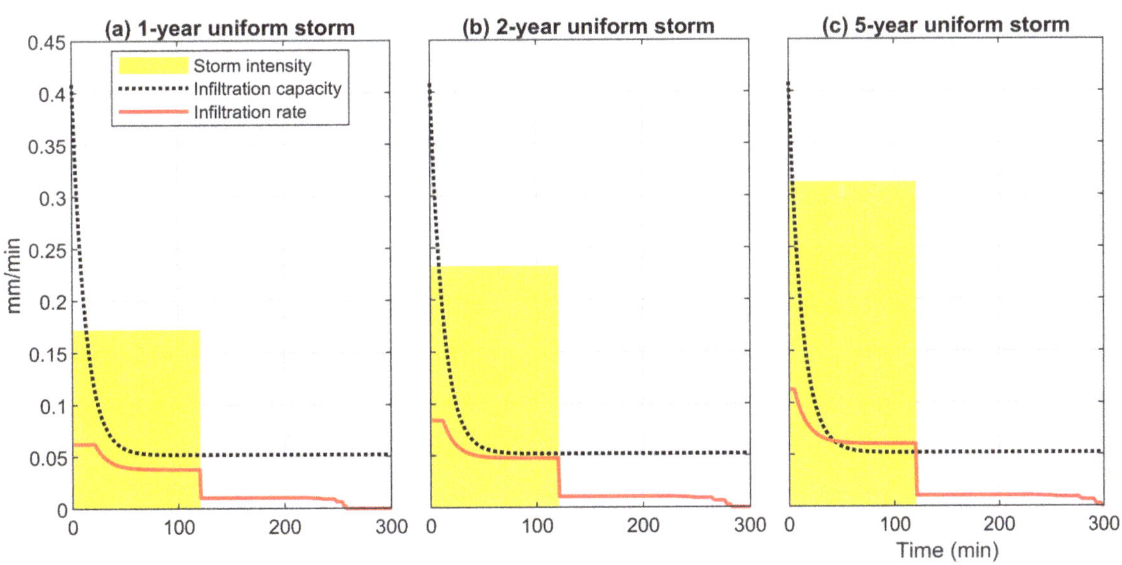

Figure 4. Catchment infiltration capacity and rate under uniform storms with 120 min duration and 1-, 2-, or 5-year return period for porous pavements scheme.

The catchment's infiltration capacities during the three uniform storms (black dotted lines in Figure 4) are theoretical values calculated using the Horton method, assuming sufficient water for infiltration. According to the Horton infiltration theory, when water availability is limited, the actual infiltrability may be less than the infiltration capacity at a given time and for a specific soil. In other words, the infiltration process is either supply-controlled or profile-controlled. Remarkably, under each uniform storm, the process initially follows a supply-controlled pattern, transitioning to a profile-controlled state before returning to a supply-controlled mode.

The catchment's infiltration rates (red solid lines in Figure 4) offer the following insights: At the onset of each storm event, the infiltration rates remained constant (0.0622, 0.0840, and 0.1128 mm/min). However, the duration of this steady state was shorter under more intense storms; knee points were observed on the infiltration rate curves at 21, 11, and 4 min for the respective storms, indicating that the increased storm intensity led to a faster filling of soil pores during the initial stages of the infiltration process.

The catchment infiltration volume, determined by applying the definite integral method to the infiltration rate time series, is illustrated in Figure 5. A linear correlation emerged between the infiltration volume and the uniform storm intensity. Furthermore, a precise linear equation was derived to represent this relationship accurately. The observed pattern can be attributed to higher storm intensities resulting in larger infiltration rate time series, leading to greater infiltration volumes, as represented by the enclosed area under the infiltration rate curve. This finding aligns with the observation that cumulative infiltration exhibited significant variations [3].

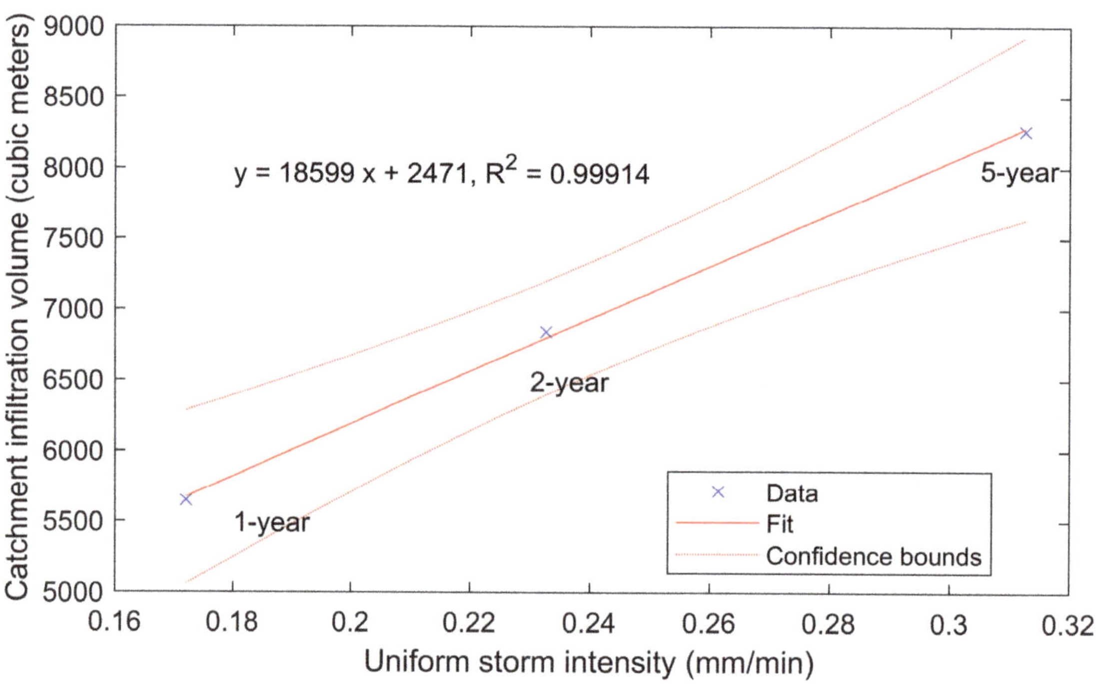

Figure 5. Catchment infiltration volumes under uniform storms with 120 min duration and 1-, 2-, or 5-year return period for porous pavements scheme.

4.3.2. Subcatchment Scale

The infiltration rates within subcatchments, including pervious areas and porous pavements, were analyzed under uniform storm conditions. Interestingly, the infiltration rates of pervious areas remained consistent within each subcatchment, regardless of whether porous pavements were present. Additionally, similar patterns in infiltration rates were observed across all subcatchments. To illustrate this, we present an example using subcatchment s7, depicting the infiltration rates of its subareas under uniform storms with return periods of 1, 2, or 5 years for the porous pavements scheme, as shown in Figure 6.

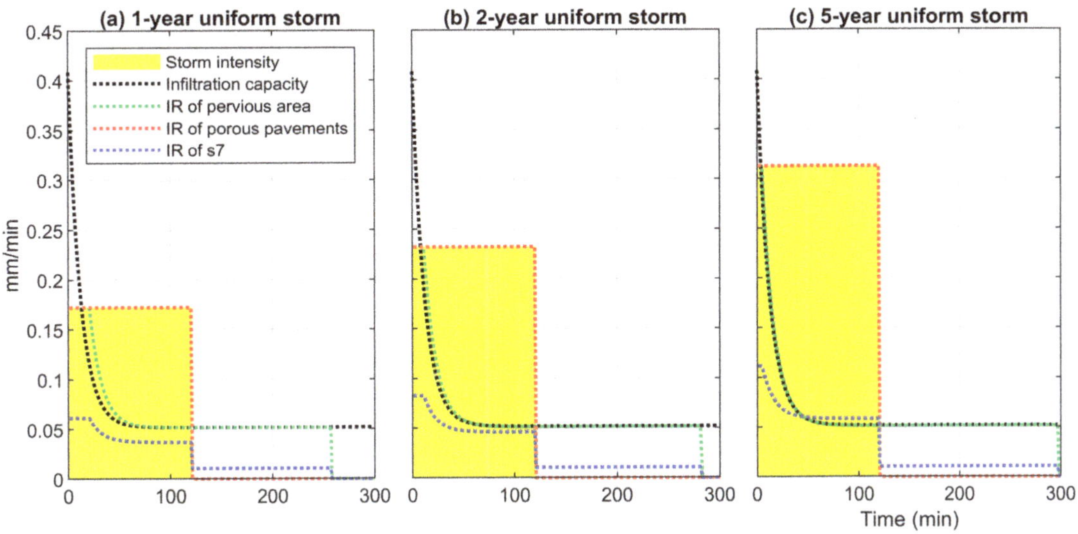

Figure 6. Infiltration rates (IRs) of pervious area, porous pavements, and subcatchment s7 under uniform storms with 120 min duration and 1-, 2-, or 5-year return period for porous pavements scheme.

The analysis of infiltration rates of pervious areas in subcatchment s7 under uniform storms (green dotted lines in Figure 6) revealed the following: (1) Initially, the infiltration rate remained constant, then gradually decreased, reached a state of equilibrium (equivalent to the minimum infiltration capacity), and eventually diminished to zero due to the absence of available rainwater for infiltration. (2) As the return period increased, the initial values of the infiltration rate rose and aligned with the corresponding storm intensities (0.1722, 0.2327, and 0.3126 mm/min). This behavior can be attributed to the Horton method, where the storm intensities were less than the maximum infiltration capacity (0.4083 mm/min, equivalent to 24.5 mm/h), resulting in infiltration rates equal to the storm intensities. (3) Higher storm intensities led to faster filling of soil pores, resulting in shorter durations (20, 11, or 4 min) of constant infiltration rates. (4) The actual infiltration rates may exceed the infiltration capacities on the Horton curve, as indicated by the green line surpassing the black line, as seen in Figure 6a, due to the initially inadequate amount of water available for infiltration.

Turning to the infiltration rates of porous pavements in subcatchment s7 under uniform storms (red dotted lines in Figure 6), it was evident that these rates remained constant throughout the storm and were equal to the storm intensities (0.1722, 0.2327, or 0.3126 mm/min). Subsequently, infiltration rates promptly dropped to zero upon the storm's cessation. This outcome can be attributed to porous pavements controlling the stormwater that falls on their surfaces and having sufficient infiltrability to filtrate the rainfall fully, thus aligning the infiltration rates with the storm intensities.

Moreover, the infiltration rates of subcatchment s7 (blue dotted lines in Figure 6) were examined, demonstrating similar patterns across different return periods, with higher return periods resulting in increased infiltration rates. Notably, the infiltration rates of s7 were significantly influenced by the infiltration rates of porous pavements compared with those of the pervious area. This finding underscores the impact of porous pavements on overall infiltration dynamics within the subcatchment.

Focusing on the infiltration rates of the catchment (red solid lines in Figure 4), pervious area (green dotted lines in Figure 6), and s7 (blue dotted lines in Figure 6), we observed that the rate of descent in the infiltration rate curve increased with the higher storm return period. This pattern is consistent with the findings of Mu et al. [27], who reported that the infiltration rate curve became steeper with increasing rainfall intensity.

4.4. Chicago Storm Parameter Impact on Infiltration

4.4.1. Catchment Scale

Nine Chicago storms, each lasting 120 min and with return periods of 1, 2, or 5 years, and time-to-peak coefficients of 0.2, 0.4, or 0.6, were utilized to calculate the infiltration rates in SWMM. The catchment infiltrations are presented in Figure 7, revealing the following observations:

(1) The infiltration rates (solid lines) peaked simultaneously with the Chicago storms. When the storm intensities exceeded the soil infiltrabilities, the infiltration rates equaled the infiltrabilities. However, at the onset of the storms, the soil infiltrability was not fully satisfied with low storm intensities, leading to gradual increases in the infiltration rates until they reached their maximum values during the storm peak.

(2) The peak infiltration rate exhibits a weak positive correlation with the return period and a weak negative correlation with the time-to-peak coefficient. Specifically, under the Chicago storm with a time-to-peak coefficient of 0.2, 0.4, or 0.6, the correlation coefficients and *p*-values of the peak infiltration rate concerning the return period are 0.9814, 0.9816, or 0.9810 and 0.1230, 0.1222, or 0.1224, respectively. Conversely, under the Chicago storm with a return period of 1 year, 2 years, or 5 years, the correlation coefficients and *p*-values of the peak infiltration rate regarding the time-to-peak coefficient are −0.9550, −0.9384, or −0.9212 and 0.1918, 0.2247, or 0.2544, respectively. This can be attributed to storms peaking later, resulting in higher soil moisture content at the storm's peak, leading to reduced infiltration rates at that specific moment. However, all *p*-values exceed 0.05 (i.e., confidence level of 95%), indicating that the observed correlations lack statistical significance. Notably, the peak infiltration rates exhibited only minor changes, consistent with the findings of Fu et al. (2023), who reported that the maximum infiltration rate remained largely consistent [3].

(3) The infiltration volumes were calculated, revealing a weak positive correlation with the return period. Specifically, under the Chicago storm with a time-to-peak coefficient of 0.2, 0.4, or 0.6, the correlation coefficients and *p*-values of infiltration volume regarding the return period are 0.9753, 0.9751, or 0.9747, and 0.1418, 0.1423, or 0.1434, respectively. Under the Chicago storm with a return period of 1 year, 2 years, or 5 years, the correlation coefficients and *p*-values of infiltration volume concerning the time-to-peak coefficient are −0.9350, −0.4647, or 0.7040, and 0.2307, 0.6923, or 0.5027, respectively. Significantly, these *p*-values exceed 0.05, indicating a lack of statistical significance in the observed correlations.

4.4.2. Subcatchment Scale

The infiltration rates of the pervious area under Chicago storms for the no-PP and PP schemes were identical. Thus, the infiltration processes in subcatchment s7 under Chicago storms with different return periods and peak-to-time coefficients were examined as an illustrative example. Figure 8 presents the infiltration rates of the pervious area, porous pavements, and subcatchment s7 under Chicago storms with 1-, 2-, or 5-year return periods and a peak-to-time coefficient of 0.4 for the porous pavement scheme.

The infiltration rates of the pervious area in s7 (green dotted lines in Figure 8) demonstrated that: (1) Initially, the infiltration rate increased and then decreased. During the early stages of the storms, the soil infiltrabilities exceeded the storm intensities, resulting in the infiltration rates being equal to the storm intensities. As the storm intensities increased and the infiltrabilities decreased, the infiltration rates peaked when these two values became equal. Subsequently, as the storm intensities continued to rise and surpass the infiltrabilities, the infiltration rates became equal to the infiltrabilities. As the infiltrabilities decreased further, the infiltration rates equaled the minimum infiltration capacity until the water-input rates reached zero, resulting in an infiltration rate of zero. (2) With increasing return periods, the peak infiltration rates varied (0.2077, 0.1966, 0.1974 mm/min), and the timing occurred earlier (38, 34, 30 min). These peak infiltration moments were earlier than the peak storm (48 min). A non-significant negative correlation was observed between peak infiltration rates and the return period (correlation coefficient = −0.6426, *p*-value = 0.5557). Similarly, a weak negative correlation was identified between peak infiltration moments and the return period (correlation coefficient = −0.9608, *p*-value = 0.1789).

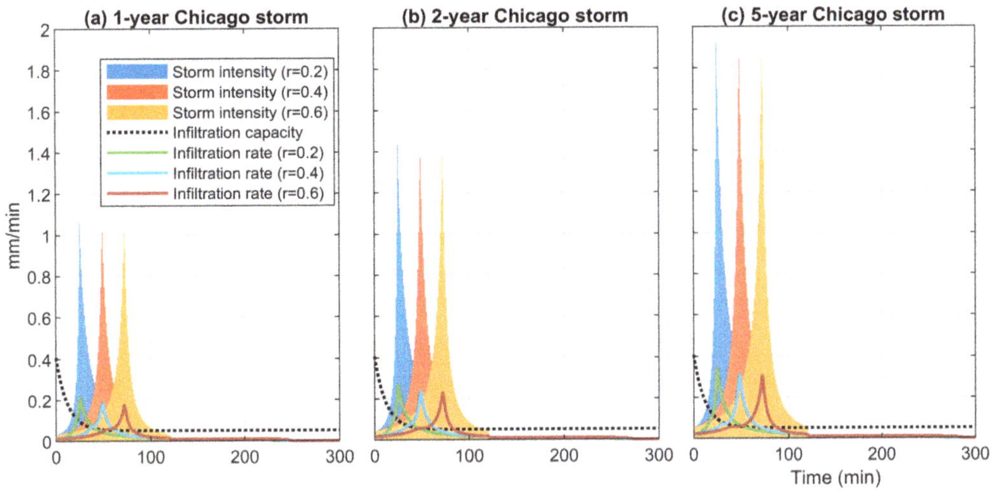

Figure 7. Catchment infiltration rates under Chicago storms with 120 min duration, 1-, 2-, or 5-year return period, and 0.2, 0.4, or 0.6 time-to-peak coefficient (denote as *r*) for porous pavement scheme.

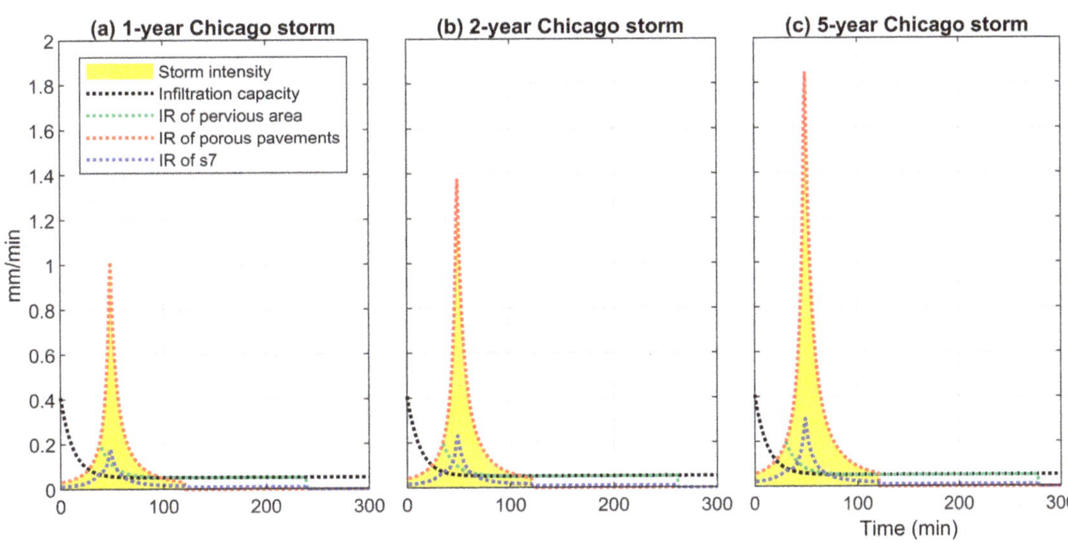

Figure 8. Infiltration rates (IR) of pervious area, porous pavements, and subcatchment s7 under Chicago storm with 120 min duration, 1-, 2-, or 5-year return period, and a time-to-peak coefficient of 0.4 for porous pavement scheme.

The infiltration rates of porous pavements in s7 (red dotted lines in Figure 8) provided the following insights: The infiltration rates equaled the storm intensities at any given time and immediately dropped to zero at the storm's end.

In addition, the infiltration rates of subcatchment s7 (blue dotted lines) generally followed the patterns of storm intensities, initially increasing and then decreasing. They may exceeded those of the pervious area when the infiltration rates of porous pavements were significant, resulting in larger catchment infiltration rates after area-weighted averaging. As the return period increased, the peak infiltration rates varied (0.1794, 0.2303, 0.2998 mm/min). A weak positive correlation was identified between the peak infiltration rates and the return period (correlation coefficient = 0.9817, p-value = 0.12).

Figure 9 illustrates the infiltration rates of the pervious area, porous pavements, and subcatchment s7 under Chicago storms, with a duration of 120 min, a 5-year return period, and time-to-peak coefficients of 0.2, 0.4, or 0.6 for the porous pavement scheme. The 5-year return period was chosen for analysis because it exhibited similar patterns to other return periods.

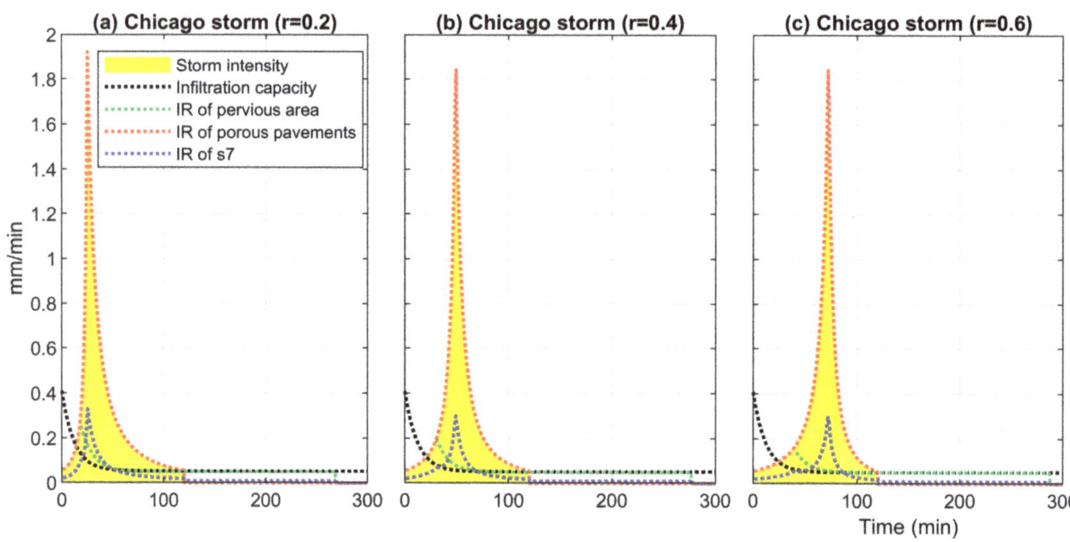

Figure 9. Infiltration rates (IR) of pervious area, porous pavements, and subcatchment s7 under Chicago storm with 120 min duration, 5-year return period, and 0.2, 0.4, or 0.6 time-to-peak coefficient (denote as r) for porous pavement scheme.

Our findings regarding the infiltrations of the pervious area (green dotted lines) reveal that a larger time-to-peak coefficient (0.2, 0.4, 0.6) resulted in a smaller peak infiltration rate (0.2593, 0.1974, 0.1585 mm/min) and a longer time to reach the peak infiltration rate (18, 30, 40 min). This observation can be attributed to the fact that with a larger time-to-peak coefficient, the cumulative precipitation at any given time before 72 min (time-to-peak coefficient of 0.6) was smaller, leading to smaller soil water content. A weak negative correlation was observed between the peak infiltration rates and the time-to-peak coefficient (correlation coefficient = −0.9914, p-value = 0.08). Conversely, there is a positive correlation between the peak infiltration moments and the time-to-peak coefficient (correlation coefficient = 0.9986, p-value = 0.0334).

On the other hand, the infiltration rates of subcatchment s7 (blue dotted lines) reached their peaks (0.3398, 0.3099, 0.3053 mm/min) at 25, 49, and 72 min, respectively. The peak moments of infiltrations aligned with or were close to those of the storms. There is a weak negative correlation between the peak infiltration rates and time-to-peak coefficient (correlation coefficient = −0.9220 and p-value = 0.25).

The Horton Model is sensitive to rainfall intensity throughout the simulation, and storm intensity and temporal distribution are crucial for accurate runoff prediction [36]. Our findings show that parameters like the Chicago storm's return periods and time-to-peak coefficient significantly impact infiltration simulation results in SWMM for sponge cities. These observations support previous research by [35].

5. Conclusions

This research examined storm parameters' impact on infiltration within a sponge city, particularly the return period and time-to-peak coefficient. Within the SWMM framework, the Horton and Green–Ampt infiltration models were employed for the pervious areas of subcatchments and porous pavements, respectively. We concluded that:

(1) Increasing the area of porous pavements results in proportional reductions in the impervious area without depression storage, the impervious area with depression storage, and the pervious area based on their initial area ratios. The infiltration rates of porous pavements under uniform and Chicago storms were supply-controlled.
(2) The infiltration rates of the pervious areas, subcatchments, and catchment under uniform storms exhibit a consistent initial stage, with the duration of this steady state becoming shorter as the return period increases. The catchment infiltration volumes demonstrate a linear growth trend with higher uniform storm intensities.
(3) The peak infiltration rate within pervious areas exhibits a non-significant negative correlation with the return period, while those within subcatchments and the overall catchment display non-significant positive correlations with the return period. The peak infiltration rate for pervious areas, subcatchments, and the catchment demonstrates non-significant negative correlations with the time-to-peak coefficient.
(4) The peak infiltration moments within pervious areas show non-significant negative correlations with the return period and non-significant positive correlations with the time-to-peak coefficient. Infiltration rates of porous pavements, subcatchments, and the overall catchment peak simultaneously to Chicago storms.

Our findings significantly advance the understanding and prediction of soil infiltration rates within sponge cities. Notably, our results underscore the critical importance of integrating considerations related to return periods and time-to-peak coefficients into infiltration analyses and the planning of infiltration-based facilities. We strongly recommend the implementation of porous pavements alongside impervious surfaces to facilitate the infiltration of runoff. It is imperative to recognize the diverse infiltration patterns that manifest under different storm scenarios, as they should inform the adaptive design, planning, and management of porous pavements. The effectiveness of these systems is substantially influenced by the characteristics of the rainfall events they encounter. Therefore, optimizing porous pavement locations and properties should be tailored to the local rainfall characteristics. Furthermore, it is worth noting that porous pavements exhibit enhanced performance when dealing with rainfall events characterized by larger time-to-peak coefficients. Consequently, retention-based solutions should be emphasized as an alternative strategy to mitigate the impacts of such rainfall events.

Nonetheless, it is crucial to recognize the limitations of our study. Specifically, despite its widespread application, the Horton model does not account for the effect of cumulative water layer depth on infiltration intensity, a consideration addressed by the Green–Ampt model. Additional advances are required for greater applicability, especially facilitating continuous simulations that include ponding and non-ponding conditions. Also, we advocate examining the dynamic connections between soil properties, storm events, runoff dynamics, and the effect of vegetation coverage in light of future research priorities. Furthermore, invaluable would be an investigation of the Hortonian overland flow mechanism and extensive field measurements to investigate the spatiotemporal heterogeneity of infiltration capacity and intensity across the catchment. These future attempts are anticipated to yield a thorough understanding of infiltration mechanisms, enabling the design of sustainable urban infrastructure that effectively manages stormwater, reduces flood risks, and encourages water conservation.

Author Contributions: Conceptualization, Y.Y.; methodology, Z.S. and Y.Y.; software, Z.S. and Y.Y.; validation, Y.Y. and X.X.; formal analysis, Z.S. and Y.Y.; investigation, Z.S. and Y.Y.; resources, Y.Y.; data curation, Z.S.; writing—original draft preparation, Y.Y. and Z.S.; writing—review and editing, Y.Y.; visualization, Z.S. and Y.Y.; supervision, Y.Y. and D.L.; project administration, Y.Y.; funding acquisition, Y.Y. All authors have read and agreed to the published version of the manuscript.

Funding: This research was funded by the National Natural Science Foundation of China, grant number 52009099 and 52279025. The APC was funded by the China Postdoctoral Science Foundation Funded Project (2019M653882XB) and the Joint Institute of the Internet of Water and Digital Water Governance (sklhse-2019-Iow06).

Data Availability Statement: Not applicable.

Acknowledgments: We thank the editors and reviewers for their helpful comments and suggestions.

Conflicts of Interest: The authors declare no conflict of interest.

References

1. Wang, W.; Liu, J.; Xu, B.; Li, C.Z.; Liu, Y.C.; Yu, F.L. A WRF/WRF-Hydro coupling system with an improved structure for rainfall-runoff simulation with mixed runoff generation mechanism. *J. Hydrol.* **2022**, *612*, 128049. [CrossRef]
2. Niyazi, B.; Masoud, M.; Elfeki, A.; Rajmohan, N.; Alqarawy, A.; Rashed, M. A Comparative Analysis of Infiltration Models for Groundwater Recharge from Ephemeral Stream Beds: A Case Study in Al Madinah Al Munawarah Province, Saudi Arabia. *Water* **2022**, *14*, 1686. [CrossRef]
3. Fu, X.; Liu, J.; Wang, Z.; Wang, D.; Shao, W.; Mei, C.; Wang, J.; Sang, Y.-F. Quantifying and assessing the infiltration potential of green infrastructure in urban areas using a layered hydrological model. *J. Hydrol.* **2023**, *618*, 128626. [CrossRef]
4. Zaqout, T.; Andradottir, H.O.; Arnalds, O. Infiltration capacity in urban areas undergoing frequent snow and freeze-thaw cycles: Implications on sustainable urban drainage systems. *J. Hydrol.* **2022**, *607*, 127495. [CrossRef]
5. Morbidelli, R.; Corradini, C.; Saltalippi, C.; Flammini, A.; Dari, J.; Govindaraju, R.S. Rainfall Infiltration Modeling: A Review. *Water* **2018**, *10*, 1873. [CrossRef]
6. Zakwan, M. Comparative analysis of the novel infiltration model with other infiltration models. *Water Environ. J.* **2019**, *33*, 620–632. [CrossRef]
7. Zakwan, M.; Niazkar, M. A Comparative Analysis of Data-Driven Empirical and Artificial Intelligence Models for Estimating Infiltration Rates. *Complexity* **2021**, *2021*, 9945218. [CrossRef]
8. Horton, R.E. An Approach Toward a Physical Interpretation of Infiltration-Capacity. *Soil Sci. Soc. Am. J.* **1941**, *5*, 399–417. [CrossRef]
9. Shiraki, S.; Thu, A.K.; Matsuno, Y.; Shinogi, Y. Evaluation of infiltration models and field-saturated hydraulic conductivity in situ infiltration tests during the dry season. *Paddy Water Environ.* **2019**, *17*, 619–632. [CrossRef]
10. Gu, Y.; Peng, D.Z.; Deng, C.N.; Zhao, K.K.; Pang, B.; Zuo, D.P. Atmospheric-hydrological modeling for Beijing's sub-center based on WRF and SWMM. *Urban Clim.* **2022**, *41*, 101066. [CrossRef]
11. Huang, Q.Z.; Hilpert, M.; Tsai, Y.Z.; Hsu, S.Y. Investigation and parameterization of the ponding height effect on dynamic suction head estimation in the Green-Ampt model. *J. Hydrol.* **2023**, *621*, 129524. [CrossRef]
12. Podeh, H.T.; Parsaie, A.; Shahinejad, B.; Arshia, A.; Shamsi, Z. Development and uncertainty analysis of infiltration models using PSO and Monte Carlo method. *Irrig. Drain.* **2023**, *72*, 38–47. [CrossRef]
13. Kim, S.; Karahan, G.; Sharma, M.; Pachepsky, Y. The site-specific selection of the infiltration model based on the global dataset and random forest algorithm. *Vadose Zone J.* **2021**, *20*, e20125. [CrossRef]
14. Parchami-Araghi, F.; Mirlatifi, S.M.; Dashtaki, S.G.; Mahdian, M.H. Point estimation of soil water infiltration process using Artificial Neural Networks for some calcareous soils. *J. Hydrol.* **2013**, *481*, 35–47. [CrossRef]
15. Molayem, M.; Abtahi, S.A.; Jafarinia, M.; Yasrebi, J. Improving infiltration prediction by point-based PTFs for semiarid soils in southern of Iran. *Environ. Earth Sci.* **2021**, *80*, 794. [CrossRef]
16. Bergeson, C.B.; Martin, K.L.; Doll, B.; Cutts, B.B. Soil infiltration rates are underestimated by models in an urban watershed in central North Carolina, USA. *J. Environ. Manag.* **2022**, *313*, 115004. [CrossRef] [PubMed]
17. Rasool, T.; Dar, A.Q.; Wani, M.A. Development of a Predictive Equation for Modelling the Infiltration Process Using Gene Expression Programming. *Water Resour. Manag.* **2021**, *35*, 1871–1888. [CrossRef]
18. Storm Water Management Model (SWMM). Available online: www.epa.gov/water-research/storm-water-management-model-swmm (accessed on 1 August 2023).
19. Singh, B.; Sihag, P.; Singh, K. Comparison of infiltration models in NIT Kurukshetra campus. *Appl. Water Sci.* **2018**, *8*, 63. [CrossRef]

20. Albalasmeh, A.A.; Alghzawi, M.Z.; Gharaibeh, M.A.; Mohawesh, O. Assessment of the Effect of Irrigation with Treated Wastewater on Soil Properties and on the Performance of Infiltration Models. *Water* **2022**, *14*, 1520. [CrossRef]
21. Rasool, T.; Dar, A.Q.; Wani, M.A. Comparative Evaluation of Infiltration Models under Different Land Covers. *Water Resour.* **2021**, *48*, 624–634. [CrossRef]
22. Duan, R.; Fedler, C.B.; Borrelli, J. Field evaluation of infiltration models in lawn soils. *Irrig. Sci.* **2011**, *29*, 379–389. [CrossRef]
23. Wang, N.; Chu, X.F. Revised Horton model for event and continuous simulations of infiltration. *J. Hydrol.* **2020**, *589*, 125215. [CrossRef]
24. Wang, J.L.; Song, J.Y.; Lin, H.J.; Peng, L.W.; Li, K.; Wang, Z.X. Comparison of infiltration models to describe infiltration characteristics of bioretention. *J. Hydro-Environ. Res.* **2021**, *38*, 35–43. [CrossRef]
25. Dahak, A.; Boutaghane, H.; Merabtene, T. Parameter Estimation and Assessment of Infiltration Models for Madjez Ressoul Catchment, Algeria. *Water* **2022**, *14*, 1185. [CrossRef]
26. Song, J.; Wang, J.; Wang, W.; Peng, L.; Li, H.; Zhang, C.; Fang, X. Comparison between different infiltration models to describe the infiltration of permeable brick pavement system via a laboratory-scale experiment. *Water Sci. Technol.* **2021**, *84*, 2214–2227. [CrossRef]
27. Mu, W.B.; Yu, F.L.; Li, C.Z.; Xie, Y.B.; Tian, J.Y.; Liu, J.; Zhao, N.N. Effects of Rainfall Intensity and Slope Gradient on Runoff and Soil Moisture Content on Different Growing Stages of Spring Maize. *Water* **2015**, *7*, 2990–3008. [CrossRef]
28. Kidron, G.J. Comparing overland flow processes between semiarid and humid regions: Does saturation overland flow take place in semiarid regions? *J. Hydrol.* **2021**, *593*, 125624. [CrossRef]
29. Davidsen, S.; Lowe, R.; Ravn, N.H.; Jensen, L.N.; Arnbjerg-Nielsen, K. Initial conditions of urban permeable surfaces in rainfall-runoff models using Horton's infiltration. *Water Sci. Technol.* **2018**, *77*, 662–669. [CrossRef]
30. Yang, M.Y.; Zhang, Y.Y.; Pan, X.Y. Improving the Horton infiltration equation by considering soil moisture variation. *J. Hydrol.* **2020**, *586*, 124864. [CrossRef]
31. Wei, L.; Yang, M.Y.; Li, Z.; Shao, J.L.; Li, L.Q.; Chen, P.; Li, S.; Zhao, R.B. Experimental Investigation of Relationship between Infiltration Rate and Soil Moisture under Rainfall Conditions. *Water* **2022**, *14*, 1347. [CrossRef]
32. Zhang, J.; Liu, J.T.; Han, X.L.; Shen, X.H.; Liang, Z.M.; Wang, S.H. Variable storage behavior controlled by rainfall intensity and profile structure upon saturation excess overland flow generation. *J. Hydrol.* **2022**, *610*, 127860. [CrossRef]
33. Amatya, D.M.; Walega, A.; Callahan, T.J.; Morrison, A.; Vulava, V.; Hitchcock, D.R.; Williams, T.M.; Epps, T. Storm event analysis of four forested catchments on the Atlantic coastal plain using a modified SCS-CN rainfall-runoff model. *J. Hydrol.* **2022**, *608*, 127772. [CrossRef]
34. Ren, X.W.; Hong, N.; Li, L.F.; Kang, J.Y.; Li, J.J. Effect of infiltration rate changes in urban soils on stormwater runoff process. *Geoderma* **2020**, *363*, 114158. [CrossRef]
35. Schoener, G.; Stone, M.C.; Thomas, C. Comparison of seven simple loss models for runoff prediction at the plot, hillslope and catchment scale in the semiarid southwestern US. *J. Hydrol.* **2021**, *598*, 126490. [CrossRef]
36. Chen, L.; Sela, S.; Svoray, T.; Assouline, S. Scale dependence of Hortonian rainfall-runoff processes in a semiarid environment. *Water Resour. Res.* **2016**, *52*, 5149–5166. [CrossRef]
37. Yang, Y.Y.; Li, J.; Huang, Q.; Xia, J.; Li, J.K.; Liu, D.F.; Tan, Q.T. Performance assessment of sponge city infrastructure on stormwater outflows using isochrone and SWMM models. *J. Hydrol.* **2021**, *597*, 126151. [CrossRef]
38. Yang, Y.Y.; Xu, X.Y.; Liu, D.F. An Event-Based Stochastic Parametric Rainfall Simulator (ESPRS) for Urban Stormwater Simulation and Performance in a Sponge City. *Water* **2023**, *15*, 1561. [CrossRef]
39. Rossman, L.A. *Storm Water Management Model Reference Manual*; Volume I—Hydrology (Revised); National Risk Management Research Laboratory, Office of Research and Development, US Environmental Protection Agency: Cincinnati, OH, USA, 2016.
40. Fengxi Sponge City (China). Available online: http://ecohydrology-ihp.org/demosites/view/1220 (accessed on 1 August 2023).
41. Yang, Y.Y.; Xin, Y.F.; Li, J.K. Surrogate-Based Multiobjective Optimization of Detention Pond Volume in Sponge City. *Water* **2023**, *15*, 2705. [CrossRef]
42. Wang, Y.; Liu, Z.; Wang, G.; Xue, W. Cellular automata based framework for evaluating mitigation strategies of sponge city. *Sci. Total Environ.* **2021**, *796*, 148991. [CrossRef]
43. Rosenberger, L.; Leandro, J.; Pauleit, S.; Erlwein, S. Sustainable stormwater management under the impact of climate change and urban densification. *J. Hydrol.* **2021**, *596*, 126137. [CrossRef]
44. Fengxi Management Committee. Available online: http://fxxc.xixianxinqu.gov.cn (accessed on 1 August 2023).
45. Yang, Y.Y.; Li, Y.B.; Huang, Q.; Xia, J.; Li, J.K. Surrogate-based multiobjective optimization to rapidly size low impact development practices for outflow capture. *J. Hydrol.* **2023**, *616*, 128848. [CrossRef]
46. Shahed Behrouz, M.; Zhu, Z.; Matott, L.S.; Rabideau, A.J. A new tool for automatic calibration of the Storm Water Management Model (SWMM). *J. Hydrol.* **2020**, *581*, 124436. [CrossRef]
47. Warwick, J.J.; Tadepalli, P. Efficacy of SWMM Application. *J. Water Res. Plan. Manag.* **1991**, *117*, 352–366. [CrossRef]
48. Hu, C.; Xia, J.; She, D.X.; Song, Z.H.; Zhang, Y.; Hong, S. A new urban hydrological model considering various land covers for flood simulation. *J. Hydrol.* **2021**, *603*, 126833. [CrossRef]
49. Fengxi New City Management Committee. *Preliminary Report of Numerical Simulation of Key Indicators of Sponge City Construction in Fengxi New City*; Fengxi New City Management Committee: Xi'an, China, 2017.

50. Yu, Y.; Zhou, Y.; Guo, Z.; van Duin, B.; Zhang, W. A new LID spatial allocation optimization system at neighborhood scale: Integrated SWMM with PICEA-g using MATLAB as the platform. *Sci. Total Environ.* **2022**, *831*, 154843. [CrossRef] [PubMed]
51. Luo, X.; Liu, P.; Cheng, L.; Liu, W.; Cheng, Q.; Zhou, C. Optimization of in-pipe storage capacity use in urban drainage systems with improved DP considering the time lag of flow routing. *Water Res.* **2022**, *227*, 119350. [CrossRef]
52. Mobilia, M.; Longobardi, A.; Amitrano, D.; Ruello, G. Land use and damaging hydrological events causing temporal changes in the Sarno River basin: Potential for green technologies mitigation by remote sensing analysis. *Hydrol. Res.* **2023**, *54*, 277–302. [CrossRef]
53. Koo, Y.M.; Seo, D. Parameter estimations to improve urban planning area runoff prediction accuracy using Stormwater Management Model (SWMM). *J. Korea Water Resour. Assoc.* **2017**, *50*, 303–313.
54. Beven, K. The era of infiltration. *Hydrol. Earth Syst. Sci.* **2021**, *25*, 851–866. [CrossRef]
55. Parnas, F.E.A.; Abdalla, E.M.H.; Muthanna, T.M. Evaluating three commonly used infiltration methods for permeable surfaces in urban areas using the SWMM and STORM. *Hydrol. Res.* **2021**, *52*, 160–175. [CrossRef]

water

MDPI

Article

Comparisons of Retention and Lag Characteristics of Rainfall–Runoff under Different Rainfall Scenarios in Low-Impact Development Combination: A Case Study in Lingang New City, Shanghai

Chen Zhang [1], Yongpeng Lv [1], Jian Chen [1], Tao Chen [1], Jinqiao Liu [1], Lei Ding [1], Nan Zhang [1] and Qiang Gao [1,2,*]

1 Shanghai Municipal Engineering Design Institute (Group) Co., Ltd., Shanghai 200092, China; zhangchen@smedi.com (C.Z.); yongpenglv@foxmail.com (Y.L.); chenjian@smedi.com (J.C.); chentao5@smedi.com (T.C.); echobadn@gmail.com (J.L.); dinglei@smedi.com (L.D.); zhangnan@smedi.com (N.Z.)
2 Shanghai Shenhuan Environmental Engineering Co., Ltd., Shanghai 200092, China
* Correspondence: gaoqiang81@gmail.com; Tel.: +86-21-5500-9961

Abstract: An increasing focus has been given to stormwater management using low-impact development (LID), which is regarded as a "near-nature" concept and is utilized to manage and reduce surface runoff during the rainfall–runoff process. According to the hydrological monitoring data, we evaluated the retention and lag characteristics of rainfall–runoff in LID combination under three rainfall-intensity scenarios (light–moderate, heavy, and torrential rainfall) in Lingang New City in Shanghai. LID facilities have been constructed for three years in the target study area, including rain gardens, retention ponds, green parking, porous pavement, and grass swales. The average runoff retention was 10.6 mm, 21.3 mm, and 41.6 mm under light–moderate, heavy, and torrential rainfall scenarios, respectively, and the corresponding runoff retention rate was 72.9%, 64.7%, and 76.1% during the study period. By comparing rainfall, runoff retention, runoff retention rate, cumulative rainfall, and lag times, it becomes evident that the ability to retain runoff can be greatly improved in the LID combination. The average runoff retention was significantly enhanced by nearly two times and four times under the heavy and torrential rainfall scenarios compared to the conditions under the light–moderate rainfall scenario. Furthermore, the lag time from the end of rainfall to the end of runoff (t_2) and the lag time between the centroid of rainfall and the centroid of runoff (t_3) showed a significantly negative correlation with rainfall intensity. Meanwhile, t_3 presented an incredibly positive correlation with rainfall duration. In this study, the LID combination demonstrated superior benefits in extending the duration of runoff in rainfall events with lower rainfall amounts, and demonstrated significant overall lag effects in rainfall events with longer durations and lower rainfall amounts. These results confirmed the vital role of the LID combination in stormwater management and the hydrologic impact of the LID combination on rainfall-induced runoff retention and lag effects. This work has provided valuable insights into utilizing LID facilities and can contribute to a better understanding of how runoff retention and lag characteristics respond to different rainfall intensity scenarios.

Keywords: Sponge City; low-impact development; stormwater management; retention time; lag time; Lingang New City

Citation: Zhang, C.; Lv, Y.; Chen, J.; Chen, T.; Liu, J.; Ding, L.; Zhang, N.; Gao, Q. Comparisons of Retention and Lag Characteristics of Rainfall–Runoff under Different Rainfall Scenarios in Low-Impact Development Combination: A Case Study in Lingang New City, Shanghai. *Water* **2023**, *15*, 3106. https://doi.org/10.3390/w15173106

Academic Editor: Marco Franchini

Received: 25 July 2023
Revised: 21 August 2023
Accepted: 25 August 2023
Published: 30 August 2023

1. Introduction

Urban development has modified land use types and occupied large areas of natural green lands, which leads to the rise of impervious surfaces [1] and the rapid increase in runoff volume [2,3]. Increasing focus has been given to stormwater management using low-impact development (LID), which is regarded as a "near-nature" concept [4] and aimed to manage stormwater runoff and reduce surface runoff [5–7]. Common LID practices mainly include bioretention, green roofing, permeable pavement, bioswales, rain barrels, rain gardens, and so on [8], which manage stormwater through infiltration, detention, storage, and purification [9,10]. In particular, LID practices play a vital role in reducing rainfall runoff volumes and peaks [5,11].

The LID practices have been evaluated in various geographical regions [7,12]. However, the construction and operation of LID have been widely applied in developed countries, although there are insufficient practices and studies in developing countries [7,13,14]. Therefore, there is an immediate requirement for experimental data on LID that accurately represent the specific local and environmental conditions in developing nations [15,16]. A "Sponge City" is a concept that utilizes nature as a sponge, intending to enhance the capacity of LID to improve the effective control of urban peak runoff and increase the effectiveness of stormwater management [17]. Since 2015, the Chinese government has carried out "Sponge City" construction projects [18], with LID as one of the critical approaches to stormwater management [19]. Effectively reducing runoff volume by using LID has become one of the primary goals of "Sponge City" [20], and has been piloted in 30 large-scale Sponge Cities across the whole country [21].

Numerous studies have evaluated the retention effect of certain individual LID practices [7,22–24]. For example, rain gardens could mitigate direct runoffs by 23.6–98.4% [22], detention ponds could reach a total runoff reduction by 45% [23], and porous pavement parking could reduce flood volume by 93% compared to asphalt parking [24]. Meanwhile, there is a lag effect of LID practices on urban flooding events, and LID practices can reduce the risk of urban flooding [25,26]. LID could minimize peak discharge depths, runoff coefficients, and discharge volumes and increase lag times and runoff thresholds compared with traditional residential development [26]. For example, Davis [25] found that the peak time of runoff in retention ponds can be lagged twice or even more compared with the rainfall process. Xia [27] showed that the outflow peaks in bioretention were delayed for at least 13 min and lowered at least 52% under high, medium, and low inflow rate conditions. The magnitude of the time delay and flood detention of peak flow using green roofing could be enhanced by 22–70% [28].

But even here, many studies have indicated that LID combinations with various characteristics may provide more effective performance than certain individual LID practices [29–31]. For instance, the simulation of individual LID practices led to a 3–40% reduction in average annual flood volumes, whereas LID combinations could reduce annual flood volumes by 16–47% [31]. Overall, green infrastructures could lessen total rainfall runoffs by 85–100% and decrease peak flows by 92–100% [30]. In contrast, the reduced flood capacity of a single infrastructure was limited [30]. However, there is still a lack of sufficient studies on the lag effect of LID practices in China. This topic holds significant importance and requires in-depth investigation, particularly concerning the study on the lag effect in combination of LID practices [16,19,30].

Shanghai is located at the intersection of the coastline and the Yangtze River Estuary (Figure 1), with a mean elevation of approximately four meters. There is continuous rainy weather annually in the Yangtze River Delta from late spring to early summer, commonly called the plum rain season. The plum rain season usually begins in mid-June and ends at the beginning of July, lasting approximately 20-plus days [32]. Shanghai is among the 30 pilot cities of China's second round of the national "Sponge Cities" project [22]. Due to rapid urbanization, a large population, and climate change, the high probability of heavy rainfall risk is rising in Shanghai, increasing the potential dangers of pluvial flooding events in Shanghai [22,33–35]. However, how retention and lag characteristics in the LID combination respond to different rainfall scenarios is still not well known and needs to be stressed.

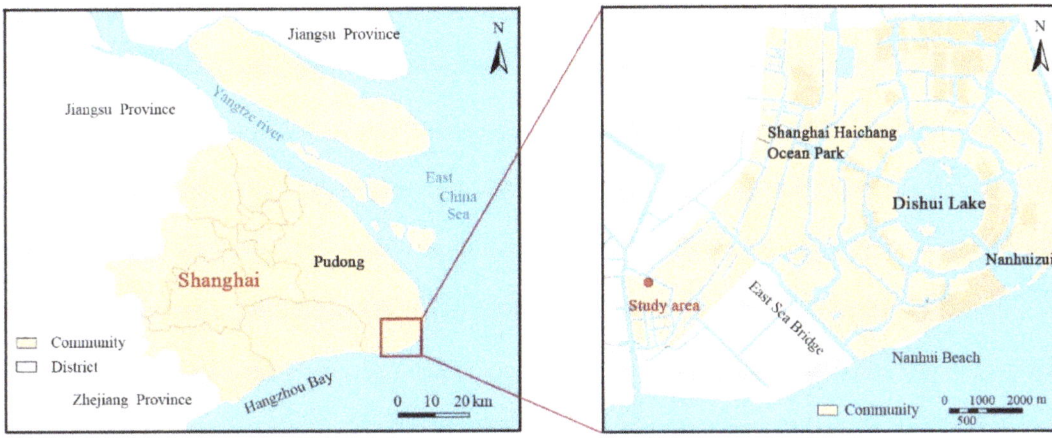

Figure 1. Location of the study area.

In this study, based on the continuous hydrological monitoring data, we selected a residential area of Lingang New City in Shanghai with a LID combination, which has been constructed for three years, and specifically addressed the following aims: (1) compare retention and lag characteristics of runoff in LID combination; (2) stress the different retention and lag characteristics of runoff in LID combination under different rainfall intensity scenarios; (3) discuss the correlation between rainfall characteristics and retention or lag characteristics during the rainfall–runoff process.

2. Materials and Methods

2.1. Study Area

Lingang New City, situated at the southeastern tip of Shanghai, is the largest "Sponge City" pilot area among the 30 pilot cities in China, spanning a pilot area of 79 km^2 (Figure 1). It is approximately 50 km away from the urban area of Shanghai [36]. Lingang New City has been focused on developing and expanding advanced manufacturing industries for nearly 20 years, contributing to its economic growth [37].

There have been considerable changes in the landscape because of sediment deposition, erosion, sand excavation, dam construction, and land reclamation in the history of Lingang New City [38]. Due to the low elevation and extensive land use of tidal flats in this area, Lingang New City is frequently threatened by pluvial flooding and freshwater shortages [39].

The land terrain in Lingang New City is predominantly flat, as it is located within the impact plain of the Yangtze River Delta. The land, for the most part, has been formed through beach reclamation [40]. The area has a stable and modern Quaternary sedimentary structure, with no occurrences of geological disasters such as new active faults or landslides in the pilot area. The foundation soil layer mainly consists of clay, silty, and sandy soil compositions [40]. The surface layer of the soil is composed of blown fill soil with a thickness ranging from 0.5 to 3.5 m. This layer primarily consists of clay, silt, and other similar materials. The soil in the surface layer is characterized by its uneven nature, loose structure, and poor permeability. Beneath the surface layer lies a layer of sandy silt, which exhibits more favorable soil properties. This bottom layer possesses relatively higher bearing capacity and permeability capabilities than the surface layer [40].

The soil permeability coefficient in Lingang New City can be found in Table 1 [40]. In general, the undisturbed soil in the pilot area is characterized by low permeability, high salinity–alkalinity, and soil depletion. These soil properties can affect water infiltration and drainage capabilities in the area. Additionally, Lingang New City experiences an average annual rainfall of 1228.1 mm. A significant portion of this rainfall occurs from May to

October, accounting for more than 70% of the annual rainfall. During this period, there is a higher occurrence of typhoons accompanied by heavy rainfall and high tide phenomena. These weather conditions can contribute to increased precipitation levels during these months [40].

Table 1. Soil permeability coefficient in Lingang New City.

Soil Type	Permeability Coefficient *k* (cm/s)
Sandy silt	1.54×10^{-4}~1.83×10^{-4}
Silt	2.63×10^{-4}~2.77×10^{-4}
Mucky clay	1.54×10^{-5}~1.55×10^{-5}
Clay	1.38×10^{-5}~1.54×10^{-5}

The target study area (Xinluyuan F residential district) is a resettlement community built in 2006. In 2021, the annual rainfall in the target study area was recorded as 1948.6 mm. The area has recently been constructed and equipped with LID facilities since late 2017. The target study area is in the southwestern part of Lingang New City, as shown in Figure 1. The study area encompasses a total land area of 3.36 hm^2 with a designated green area covering 1.33 hm^2. The greening rate of the study area is approximately 40%.

In this study area, the comprehensive runoff coefficient for the underlying surface of the land parcel is about 0.65. The design recurrence interval for the rainwater pipe network is once every 5 years. Surface runoff from rainwater within this target residential community is collected through various LID facilities and directed into the on-site rainwater pipe network. The rainwater pipe network is designed to discharge into the municipal rainwater pipe network on the east side of Chao Le Road. The outflow pipe has a diameter of DN800, and the bottom elevation inside the pipe is at an absolute height of 1.92 m (using the Wu Song elevation system).

The primary purpose of this sponge engineering construction is to reduce emissions at the source, implement total stormwater runoff control from the source of runoff production and confluence, reduce the peak and flow of runoff, delay the runoff time, and improve the drainage capacity of the original drainage facilities in the target study area of Lingang New City. Five LID facilities were selected considering the Sponge City's goals and the study area's characteristics, including retention ponds, rain gardens, green parking, grass swales, and porous pavement (Figure 2). LID facilities were set up in an independent and parallel manner, and the characteristics of different LID facilities in the study area are shown in Table 2.

Table 2. Characteristics of different LID facilities in the study area.

LID	Area	Number
Retention pond	25 m^3	25
Rain garden	773.3 m^2	75
Green parking	2631.6 m^2	58
Grass swale	50.12 m^2	47
Porous pavement	158 m^2	2

Figure 2. The layout of LID facilities and their combination in the study area.

2.2. Data Acquisition and Analysis

In this study, the rainfall and runoff data were collected from the Lingang Sponge City Management Platform, established in 2019. Through the Lingang Sponge City Management Platform, total rainfall and runoff data were continuously recorded and uploaded in the study area. Rainfall data were collected using a tipping bucket rain gauge of type L99-YL with a precision of ±2% to ±4% mm. The rain gauge was strategically positioned near the center of the study area (Figure 2). On the other hand, runoff data were collected using an acoustic doppler flowmeter of type ISCO 2150 with a precision of ±0.03 m/s (−1.5 m to 1.5 m/s). The runoff meter was strategically positioned near the main outlet in the study area (Figure 2). The data were recorded at intervals of 15 min for rainfall and 10 min for runoff.

The rainfall and runoff data used in this study were collected from fifteen valid rainfall events during 2021. The criterion for distinguishing a new valid rainfall event was based on two conditions: (1) no rainfall observed for at least 30 min prior to the commencement of rain; (2) no runoff outflow detected from the LID facilities during this period as well [16]. The rainfall data can be divided into four groups based on the intensity of rainfall [34]. These groups are determined as follows: (i) light rainfall: cumulative rainfall less than 10 mm/day ($r < 10$ mm/day); (ii) moderate rainfall: cumulative rainfall between 10 mm/day and 25 mm/day ($10 \leq r < 25$ mm/day); (iii) heavy rainfall: cumulative rainfall between 25 mm/day and 50 mm/day ($25 \leq r < 50$ mm/day); and (iv) torrential rainfall: cumulative rainfall exceeding 50 mm/day ($r > 50$ mm/day). Additionally, for ease of interpretation, the categories of light and moderate rainfall have been combined into a single group referred to as "light-moderate rainfall".

In the study, the retention effect was analyzed using indicators such as total retention, retention rate, cumulative rainfall (A_0), and retention time (t_1), along with their changes under different rainfall intensity scenarios. Among these indicators, runoff retention and retention rate were used to assess the regulation effect of the LID combination on runoff. Runoff retention represents the difference between rainfall and runoff, while retention rate is the retention ratio to rainfall. A_0 was measured to indicate the cumulative rainfall from the beginning of rainfall to the beginning of runoff.

T_1 represents the time interval from the beginning of rainfall (t_{p0}) to the beginning of runoff (t_{r0}) (as shown in Figure 3). In addition, our study investigated three other indices to represent the lag characteristics during the process of rainfall and runoff (Figure 3): (1) t_2: the lag time from rainfall end (t_{p3}) to the runoff end (t_{r3}); (2) t_3: the lag time from the centroid of rainfall (t_{p2}) to the centroid of runoff (t_{r2}); and (3) t_4: the time from the peak rainfall intensity (t_{p1}) to the peak runoff (t_{r1}). In the study, we collected data on the total lag times of t_2, t_3, and t_4, as well as their changes, under various rainfall intensity scenarios.

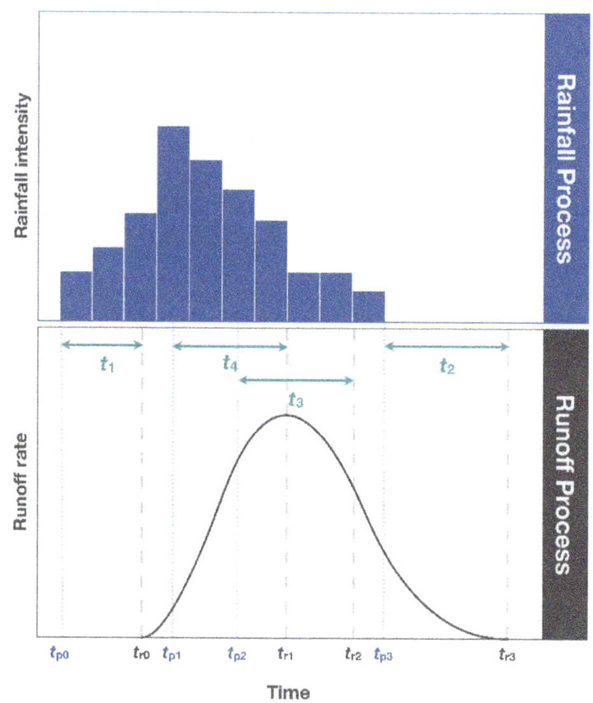

Figure 3. Schematic diagram illustrating indices of retention and lag effects for rainfall and runoff (Based and modified on Hood et al., 2007 [26] and Yin et al., 2022 [16]).

The centroid of rainfall and runoff were calculated as follows, respectively [26]:

$$t_{p2} = \frac{\sum_{i=1}^{n} W_i \times t_i}{\sum_{i=1}^{n} W_i} \quad (1)$$

where t_{p2} = centroid of rainfall, W_i = rainfall for period i, and t_i = time for period i.

$$t_{r2} = \frac{\sum_{i=1}^{n} A_i \times t_i}{\sum_{i=1}^{n} A_i} \quad (2)$$

where t_{r2} = centroid of runoff, A_i = runoff for period i, and t_i = time for period i.

The differences in runoff retention, runoff retention rate, A_0, and runoff retention and lag time among different rainfall intensity scenarios were analyzed using non-parametric comparison (Kruskal–Wallis test) for pairwise multiple comparisons. Spearman's test was used to examine the relationship between characteristics of rainfall events and indicators representing the runoff retention and lag effects.

All data analyses were performed in R (version 4.0.2) through RStudio (version: 2022.12.0+353; https://posit.co/download/rstudio-desktop/, accessed on 31 January 2023). The Kruskal–Wallis non-parametric comparison procedure was performed using the R package 'PMCMRplus' [41]. Statistically significant differences were identified when $p < 0.05$, unless otherwise stated in this study.

3. Results

3.1. Rainfall Runoff Retention Effect

Among the fifteen rainfall events studied, the total rainfall and runoff retention were 427.4 mm and 296.2 mm, respectively. Specifically, the average rainfall amounts were 14.6 mm, 33 mm, and 54.6 mm under the light–moderate, heavy, and torrential rainfall scenarios, respectively. The average rainfall under the heavy and torrential rainfall scenarios was 2.3 and 3.7 times higher, respectively, compared to the light–moderate rainfall scenario. The maximum recorded rainfall was 21 mm, 41 mm, and 55 mm under the light–moderate, heavy, and torrential rainfall scenarios, respectively.

The average runoff retention values were 10.6 mm, 21.3 mm, and 41.6 mm under the light–moderate, heavy, and torrential rainfall scenarios, respectively (Figure 4). The maximum runoff retention values recorded were 15.8 mm, 35.1 mm, and 42 mm under the light–moderate, heavy, and torrential rainfall scenarios, respectively. The retention showed a significant increase of approximately two and four times under the heavy and torrential rainfall scenarios, respectively, in comparison with the light-moderate rainfall scenario (p = 0.026 and 0.003, respectively) (Figure 4). However, no significant difference in runoff retention was observed between heavy and torrential rainfall scenarios (p = 0.108) (Figure 4).

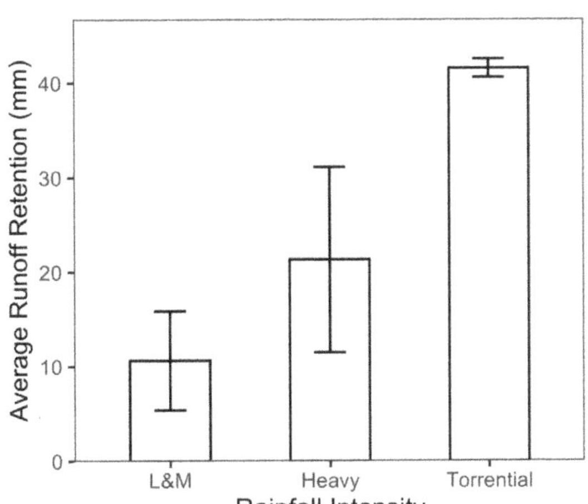

Figure 4. The average runoff retention of LID combination under three different rainfall intensity scenarios. The data in the bar plot represent the mean ± standard deviation for each rainfall intensity scenario. Abbreviations: L&M—light and moderate rainfall.

As described in Figure 5, the rainfall and retention values are nearly equal under lower rainfall scenarios and closely aligned along the 1:1 diagonal line. However, as the rainfall intensity increases, the retention values become gradually lower than the corresponding rainfall values. This is evident from the increasing deviation from the 1:1 diagonal line. The observed trend indicates that the difference between rainfall and retention increases as rainfall intensity increases.

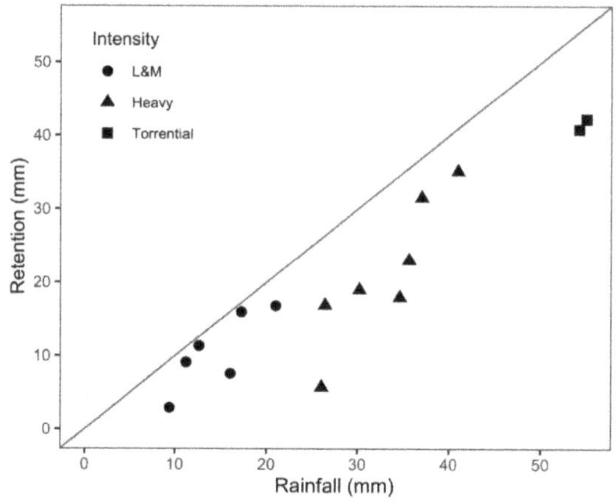

Figure 5. Discrete distribution of rainfall and runoff retention under different rainfall intensity scenarios. Abbreviations: L&M—light and moderate rainfall.

The total runoff retention rate was calculated as 69.3% among the fifteen rainfall events. The average runoff retention rates were 72.9%, 64.7%, and 76.1% under the light–moderate, heavy, and torrential rainfall scenarios, respectively (Figure 6). Despite the slight decrease in the total runoff retention rate under the heavy rainfall scenario compared to the light–moderate and torrential rainfall scenarios, no significant relationship was found between the runoff retention rates among different rainfall intensity scenarios (Figure 6). The maximum runoff retention rates recorded were 92.7%, 85.7%, and 76.8% under the light–moderate, heavy, and torrential rainfall scenarios, respectively (Figure 6).

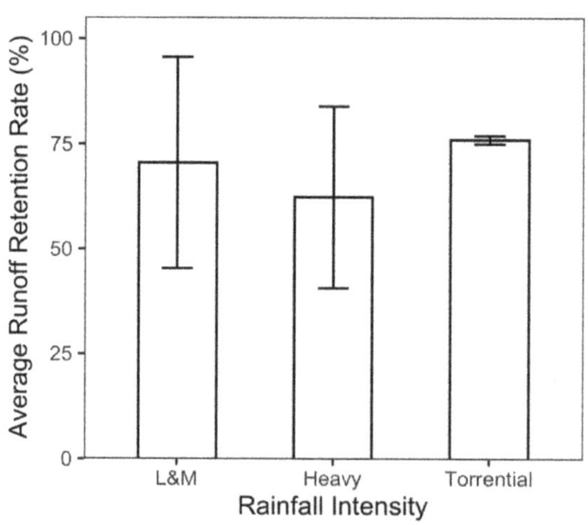

Figure 6. The retention rate of LID combinations under three different rainfall intensity scenarios. The data in the bar plot represent the mean ± standard deviation for each rainfall intensity scenario. Abbreviations: L&M—light and moderate rainfall.

The average A_0 values were 3.4 mm, 9.7 mm, and 7.0 mm under the light–moderate, heavy, and torrential rainfall scenarios, respectively (Figure 7). Among the scenarios, the average A_0 value was highest under the heavy rainfall scenario. Furthermore, the average A_0 values were nearly three and two times higher under the heavy and torrential rainfall scenarios, respectively, compared to the light–moderate rainfall scenario (Figure 7). However, no significant differences in A_0 were observed under different rainfall intensity scenarios (Figure 7). The maximum A_0 values recorded were 9.4 mm, 23.8 mm, and 10 mm under the light–moderate, heavy, and torrential rainfall scenarios, respectively (Figure 7).

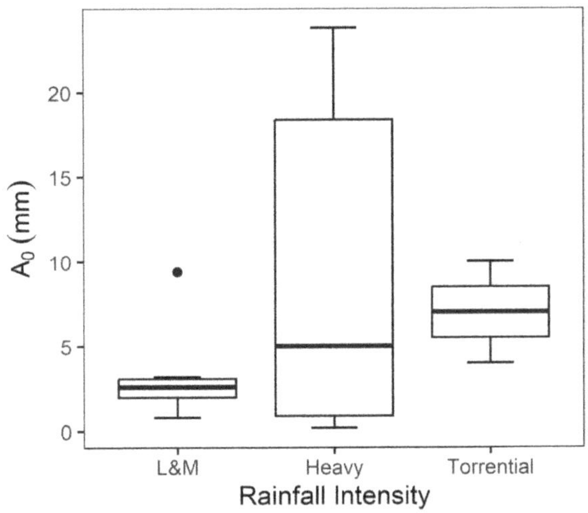

Figure 7. Distribution of cumulative rainfall (A_0) from the beginning of rainfall to runoff in LID combination. In the box plot, the horizontal line within each box represents the median value for each rainfall intensity scenario. The black dot on the plot indicates data outlier. Abbreviations: L&M—light and moderate rainfall.

3.2. Rainfall–Runoff Lag Effect

No significant differences in retention and lag time were observed under different rainfall intensity scenarios within each group, as shown in Figure 8. The average values for t_1, t_2, t_3, and t_4 were 26 min, 43 min, 34 min, and 20 min, respectively. Among the different rainfall scenarios, the range of values was 5 to 60 min for t_1. For t_2, the values varied between 10 to 115 min. For t_3, the vaues ranged from 8 to 73 min. For t_4, the values exhibited a wider range from −60 to 100 min.

Specifically, under the light–moderate rainfall scenario, the average values for t_1, t_2, t_3, and t_4 were 28 min, 57 min, 34 min, and 24 min, respectively. The range of values for t_1, t_2, t_3, and t_4 under the light–moderate rainfall scenario were 5 to 60 min, 25 to 115 min, 8 to 88 min, and 10 to 50 min, respectively. Under the heavy rainfall scenario, the average values for t_1, t_2, t_3, and t_4 were 25 min, 36 min, 34 min, and 15 min, respectively. The ranges for t_1, t_2, t_3, and t_4 under the heavy rainfall scenario were 10 to 40 min, 10 to 85 min, 9 to 70 min, and −60 to 100 min, respectively. Under the torrential rainfall scenario, the average values for t_1, t_2, t_3, and t_4 were 20 min, 30 min, 30 min, and 28 min, respectively. The ranges for t_1, t_3, and t_4 under the torrential rainfall scenario were 10 to 30 min, 19 to 40 min, and 15 to 40 min, respectively. It is noteworthy that t_2 was consistently 30 min under the torrential rainfall scenario (Figure 8).

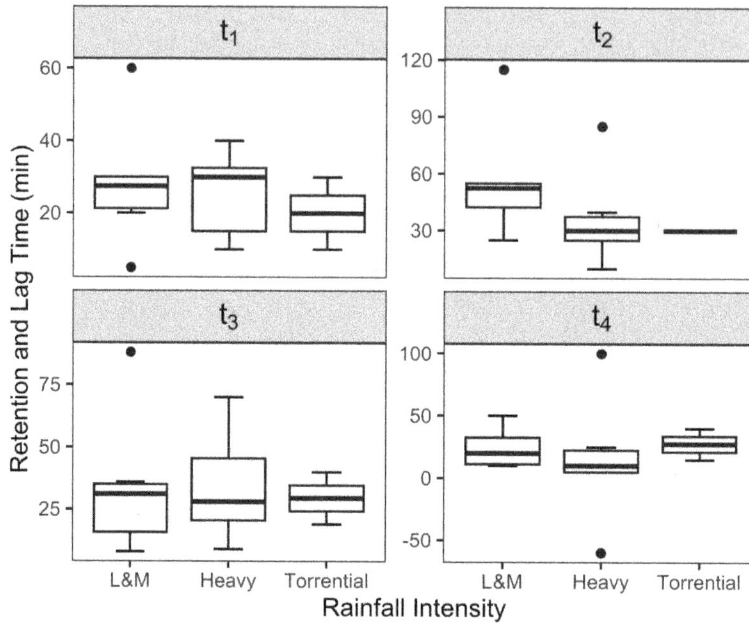

Figure 8. Patterns of retention and lag time under different rainfall intensity scenarios. t_1 represents the retention time from the beginning of rainfall to the beginning of runoff; t_2 indicates the lag time from the end of rainfall to the end of runoff; t_3 represents the lag time between the centroid of rainfall and the centroid of runoff; t_4 indicates the lag time from the peak rainfall to the peak runoff. In the box plots, the horizontal line within each box represents the median value for each rainfall intensity scenario. Black dots on the plots indicate data outliers. Abbreviations: L&M—light and moderate rainfall scenario.

3.3. *Effects of Rainfall Characteristics on Retention and Lag Effects*

The runoff retention effect exhibited a significant positive correlation with rainfall amount and intensity ($p < 0.001$ and $p = 0.005$, respectively, as shown in Table 3). Additionally, there was a significant negative correlation between the runoff retention rate and rainfall duration ($p = 0.013$, Table 3). On the other hand, the correlation between A_0 and rainfall intensity was only marginally positive ($p = 0.067$, Table 3).

Table 3. Results of Spearman's correlation tests between rainfall characteristics and retention effects. The table presents the correlation coefficients, and the corresponding *p*-values are listed in parentheses. Significant correlations are indicated in bold.

Retention Effect	Rainfall Duration	Rainfall Amount	Rainfall Intensity
Runoff retention	−0.227(0.417)	**0.932 (<0.001)**	**0.704 (0.005)**
Runoff retention rate	**−0.622 (0.013)**	0.107 (0.705)	0.446 (0.097)
A_0	−0.317 (0.250)	0.414 (0.126)	0.489 (0.067)

According to Table 4, neither t_1 nor t_4 exhibited significant correlations with rainfall duration, amount, or intensity. However, t_2 and t_3 showed significant negative correlations with rainfall intensity ($p = 0.003$ and $p = 0.011$, respectively, Table 4). Additionally, t_3 demonstrated a significant positive correlation with rainfall duration ($p = 0.022$, Table 4).

Table 4. The results of Spearman's correlation tests between rainfall characteristics and retention and lag time. The table presents the correlation coefficients, and the corresponding p-values are listed in parentheses. Significant correlations are indicated in bold.

Retention and Lag Time	Rainfall Duration	Rainfall Amount	Rainfall Intensity
t_1	0.228 (0.413)	−0.084 (0.766)	−0.301 (0.275)
t_2	0.496 (0.060)	−0.507 (0.054)	**−0.706 (0.003)**
t_3	**0.585 (0.022)**	−0.147 (0.602)	**−0.633 (0.011)**
t_4	0.141 (0.616)	−0.332 (0.226)	−0.303 (0.271)

4. Discussion

Generally, the effectiveness of LID combination tends to exceed that of individual LID facilities, as the performance of LID combination may not be equally obtained from individual LID facilities [7,29]. LID combination often exhibit varying degrees of impact in managing stormwater, and the overall retention effect of LID combination can provide more prominent advantages compared to the simple superposition of individual LID facilities [16]. During the rainfall–runoff process, the runoff retention and lag effects of LID combinations are influenced by a variety of factors [7,8], including soil type, rainfall amount, rainfall duration, timing of peak rainfall intensity, antecedent rainfall, and the conditions of the constructed area [26,42]. All these comprehensive factors collectively contribute to the overall performance of runoff retention and lag effects in LID combination.

In this study, the runoff retention effect exhibited a significant positive correlation with both rainfall amount and intensity because it is widely recognized that both rainfall amount and intensity have a considerable impact on the hydrological behavior of permeable surfaces [42]. In particular, rainfall intensity and amount can influence soil infiltration and runoff production [42–44]. It is expected that soil infiltration would increase with higher rainfall intensity, mainly due to the spatial heterogeneity in soil infiltration characteristics. When rainfall exceeds the maximum infiltration rate, soil moisture does not reach saturation, resulting in a higher soil infiltration rate [45]. Consequently, the proportion of water transitioning from rainfall to runoff would also increase with increasing rainfall intensity [45]. Additionally, the soil could stabilize the infiltration rate and sustain infiltration even after soil moisture reaches saturation under higher rainfall intensity conditions [45]. However, some other studies have reported that spatial heterogeneity in the infiltration characteristics of the soil surface may decrease with increasing rainfall intensity, even for the same duration of rainfall [45]. In particular, the land terrain in Lingang New City is predominantly flat, and the pilot area is characterized by low permeability, leading to a significant decrease in soil infiltration volume with increasing rainfall intensity scenarios. Some studies, such as those by Römkens [46] and Parsons [47], have also found a potential decrease in runoff and an increase in retention with increasing rainfall intensity.

In addition, previous studies have indicated a threshold of rainfall amount that can trigger a change in the hydrological behavior of a catchment [48–50]. However, it has been observed that rainfall intensity has a more significant influence on determining the threshold value compared to rainfall amount [42]. Primarily, we found that the retention showed a substantial increase of approximately two and four times under the heavy and torrential rainfall scenarios, respectively, in comparison with the light-moderate rainfall scenario. Moreover, the difference between rainfall intensity and retention increases as rainfall intensity increases. This suggests that the retention capacity of the LID combination tends to stabilize and gradually reach saturation as rainfall intensity rises.

The average runoff retention rate was not significantly affected by increasing rainfall intensity in the LID combination. This is probably because the changes in A_0 under different rainfall intensity scenarios reflect the site-specific soil conditions, vegetation type, the prior rainfall conditions, as well as the retention effect in the study area [8,51]. This coincides with the soil condition in the pilot area mentioned previously. The soil in the study area has low permeability, making it difficult for water to penetrate through the soil layers. Furthermore, a significant negative correlation was observed between the runoff retention rate and rainfall duration. It has been shown that LID facilities with porous surfaces are more effective in reducing floods during shorter rainfall durations [7], which chould be partly explained by the fact that A_0 does not vary with increases in rainfall duration.

Retention and lag time are considered critical indices, as they encompass various aspects of runoff generation [26,52]. This study selected four characteristic indices (t_1, t_2, t_3, and t_4) to represent retention and lag time. First, t_1 represents the lag time from the start of rainfall to the initiation of runoff. During this process, the rainfall is primarily absorbed by vegetation, infiltrated into the soil, and fills the soil macropores [44]. In this study, it did not show any significant response under different rainfall intensity scenarios and no significant correlation with rainfall duration, amount, or intensity. This could correspond to the response of A_0 mentioned earlier under the various rainfall intensity scenarios, which can be primarily attributed to the antecedent soil moisture conditions prior to the onset of rainfall [16,26]. Nevertheless, the influence of antecedent soil moisture on soil infiltration was more noticeable under relatively lower rainfall intensity. But this effect diminished gradually as the rainfall intensity increased [53]. Additionally, the impact of vegetation coverage during the rainfall–runoff process may also contribute to the response of t_1 [44].

Second, t_2 represents the lag time from the end of rainfall to the end of runoff. While the rainfall process concludes, the LID facilities continue to generate runoff during this period. It was observed that t_2 decreased with increasing rainfall intensity scenarios, and the findings of this study also revealed a significant negative correlation between t_2 and rainfall intensity. This result can be attributed to the relationship between rainfall intensity and soil infiltration, which is highly influenced by rainfall intensity [42,54,55]. The delay between the end of rainfall and the end of runoff has a buffering effect on storing rainwater and can effectively conserve urban water resources. During the study period, this delay between the rainfall and runoff could be due to the decrease in soil infiltration volume with increasing rainfall intensity scenarios, resulting in the gradual decrease in t_2 in the LID combination. Therefore, in this study, the LID combination demonstrates superior benefits in extending the duration of runoff in rainfall events with lower rainfall amounts.

Third, t_3 represents the time delay between the centroid time of rainfall and runoff. The lag effect of the centroid time can be used to assess the overall lag effect of LID combination during the rainfall and runoff production processes. In this study, a significant positive correlation between t_3 and rainfall duration indicates that the overall lag effect in the LID combination could be more pronounced with prolonged rainfall. However, t_3 showed a significant negative correlation with rainfall intensity, suggesting that the centroid time might occasionally be longer due to the extended duration of runoff generation under short-duration rainstorms. Thus, in this study, the LID combination demonstrates significant overall lag effects in rainfall events with longer durations and lower rainfall amounts.

Fourth, t_4 represents the lag time from the peak rainfall to the peak runoff, commonly used to characterize the lag effect of LID combination under different instantaneous rainfall intensity scenarios. The peak value is an immediate value, and the peak effect of rainfall in runoff generation is quickly reflected. Typically, when the rainfall reaches its peak, the corresponding peak in runoff generation will also occur rapidly in theory. In this study, we found no significant correlation between t_4 and rainfall in the various rainfall intensity scenarios. The lack of correlation and differences in t_4 can be partly attributed to the shifting positions of the peak rainfall and peak runoff. This shift often depends on the different intensities of rainfall throughout the entire process. The peak rainfall can occur either early on or after the rainfall period, leading to a delay in peak runoff. In addition, negative values for t_4 can be encountered in this study, where the peak runoff occurs before the peak rainfall. This is because, in certain rainfall events, the average rainfall intensity in the early stages is high, but the peak rainfall occurs later in the rainfall process. As a result, the peak runoff primarily responds to the earlier rainfall process rather than the later peak rainfall.

5. Conclusions

This study investigated the response of rainfall–runoff retention and lag effects in an LID combination under three different rainfall intensity scenarios in a resettlement community in Lingang New City. We compared various parameters in the LID combination along a rainfall gradient, including rainfall, runoff retention, runoff retention rate, cumulative rainfall, and different lag times in the LID combination. The results showed a gradual reduction in runoff retention within the LID combination during heavier rainfall intensity scenarios, accompanied by an increasing difference between rainfall and runoff retention. Additionally, the average runoff retention was significantly enhanced, by nearly two times and four times, under heavy and torrential rainfall events compared to light–moderate rainfall. Moreover, the runoff retention effect positively correlated with rainfall amount and intensity. However, we only found a significant negative correlation between runoff retention rates and rainfall duration, but not with rainfall and rainfall intensity.

No significant differences in A_0 were observed between different rainfall intensities. Additionally, no significant differences in the lag effect were observed under different rainfall scenarios in each group. However, t_2 and t_3 showed a significant negative correlation with rainfall intensity. On the other hand, t_3 exhibited a strong positive correlation with rainfall duration. The findings from this study provide a valuable case study using observational data to analyze rainfall retention and lag effects in the LID combination. This study contributes to the existing empirical evidence in stormwater management using the LID combination. However, due to limitations and uncertainties, it is recommended that future studies consider long-term monitoring analysis and explore other influencing factors to improve the accuracy of evaluation.

Author Contributions: Methodology and software, L.D. and N.Z.; investigation, J.C. and J.L.; data curation, T.C. and J.L.; writing—original draft preparation, C.Z., Y.L. and Q.G., writing—review and editing, C.Z., Y.L. and Q.G.; project administration, J.C. and T.C.; funding acquisition, Y.L. All authors have read and agreed to the published version of the manuscript.

Funding: This research was funded by the National Key R&D Program of China (NO. 2022YFC3800500).

Data Availability Statement: Not applicable.

Acknowledgments: We would like to thank the Lingang Sponge City Management Platform for all the data and material provided for this research. We also extend our collective thanks to the editor and the reviewers for the comments and additions.

Conflicts of Interest: The authors declare no conflict of interest.

References

1. Brown, J.N.; Peake, B.M. Sources of heavy metals and polycyclic aromatic hydrocarbons in urban stormwater runoff. *Sci. Total Environ.* **2006**, *359*, 145–155. [CrossRef] [PubMed]
2. de Macedo, M.B.; Rosa, A.; do Lago, C.A.F.; Mendiondo, E.M.; de Souza, V.C.B. Learning from the operation, pathology and maintenance of a bioretention system to optimize urban drainage practices. *J. Environ. Manag.* **2017**, *204*, 454–466. [CrossRef] [PubMed]
3. O'Sullivan, A.D.; Wicke, D.; Hengen, T.J.; Sieverding, H.L.; Stone, J.J. Life cycle Assessment modelling of stormwater treatment systems. *J. Environ. Manag.* **2015**, *149*, 236–244. [CrossRef] [PubMed]
4. Zhu, Z.H.; Chen, Z.H.; Chen, X.H.; Yu, G. An assessment of the hydrologic effectiveness of low impact development (LID) practices for managing runoff with different objectives. *J. Environ. Manag.* **2019**, *231*, 504–514. [CrossRef]
5. Tredway, J.C.; Havlick, D.G. Assessing the potential of low-impact development techniques on runoff and streamflow in the templeton gap watershed, Colorado. *Prof. Geogr.* **2017**, *69*, 372–382. [CrossRef]
6. Sohn, W.; Kim, J.H.; Li, M.H. Low-impact development for impervious surface connectivity mitigation: Assessment of directly connected impervious areas (DCIAs). *J. Environ. Plan. Manag.* **2017**, *60*, 1871–1889. [CrossRef]
7. Pour, S.H.; Wahab, A.K.A.; Shahid, S.; Asaduzzaman, M.; Dewan, A. Low impact development techniques to mitigate the impacts of climate-change-induced urban floods: Current trends, issues and challenges. *Sustain. Cities Soc.* **2020**, *62*, 102373. [CrossRef]
8. Golden, H.E.; Hoghooghi, N. Green infrastructure and its catchment-scale effects: An emerging science. *Wiley Interdiscip. Rev. Water* **2018**, *5*, e1254. [CrossRef]

9. Dong, X.; Guo, H.; Zeng, S.Y. Enhancing future resilience in urban drainage system: Green versus grey infrastructure. *Water Res.* **2017**, *124*, 280–289. [CrossRef]
10. Jia, H.F.; Wang, Z.; Zhen, X.Y.; Clar, M.; Yu, S.L. China's sponge city construction: A discussion on technical approaches. *Front. Environ. Sci. Eng.* **2017**, *11*, 18. [CrossRef]
11. Ahmed, K.; Chung, E.S.; Song, J.Y.; Shahid, S. Effective design and planning specification of low impact development practices using Water Management Analysis Module (WMAM): Case of Malaysia. *Water* **2017**, *9*, 173. [CrossRef]
12. Chang, N.B.; Lu, J.W.; Chui, T.F.M.; Hartshorn, N. Global policy analysis of low impact development for stormwater management in urban regions. *Land Use Policy* **2018**, *70*, 368–383. [CrossRef]
13. Wang, M.; Zhang, D.Q.; Su, J.; Dong, J.W.; Tan, S.K. Assessing hydrological effects and performance of low impact development practices based on future scenarios modeling. *J. Clean. Prod.* **2018**, *179*, 12–23. [CrossRef]
14. Bedan, E.S.; Clausen, J.C. Stormwater runoff quality and quantity from trational and low impact development watersheds. *J. Am. Water Resour. Assoc.* **2009**, *45*, 998–1008. [CrossRef]
15. Chen, Y.; Tan, M.; Wan, J.H.; Weise, T.; Wu, Z.Z. Effectiveness evaluation of the coupled LIDs from the watershed scale based on remote sensing image processing and SWMM simulation. *Eur. J. Remote Sens.* **2021**, *54*, 77–91. [CrossRef]
16. Yin, Y.X.; Qin, H.P.; Yu, S.Q.; Zheng, Y.Y.; He, K.M. Retention and lag effects of rainfall runoff in a low impact development area. *J. Shenzhen Univ. Sci. Eng.* **2022**, *39*, 142–151. (In Chinese) [CrossRef]
17. Song, C. Application of nature-based measures in China's sponge city initiative: Current trends and perspectives. *Nat.-Based Solut.* **2022**, *2*, 100010. [CrossRef]
18. Liu, D.S. China's sponge cities to soak up rainwater. *Nature* **2016**, *537*, 307. [CrossRef]
19. Guo, X.C.; Guo, Q.Z.; Zhou, Z.K.; Du, P.F.; Zhao, D.Q. Degrees of hydrologic restoration by low impact development practices under different runoff volume capture goals. *J. Hydrol.* **2019**, *578*, 124069. [CrossRef]
20. Ding, L.; Ren, X.Y.; Gu, R.Z.; Che, Y. Implementation of the "sponge city" development plan in China: An evaluation of public willingness to pay for the life-cycle maintenance of its facilities. *Cities* **2019**, *93*, 13–30. [CrossRef]
21. Li, Z.M.; Xu, S.Y.; Yao, L.M. A systematic literature mining of sponge city: Trends, foci and challenges standing ahead. *Sustainability* **2018**, *10*, 1182. [CrossRef]
22. Du, S.Q.; Wang, C.X.; Shen, J.; Wen, J.H.; Gao, J.; Wu, J.P.; Lin, W.P.; Xu, H. Mapping the capacity of concave green land in mitigating urban pluvial floods and its beneficiaries. *Sustain. Cities Soc.* **2019**, *44*, 774–782. [CrossRef]
23. Shannak, S.d.A.; Jaber, F.H.; Lesikar, B.J. Modeling the effect of cistern size, soil type, and irrigation scheduling on rainwater harvesting as a stormwater control measure. *Water Resour. Manag.* **2014**, *28*, 4219–4235. [CrossRef]
24. Dreelin, E.A.; Fowler, L.; Ronald Carroll, C. A test of porous pavement effectiveness on clay soils during natural storm events. *Water Res.* **2006**, *40*, 799–805. [CrossRef]
25. Davis, A.P. Field performance of bioretention: Hydrology impacts. *J. Hydrol. Eng.* **2008**, *13*, 90–95. [CrossRef]
26. Hood, M.J.; Clausen, J.C.; Warner, G.S. Comparison of stormwater lag times for low impact and traditional residential development. *J. Am. Water Resour. Assoc.* **2007**, *43*, 1036–1046. [CrossRef]
27. Xia, J.; Wang, H.P.; Stanford, R.L.; Pan, G.Y.; Yu, S.L. Hydrologic and water quality performance of a laboratory scale bioretention unit. *Front. Environ. Sci. Eng.* **2018**, *12*, 14. [CrossRef]
28. Alfredo, K.; Montalto, F.; Goldstein, A. Observed and modeled performances of prototype green roof test plots subjected to simulated low- and high-intensity precipitations in a laboratory experiment. *J. Hydrol. Eng.* **2010**, *15*, 444–457. [CrossRef]
29. Gao, J.; Wang, R.S.; Huang, J.L.; Liu, M. Application of BMP to urban runoff control using SUSTAIN model: Case study in an industrial area. *Ecol. Model.* **2015**, *318*, 177–183. [CrossRef]
30. Huang, C.L.; Hsu, N.S.; Liu, H.J.; Huang, Y.H. Optimization of low impact development layout designs for megacity flood mitigation. *J. Hydrol.* **2018**, *564*, 542–558. [CrossRef]
31. Ahiablame, L.; Shakya, R. Modeling flood reduction effects of low impact development at a watershed scale. *J. Environ. Manag.* **2016**, *171*, 81–91. [CrossRef]
32. Zhou, L.; Ye, B.; Xia, S. Assessing membrane biofouling and its gel layer of anoxic/oxic membrane bioreactor for megacity municipal wastewater treatment during plum rain season in Yangtze River Delta, China. *Water Res.* **2017**, *127*, 22–31. [CrossRef] [PubMed]
33. Huang, Q.Y.; Wang, J.; Li, M.Y.; Fei, M.L.; Dong, J.G. Modeling the influence of urbanization on urban pluvial flooding: A scenario-based case study in Shanghai, China. *Nat. Hazards* **2017**, *87*, 1035–1055. [CrossRef]
34. Yuan, Y.; Xu, Y.S.; Arulrajah, A. Sustainable measures for mitigation of flooding hazards: A case study in Shanghai, China. *Water* **2017**, *9*, 310. [CrossRef]
35. Yin, J.; Yin, Z.E.; Hu, X.M.; Xu, S.Y.; Wang, J.; Li, Z.H.; Zhong, H.D.; Gan, F.B. Multiple scenario analyses forecasting the confounding impacts of sea level rise and tides from storm induced coastal flooding in the city of Shanghai, China. *Environ. Earth Sci.* **2011**, *63*, 407–414. [CrossRef]
36. Zheng, S.J.; Meng, C.; Xue, J.H.; Wu, Y.B.; Liang, J.; Xin, L.; Zhang, L. UAV-based spatial pattern of three-dimensional green volume and its influencing factors in Lingang New City in Shanghai, China. *Front. Earth Sci.* **2021**, *15*, 543–552. [CrossRef]
37. Shen, J.; Luo, X.; Wu, F.L. Assembling mega-urban projects through state-guided governance innovation: The development of Lingang in Shanghai. *Reg. Stud.* **2020**, *54*, 1644–1654. [CrossRef]

38. Yin, J.; Zhao, Q.; Yu, D.P.; Lin, N.; Kubanek, J.L.; Ma, G.Y.; Liu, M.; Pepe, A. Long-term flood-hazard modeling for coastal areas using InSAR measurements and a hydrodynamic model: The case study of Lingang New City, Shanghai. *J. Hydrol.* **2019**, *571*, 593–604. [CrossRef]
39. Xu, H.S.; Chen, L.; Zhao, B.; Zhang, Q.Z.; Cai, Y.L. Green stormwater infrastructure eco-planning and development on the regional scale: A case study of Shanghai Lingang New City, East China. *Front. Earth Sci.* **2016**, *10*, 366–377. [CrossRef]
40. Zhang, C.; Lv, Y.P. *Research on Sponge City Construction Technology in Plain River Network Area and Case Study in Lingang New City*; China Architecture & Building Press: Beijing, China, 2022.
41. Pohlert, T. PMCMRplus: Calculate Pairwise Multiple Comparisons of Mean Rank Sums Extended. 2021. Available online: https://cran.r-project.org/web/packages/PMCMRplus/index.html (accessed on 24 August 2023).
42. Guan, M.F.; Sillanpää, N.; Koivusalo, H. Storm runoff response to rainfall pattern, magnitude and urbanization in a developing urban catchment. *Hydrol. Process.* **2016**, *30*, 543–557. [CrossRef]
43. Dunkerley, D. Effects of rainfall intensity fluctuations on infiltration and runoff: Rainfall simulation on dryland soils, Fowlers Gap, Australia. *Hydrol. Process.* **2012**, *26*, 2211–2224. [CrossRef]
44. Mu, W.B.; Yu, F.L.; Li, C.Z.; Xie, Y.B.; Tian, J.Y.; Liu, J.; Zhao, N.N. Effects of rainfall intensity and slope gradient on runoff and soil moisture content on different growing stages of Spring Maize. *Water* **2015**, *7*, 2990–3008. [CrossRef]
45. He, L.; Li, S.; Cui, C.H.; Yang, S.S.; Ding, J.; Wang, G.Y.; Bai, S.W.; Zhao, L.; Cao, G.-L.; Ren, N.Q. Runoff control simulation and comprehensive benefit evaluation of low-impact development strategies in a typical cold climate area. *Environ. Res.* **2022**, *206*, 112630. [CrossRef]
46. Römkens, M.J.M.; Prasad, S.N.; Parlange, J.Y. Surface seal development in relation to rainstorm intensity. *Catena* **1990**, *17*, 1–11.
47. Parsons, A.J.; Stone, P.M. Effects of intra-storm variations in rainfall intensity on interrill runoff and erosion. *Catena* **2006**, *67*, 68–78. [CrossRef]
48. Burton, G.A., Jr.; Pitt, R.E. *Stormwater Effects Handbook: A Toolbox for Watershed Managers, Scientists, and Engineers*, 1st ed.; CRC Press: Boca Raton, FL, USA, 2002.
49. Said, A.; Downing, H. Estimating connected impervious areas as function of rainfall depth. In Proceedings of the AWRA2010 Spring Specialty Conference, Orlando, FL, USA, 29–31 March 2010.
50. Sillanpää, N.; Koivusalo, H. Impacts of urbanization and event magnitude on runoff contributing area and runoff coefficients. In Proceedings of the 13th International Conference on Urban Drainage, Sarawak, Malaysia, 7–12 September 2014.
51. Boyd, M.J.; Bufill, M.C.; Knee, R.M. Pervious and impervious runoff in urban catchments. *Hydrol. Sci. J.* **1993**, *38*, 463–478. [CrossRef]
52. Leopold, L.B. Lag times for small drainage basins. *Catena* **1991**, *18*, 157–171. [CrossRef]
53. Liu, H.; Lei, T.W.; Zhao, J.; Yuan, C.P.; Fan, Y.T.; Qu, L.Q. Effects of rainfall intensity and antecedent soil water content on soil infiltrability under rainfall conditions using the run off-on-out method. *J. Hydrol.* **2011**, *396*, 24–32. [CrossRef]
54. Hawke, R.M.; Price, A.G.; Bryan, R.B. The effect of initial soil water content and rainfall intensity on near-surface soil hydrologic conductivity: A laboratory investigation. *Catena* **2006**, *65*, 237–246. [CrossRef]
55. Assouline, S.; Mualem, Y. Modeling the dynamics of seal formation and its effect on infiltration as related to soil and rainfall characteristics. *Water Resour. Res.* **1997**, *33*, 1527–1536. [CrossRef]

water

MDPI

Article

Simulation and Comprehensive Evaluation of the Multidimensional Environmental Benefits of Sponge Cities

Jingyu Wang, Xuehui Zhou, Shuai Wang, Lei Chen * and Zhenyao Shen

State Key Laboratory of Water Environment Simulation, School of Environment, Beijing Normal University, Beijing 100875, China; wjy6526@126.com (J.W.); zyshen@bnu.edu.cn (Z.S.)
* Correspondence: chenlei1982bnu@bnu.edu.cn

Abstract: The implementation of grey and green infrastructure is an effective means to address urban flooding and nonpoint source pollution, but due to the complexity of the process and the diversity of benefits, there is a lack of measurement of the comprehensive benefits. Adopting a typical university in Beijing as an example, this paper simulated the multidimensional benefits of the water quantity, water quality, and ecology of grey and green facility renovation by coupling the storm water management model (SWMM) and InfoWorks Integrated Catchment Management (ICM). Monetization methods and economical means were employed to characterize the comprehensive benefits. The results showed that grey and green infrastructure retrofitting reduced the number of severe overflow nodes in the study area by 54.35%, the total overflow volume by 22.17%, and the nonpoint source pollution level by approximately 80% under the heavy rain scenario and 60% under the rainstorm scenario. The annual benefits of grey and green infrastructure renovation reached CNY764,691/year: of this amount, CNY275,726/year was from hydrological regulation, CNY270,895/year was from nonpoint source pollution reduction, and CNY218,070/year was from ecological improvement. The benefits of green facilities were higher than those of grey facilities, and the combined benefits were negatively correlated with the rainfall level, with a total benefit–cost ratio of 1.19. The results provide methodological and data support for grey and green infrastructure retrofitting within the context of sponge cities.

Keywords: sponge city; grey and green infrastructure; stormwater management model; integrated environmental benefits; monetary value; stormwater use

Citation: Wang, J.; Zhou, X.; Wang, S.; Chen, L.; Shen, Z. Simulation and Comprehensive Evaluation of the Multidimensional Environmental Benefits of Sponge Cities. *Water* **2023**, *15*, 2590. https://doi.org/10.3390/w15142590

Academic Editor: Pankaj Kumar

Received: 21 June 2023
Revised: 14 July 2023
Accepted: 15 July 2023
Published: 16 July 2023

1. Introduction

The urbanization process is often accompanied by an increase in impervious underlying surfaces such as buildings and roads, which leads to difficult rainfall infiltration and, in severe cases, the formation of urban flooding, which adversely affects the normal functioning of cities as well as the lives of residents (Zhao et al., 2023; Merchán-Sanmartín et al., 2023) [1,2]. At the same time, human activities release a large number of pollutants, which settle and accumulate on the surface and cause serious nonpoint source pollution under the effect of rainfall erosion [3]. The main pollutants include reductive substances (chemical oxygen demand (COD) is usually used to measure their content) from industrial pollution emissions and vehicle exhaust emissions; suspended solids (SSs) from urban waste, building construction site stockpiles, etc.; total nitrogen (TN) and total phosphorus (TP) from agricultural pollution, leaf litter and animal manure; heavy metals and polycyclic aromatic hydrocarbons (PAHs) from road wear, tire wear, oil spills and corrosion of construction materials [4]. Runoff carrying large amounts of pollutants into the sewers leads to high concentrations of pollutants in the drainage system, which, combined with erosion, pollute both groundwater and surface water [5]. In addition to the destruction of water quality, aquatic ecosystems are degraded as a result, human health is greatly endangered and the world's biodiversity is reduced [6]. These urban water problems often occur simultaneously, which in turn increases the difficulty of their solution.

In response to the multidimensional water problems in cities, Sustainable Stormwater Management (SSM) is widely used in various countries, such as Low Impact Development (LID) in the USA, Sustainable Urban Drainage (SuDS) in the UK, Water Sensitive Urban Design (WSUD) in Australia, Best Management Practices (BMPs) in Europe, etc. [7]. In 2013, the concept of 'sponge cities' was introduced in China to address related issues [8].

Sponge cities constitute a new urban development model that uses small source-control facilities to control rainfall, reduce surface runoff, and improve the urban water quality under the premise of harmonious coexistence between humans and nature [9,10]. In recent years, the construction of sponge cities has emphasized the combination of grey and green infrastructure [11,12]; namely, green infrastructure is the main focus, supplemented by traditional grey engineering drainage facilities.

The concept of the combination of grey and green infrastructure has been widely adopted worldwide, but infrastructure construction requires high investment [13], so the multidimensional benefits provided must be fully studied to comprehensively evaluate the feasibility of construction [14].

The hydrological and nonpoint source pollution control benefits of grey and green infrastructure are the most important. In addition, green infrastructure can solve the problem of moderate or low rainfall runoff to a greater extent, whereas under high rainfall, green facilities can hardly completely dissipate rainfall, and grey infrastructure can then quickly achieve runoff evacuation, which can avoid flooding and control nonpoint source pollution to a certain extent [15].

In addition to water quantity and quality benefits, green infrastructure provides various ecological benefits: for example, plants can mitigate the greenhouse effect by absorbing carbon dioxide through photosynthesis, alleviate the urban heat island effect by absorbing heat through transpiration, reduce soil erosion through soil sequestration by plant roots, and protect urban biodiversity by restoring the ecological environment.

For example, Glick et al. [16], Abduljaleel et al. [17], and Quichimbo-Miguitama et al. [18] simulated the hydrological benefits in their study areas, among which Quichimbo-Miguitama also focused on the inundation reduction benefits. Seo et al. [19] and Deng et al. [20] conducted simulations to evaluate the hydrological and nonpoint source benefits in the study area.

In regard to ecological improvement benefits, LeBleu et al. [21] found that LID stormwater control measures would reduce the heat load of stormwater runoff and mitigate the urban heat island effect to some extent. Shen [22] simulated the mitigation of the heat island effect by green roofs. Lin et al. [23] used the life-cycle assessment method to quantify the carbon reduction in the study area.

In cost–benefit research into grey and green infrastructure, Wilbers et al. [24] divided the benefits of grey and green facilities into direct benefits (avoidance of sewage overflows and urban flooding) and co-benefits (aesthetic value, increase in house prices due to green roof installation, prevention of sewage disposal, water use, etc.) for cost–benefit accounting. Wei et al. [25], and Li et al. [26] divided the benefits of these facilities into economic, social, and environmental benefits. Raei et al. [27] and Saadatpour et al. [28] made a comprehensive decision based on construction costs and hydrological and nonpoint source benefits.

The hydrological, nonpoint source or ecological benefits for the grey and green facilities in some of these studies are shown in Appendix A.

There is an urgent need to integrate the benefits of these three aspects. Fewer previous studies on grey and green facilities have examined hydrological, nonpoint source, and ecological benefits in an integrated manner. In addition, previous studies have rarely considered construction costs, and cost–benefit accounting of the hydrological, nonpoint source, and ecological aspects of grey and green facilities is becoming increasingly complicated and must be explored by introducing methods of monetization.

This study adopted the Beijing Normal University as the study area and simulated the comprehensive benefits of hydrological regulation, nonpoint source reduction, and ecological improvement before and after the retrofitting of grey and green facilities. This study coupled the storm water management model (SWMM) and InfoWorks Integrated

Catchment Management (ICM). The combination of SWMM, with its excellent simulation of hydrology and water quality, and Infoworks ICM, with its powerful and accurate simulation of 2D flooding, provides a more comprehensive assessment of the contribution of grey and green facilities to rainfall runoff. This study also constructed a comprehensive evaluation index system for the benefits of grey and green infrastructure and monetized the benefits of the above three aspects. Finally, the benefit–cost ratio of grey and green infrastructure renovation in the study area was evaluated.

2. Materials and Methods

2.1. Simulation of the Benefits of Grey and Green Facilities

2.1.1. Simulation Index Determination

The study of the hydrological control benefits of grey and green infrastructure mainly examines their control on runoff and waterlogging [29]. The Technical Guide for Sponge City Construction, issued by the Ministry of Housing and Urban–Rural Construction in 2014, mentioned the planning control objectives for sponge city construction including the total runoff control and peak runoff control [20]. Therefore, four indicators were selected for runoff control, including the total runoff reduction, peak flow reduction, total runoff reduction rate and peak flow reduction rate [30]. The actual effect of a grey and green facility can be determined by the change in volume, but the change in volume depends to some extent on the magnitude of the rainfall that is the subject of the study. For example, when rainfall is low, the total reduction in runoff may be small, but the reduction rate may be high. Therefore, the reduction rate should be included to judge the effectiveness of the grey and green facilities. Both types of indicators need to be considered in order to evaluate the effectiveness of facilities in a comprehensive manner. Since cities focus on the environmental risk of sewer overflows [31], two indicators, namely overflow reduction and the reduction rate of overflow nodes, were selected for flooding mitigation [32].

The water quality benefit reflects the ability of grey and green infrastructure to absorb and transform pollution resulting from rainfall runoff [33]. It has been found that the pollutants commonly present at high levels in rainfall runoff include the COD, SSs, TN, TP, and heavy metal pollutants [34]. Among these, heavy metal pollutants are various and complex, which makes them difficult to be fully explored in the study. It was found that SS in road stormwater runoff from urbanized areas had a good positive correlation with most particulate-bound metals, with correlation coefficients ranging from 0.52–0.61 with heavy metals such as Cd, Cr, Cu, Zn and Pb. So SS is used to represent heavy metal contaminants in this study [35]. Therefore, the reductions in COD, SS, TN and TP levels were used as the main indicators of the simulation of water quality benefits.

There are numerous ecological improvement benefits, including groundwater replenishment, urban heat-island effect mitigation, storm water and sewage recycling, soil erosion improvement, and carbon sequestration and oxygen release from green areas [36,37]. Since the ecological benefits of different regions and different grey green infrastructures vary, the evaluation indexes for ecological benefits must be selected according to the specific study area.

2.1.2. The Storm Water Management Model (SWMM)

The SWMM is a dynamic simulation model for the calculation and prediction of surface runoff and nonpoint source pollution loads under the influence of rainfall events, simulation and optimization of stormwater management measures, and planning and design of drainage networks. The infiltration models for its flow-producing process include the Horton model, Green–Ampt model, and Soil Conservation Service (SCS) curve model, which simulate rainwater and runoff infiltration into the soil during rainfall events. The SWMM catchment process is based on the nonlinear reservoir approach. The methods used to calculate the pipe network are divided into steady flow, dynamic waves, and kinematic waves. The runoff simulation mainly simulates the moment-to-moment changes in runoff volume, infiltration, evaporation, and other processes during rainfall, as well as runoff from various nodes and outfalls. SWMM can obtain the amount of total runoff as well as peak flow, and other data required for the study.

The SWMM water quality simulation process is based on different land-use pollutant accumulation models and pollutant flushing models. The surface accumulation model includes a power function, exponential function and saturation function, and the surface flushing model includes exponential flushing, performance curve flushing and event average concentration modules [38]. The water quality simulation mainly simulates the moment-to-moment change of pollutant concentrations during rainfall and the discharge of pollutants at each node and outfall. SWMM can obtain the average concentration of pollutants, total amount of pollutants and other data for each rainfall.

The choice of model is determined by the actual conditions in the study area. Specific modelling results in this paper are presented later.

2.1.3. The InfoWorks Integrated Catchment Management (ICM)

InfoWorks ICM is a powerful two-dimensional (2D) flood simulation tool that provides a more realistic simulation of the interaction between the pipe network system and surface water in order to simulate the process of surface runoff movement and the occurrence of flooding, hence the introduction of InfoWorks ICM in this study to assess urban flooding and the ability of the urban pipe network system. In addition, ICM provides powerful pre-processing and post-processing data capabilities, and it is compatible with the SWMM network, allowing statistical analysis of simulation results based on this platform [39]. The 2D flooding simulation requires the input of digital elevation model (DEM) data to create a ground irregular triangular network (TIN) model. A 2D simulation polygon is then created within the TIN model, which is used as the basis for the flooding calculations. With Infoworks ICM, it is possible to obtain a range of data such as the time curve of the flooded area and the flooded points during the rainfall in the study area. The modelling process and the relevant input data are described in detail later.

2.2. Evaluation of the Benefits of Grey and Green Infrastructure

Figure 1 shows the aspects covered in the cost–benefit evaluation of grey and green infrastructure. Specific calculations should be screened and adjusted to the actual situation in the study area.

2.2.1. Runoff Control Benefits

The discharge of stormwater runoff increases urban construction costs, such as the maintenance and refurbishment costs of impervious and permeable underlying surfaces, construction and operation and maintenance costs of drainage networks, as well as rainwater-saving facilities, and energy use costs [40]. The construction of grey and green infrastructure reduces runoff discharge and lowers these costs to a certain extent, yielding economic benefits. Because of the wide variety of factors, the cost statistics are highly complex. There is no uniform domestic fee standard in China, so this study referred to the stormwater drainage fees levied on stormwater in other countries to obtain the economic value [41], which can be calculated as follows:

$$V_{ro} = \alpha R P_{ro} S / 1000 \tag{1}$$

where V_{ro} is the runoff control benefit, [CNY]; R is the precipitation in the study area, [mm]; P_{ro} is the discharge cost of rainfall runoff, [CNY/m^3]; S is the catchment area of the study area, [m^2]; and α is the total runoff reduction rate.

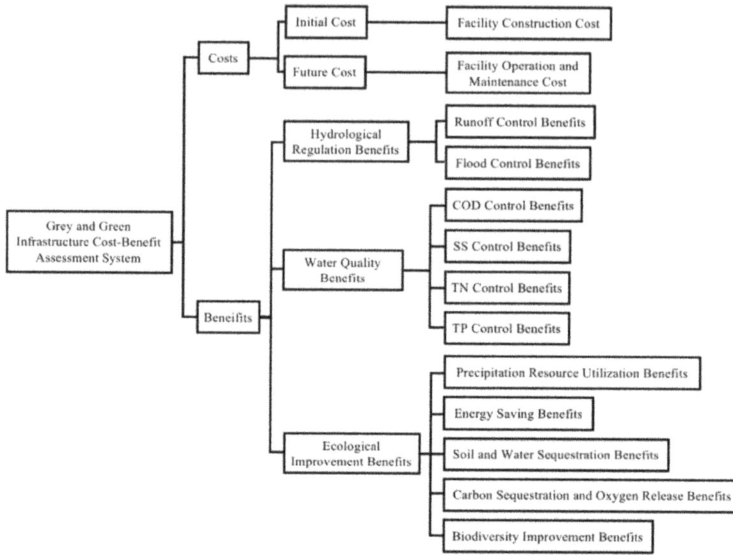

Figure 1. Grey and green infrastructure cost–benefit evaluation system.

The cost of discharging rainfall runoff is referenced to the levy in countries with more established systems for collecting stormwater drainage fees; the catchment area is the projected area of the catchment area in the study area; and the total runoff reduction rate is the reduction rate of the total runoff before and after the facility modification for the same rainfall level. The amount of rainfall in the study area is selected according to the actual measurement needs, such as annual/daily/field rainfall. In this paper, the benefits of runoff control are calculated separately for different rainfall levels, as the amount of rainfall varies considerably between different rainfall levels. Precipitation in the study area is averaged over the eight rainfall events measured in the field for the different rainfall levels.

2.2.2. Flood Control Benefits

Under large rainfall, the flooding problem can be effectively alleviated and the economic and social losses can be reduced through the implementation of permeable paving, water storage ponds and upgraded pipe network systems. Using the shadow engineering method to estimate the economic losses caused by flooding by assuming the cost of manually constructing a flood control reservoir, the monetization of the flood control benefits can be calculated as follows:

$$V_f = V_{vf} P_f \tag{2}$$

where V_f is the flood control benefit, [CNY]; V_{vf} is the flood control volume, [m^3]; and P_f is the flood control reservoir cost per unit volume, [CNY/m^3].

Among these, the flood control volume is the reduction in the overflow volume after the installation of additional renovation facilities. The cost of the flood control reservoir is determined with reference to the standard of relevant documents on construction in the study area or the average market cost after market research.

2.2.3. Water Quality Benefits

If a combined drainage system is adopted, rainwater and domestic sewage, industrial wastewater, etc., are sent together to the sewage treatment plant for treatment, which increases the treatment cost of the sewage treatment plant. If a divided drainage system is adopted, rainfall runoff is discharged directly into the water without treatment, causing pollution to the water. Residential areas and campuses, etc., mainly adopt a separate drainage system [42]. Green infrastructure can effectively dissipate various pollutants, such as COD, SS, TN, and TP, and decrease the concentration of pollutants in runoff, which can be converted into a reduction in treatment costs using the opportunity cost approach in the case of diversion systems. In addition, the reduction in pollutant inflow to the water will reduce the negative impacts of eutrophication, non-carcinogenic toxicity and ecotoxicity of the water, resulting in certain nonpoint source reduction benefits [42]. Therefore, the monetization of water quality benefits can be seen as the sum of the reduction in treatment operation costs and the reduction in the negative impacts of the receiving water, which can be calculated as follows:

$$V_{NPS} = P_{1-i}M_{1-i} + P_{2-i}M_{2-i} \tag{3}$$

where V_{NPS} is the water quality benefit, [CNY]; P_{1-i} is the unit cost of pollutant treatment, [CNY/t]; P_{2-i} is the economic benefit of reducing the negative impact of water per unit of pollutant treated, [CNY/t]; and M_{1-i} together with M_{2-i} are the amount of pollutants abated, [kg].

Where P_{1-i} and P_{2-i} varies by region, calculations are based on relevant regional studies or public financial data from regional governments. M_{1-i} and M_{2-i} can be fitted using the SWMM, which has different modelling equations for different LID facilities and will fit the amount of pollutants from the additional LID facilities to the corresponding equations based on the amount of monitored rainfall runoff pollution.

2.2.4. Benefit of Hydrological Regulation and Water Quality

As the benefits calculated in this study area are for a single rainfall event of different rainfall classes, in order to obtain data that can be compared, they should be converted into annual benefits and discounted according to the proportion of annual rainfall classes in the study area, which can be calculated as follows:

$$V_{i,a} = \sum V_i M_i \tag{4}$$

where $V_{i,a}$ is the average annual benefit of hydrological regulation and water quality, [CNY]; V_i is the economic benefit of a single rainfall event for a given rainfall level, [CNY/field]; and M_i is the annual average of the number of rainfall events at a given level.

Here, M_i is discounted based on the total number of years of rainfall at different levels divided by the number of years in the study area. The rainfall levels in this paper follows the method currently practiced in China, which is based on the amount of rainfall received in a 24-h period. The exact classification is described in detail later.

2.2.5. Resource Utilization Benefits

LID facilities increase the amount of available water resources through the retention of rainwater. The storage LID facilities collect and store rainwater, which can be used for urban green-space irrigation, road cleaning, fire fighting, etc., saving water costs [43], calculated as in (5) [44]. The infiltration LID facilities increase rainwater infiltration to replenish groundwater resources, alleviating groundwater overdraft to a certain extent, with benefits calculated as in (6) [45]:

$$V_{rs} = R_a S_{LID} \delta V_{r,v} / 1000 \tag{5}$$

$$V_{ri} = \mu V_v \tag{6}$$

where V_{rs} is the precipitation resource storage benefit, [CNY]; R_a is the average annual precipitation in the study area, [mm]; S_{LID} is the construction area of LID facilities, [m^2]; δ is the runoff coefficient; and $V_{r,v}$ is the economic benefit per unit of the rainwater volume, [CNY/m^3]; V_{ri} is the infiltration benefit of precipitation resources, [CNY]; μ is the infiltration benefit coefficient; and V_v is the controlled stormwater volume, [m^3].

The calculation is based on the shadow price of water resources in the study area and the saturated water content of the soil (28.90%) [45].

2.2.6. Energy Saving Benefits

Green roofs can increase the area of urban green space and alleviate the heat island effect. In addition, they also have a regulating effect on the temperature of the building's roof and interior. As the indoor temperature decreases in summer by using green roofs, the frequency and duration of the use of cooling equipment decreases, which results in the reduction in the energy consumption of air conditioners and electric fans [46,47]. This benefit can be calculated as follows:

$$V_{UHI} = Q_{elec} S_{gr} P_{elec} \tag{7}$$

where V_{UHI} is the energy saving benefit, [CNY]; Q_{elec} is the reduction in electricity consumption by green roofs, [kWh/m^2]; S_{gr} is the total green roof area, [m^2]; and P_{elec} is the electricity price, [CNY/kWh].

A hectare of green space can absorb 8.1×10^4 kJ of heat in the surrounding environment in summer, and its cooling effect is the same as 189 air conditioners in a full day. The total amount of heat absorbed is calculated based on the area of green roof construction, and the electricity consumption of cooling equipment required to achieve the same cooling effect is measured as Q_{elec} [48].

2.2.7. Soil and Water Sequestration Benefits

Soil erosion leads to a significant loss of soil nutrients, especially nutrients such as nitrogen (*N*), phosphorus (*P*) and potassium (*K*). Green infrastructure can effectively reduce the flow and speed of rainwater runoff, thus weakening soil erosion runoff, weakening urban soil erosion to a certain extent, and maintaining the original soil fertility. The economic value of soil fertility maintenance is assessed using the opportunity cost approach and is calculated as follows [48]:

$$V_s = \sum_i Q_{sr} C_i P_i / 10,000 \tag{8}$$

$$Q_{sr} = RKLS(1 - C) \tag{9}$$

where V_s denotes the soil and water sequestration benefits, [CNY]; Q_{sr} is soil retention, [t/a]; i refers to N, P and K nutrients, respectively; C_i is the percentage of pure content of N, P and K in the soil; P_i is the average price of N, P and K fertilizers, [CNY/t]; R is the rainfall erosion force factor; K is the soil erodibility factor; L is the slope length factor; S is the slope factor; and C is the vegetation cover factor.

Sampling points were laid out for soil sampling based on factors such as land use patterns, crop types, fertilizer application methods, farming history, management systems, etc. in the study area. After measuring the content of N, P and K in the soil samples, they were imported into the GIS to calculate C_i; the calculation of P_i is based on the average value of the market in different regions after market research; the calculation of R, K, L, S and C is based on the study of Ouyang et al. [49] and 'The Technical Specification for Investigation and Assessment of National Environmental Standards' in China.

2.2.8. Carbon Sequestration and Oxygen Release Benefits

Green roofs, constructed wetlands, and concave herbaceous fields are planted with a large number of plants, which are highly valuable in carbon sequestration and oxygen release and very important for urban carbon emission reduction. Therefore, we focused on quantifying the carbon sequestration and oxygen release benefits of green infrastructure as follows:

$$V_{cfor} = V_{CO_2} + V_{O_2} \tag{10}$$

$$V_{CO_2} = M_{CO_2} P_{CO_2} \tag{11}$$

$$M_{CO_2} = S_{ng}F_{CO_2} \tag{12}$$

where V_{cfor} denotes the carbon sequestration and oxygen release benefits, [CNY]; V_{CO_2} and V_{O_2} are the new green space carbon sequestration and oxygen release benefits, respectively, [CNY]; M_{CO_2} is the new green space carbon dioxide fixation amount, [t]; P_{CO_2} is the carbon sequestration price, [CNY/t]; S_{ng} is the new green space area, [m^2]; and F_{CO_2} is the green space carbon dioxide fixation amount, [kg/(m^2·d)].

In addition, due to different climatic conditions such as temperature and humidity in different regions, different topography, and different types of plants grown, the amount of CO_2 fixed per unit area per unit time varies and is determined according to the specific study area or relevant studies in areas with similar conditions. The calculation uses the internationally accepted carbon tax method, and the price of carbon sequestration is determined according to the carbon tax rate in the specific study area, with reference to carbon tax rates in other countries where carbon taxes are not implemented or where the carbon tax market mechanism is less mature.

$$V_{O_2} = M_{O_2}P_{O_2} \tag{13}$$

$$M_{O_2} = S_{ng}F_{O_2} \tag{14}$$

where M_{O_2} is the amount of oxygen released from the new green space, [t]; P_{O_2} is the price of industrial oxygen production, [CNY/t]; and F_{O_2} is the amount of oxygen released from the green space, [kg/(m^2·d)].

Here, F_{O_2} is again determined based on specific study areas or relevant studies in areas with similar conditions.

2.2.9. Biodiversity Benefits

Concave herbaceous fields, green roofs, and constructed wetlands can effectively increase urban plant diversity through vegetation planting and improve the urban ecological environment while also attracting various animals, such as insects and birds, thus contributing to biodiversity improvement [50]. Since it is difficult to quantify the changes in the number of plant and animal species resulting from the application of these green infrastructures in detail, the biodiversity improvement benefits were calculated using the results of the study of Xie et al. [51] on the ecological service value equivalent per unit area of different ecosystems, as follows:

$$V_{BIO} = EQ_{BIO} \tag{15}$$

where V_{BIO} denotes the biodiversity improvement benefits, [CNY]; E is the amount of economic value of an ecological service value equivalent factor, [CNY/hm^2]; and Q_{BIO} is the equivalent value of biodiversity maintenance.

2.3. *Accounting for the Benefit–Cost Ratio of Grey and Green Infrastructure*

2.3.1. Benefit Accounting in the Life Cycle

Benefit accounting can be expressed as follows:

$$V_B = \frac{(1+i)^n - 1}{i(1+i)^n} V_{B,a} \tag{16}$$

where V_B is the present value of the total benefits of the facility over the life cycle, [CNY]; and $V_{B,a}$ is the average annual total benefits of the facility, [CNY/a].

2.3.2. Cost Accounting in the Life Cycle

The cost includes the construction cost and operation and maintenance cost of the project, which can more comprehensively reflect the life cycle cost of grey and green infrastructure and can be calculated as follows:

$$V_C = V_{IC} + \frac{(1+i)^n - 1}{i(1+i)^n} V_{C,a} \tag{17}$$

$$V_{IC,GI} = C_{GI} \cdot S_{GI} \tag{18}$$

$$V_{IC,R} = C_R \cdot V_{Re} \tag{19}$$

$$V_{IC,DP} = C_{DP} \cdot L_{DP} \tag{20}$$

where V_C is the present value of the life cycle facility engineering cost, [CNY]; V_{IC} is the facility construction cost, [CNY]; $V_{C,a}$ is the facility operation and maintenance cost, [CNY/a]; n is the facility design life, [a]; i is the discount rate; $V_{IC,GI}$, $V_{IC,R}$, and $V_{IC,DP}$ are the green infrastructure, water storage facility, and drainage network system construction costs, respectively, [CNY]; C_{GI} is the construction cost of green infrastructure per unit area, [CNY/m^2]; S_{GI} is the construction area of green infrastructure, [m^2]; C_R is the construction cost of water storage facilities per unit volume, [CNY/m^3]; V_{Re} is the construction volume of water storage facilities, [m^3]; C_{DP} is the construction cost of the drainage pipe network per unit length, [CNY/m]; and L_{DP} is the construction length of the drainage pipe network, [m].

Among these parameters, existing studies usually set the operation and maintenance cost as a percentage of the initial cost, and this study followed this method and set the operation and maintenance cost to 3% of the facility construction cost [52,53].

Figure 2 shows the estimated unit cost data for selected low-impact development of individual facilities in the Technical Guide for Sponge City Construction.

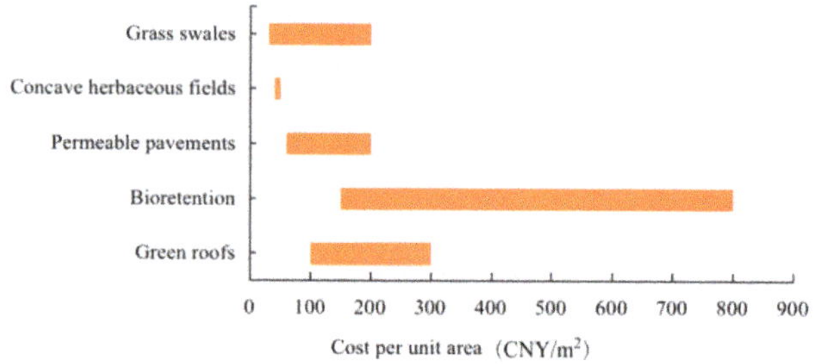

Figure 2. Construction cost of major green infrastructure project.

2.3.3. Benefit–Cost Ratio Accounting

Cost–benefit accounting is conducive to deepening the understanding of the investment, operation, and maintenance of grey and green facilities. The benefit–cost ratio, which can measure the economic effectiveness, is generally used for systematic assessment, and can be calculated as follows:

$$B/C = \frac{V_B}{V_C} \tag{21}$$

where B/C is the benefit–cost ratio, and the higher the B/C value is, the higher the effectiveness under the same investment conditions.

3. Study Case

3.1. Overview of the Study Area

The Beijing Normal University selected in this study covers an area of 58.3 ha, with nine types of underlying surfaces, including roads, sidewalks, roofs, green areas, mixed land, artificial grass sports fields, real grass sports fields, permeable pavements, and asphalt pavements. According to the type of underlying surface, the study area can be divided into 748 catchment areas, as shown in Figure 3. The stormwater pipe network in the study area contains a total of 5 outfalls, of which the catchment area controlled by outfall 3 accounts for more than 80% of the total study area, so this outfall was used as the flow and water quality monitoring object. Due to the low construction standard of the stormwater pipe network system in the study area, it is difficult to drain water in a timely manner, and the area is therefore prone to flooding, while the stormwater runoff pollution problem is very prominent. At the same time, the campus exhibits a high population density and complex functions, and the total amount of pollution discharge is large [54]. Therefore, Beijing Normal University was purposefully selected as the case study area.

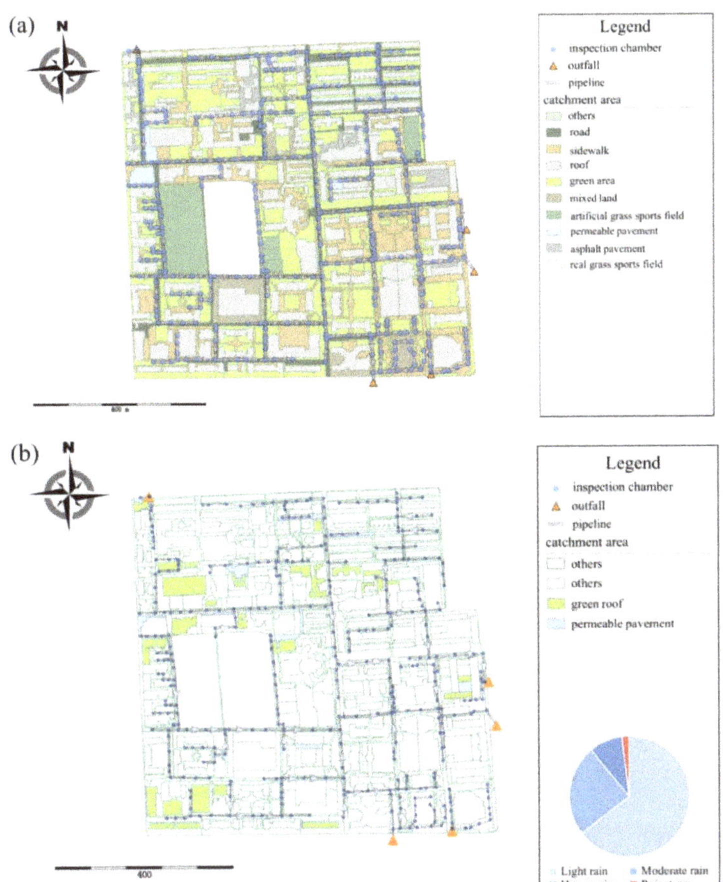

Figure 3. (**a**) Distribution of land use types and layout of pipe networks in the study area; (**b**) Design location of grey and green infrastructure renovation in the study area and probability of rainfall occurrence by level in Beijing from 2010 to 2019 (at the bottom right).

Since the study area belongs to a relatively mature community, it is difficult to renovate grey facilities, so the renovation of facilities in the study area is based on green infrastructure. However, due to the high building density and limited surface space in the study area, the addition of green facilities alone cannot completely solve the water quantity and quality problems, so they should be supplemented with grey facilities. In terms of green infrastructure, flat roofs, nonmain roads and asphalt pavements were selected for renovation. Impervious pavements were replaced with permeable pavements, and flat roofs were transformed into green roofs, with a total renovation area of 33,000 m^2, occupying approximately 5.52% of the total study area, of which 27,435 m^2 comprised green roofs. Regarding grey facilities, due to the obvious lack of drainage capacity of the current pipe network, the existing system should be renovated and upgraded. When the ratio of the backflow volume of the downstream pipe section to the incoming volume of the upstream pipe section is less than 0.5, it means that excessive upstream incoming flow is the main cause of nodal overflow. Such a pipe section should be retrofitted by increasing the pipe diameter of the downstream pipe. The locations of green and grey infrastructure renovation and design are shown in Figure 3. The green facility retrofit design parameters are shown in Table 1.

Table 1. Design parameters for green infrastructure retrofit.

Green Roof	Indexes	Value	Permeable Pavement	Indexes	Value
	Height of the berm/mm	250		Height of the berm/mm	20
Surface	Vegetation coverage	0.9	Surface	Vegetation coverage	0.15
	Surface roughness	0.1		Surface roughness	0.02
	Surface slope	1		Surface slope	1
	Thickness/mm	100		Thickness/mm	150
	Porosity	0.463	Pavement	Voids ratio	0.21
	Actual water content volume	0.232		Permeability/(mm·h^{-1})	2000
Soil	Withering point	0.116		Blockage coefficient	83
	Conductivity/(mm·h^{-1})	3.6		Thickness/mm	100
	Conductivity slope	10		Porosity	0.463
	Suction head/mm	88.9		Actual water content volume	0.232
	Thickness/mm	100	Soil	Withering point	0.116
Drainage mat	Voids ratio	0.5		Conductivity/(mm·h^{-1})	3.6
	Manning roughness	0.02		Conductivity slope	10
				Suction head/mm	88.9
			Storage	Thickness/mm	300

3.2. *Rainfall Data Collection*

3.2.1. Rainfall in Beijing

Based on the hourly rainfall data retrieved from the National Basic Weather Station Beijing Nanjiao Observatory (54,511) from 2010 to 2019 and considering the classification guideline of recognizing the next rain event if no rainfall has occurred for more than 10 h [55], a total of 631 rainfall events were obtained, and the average number of rainfall events per year in Beijing was chosen as 63.1. Unlike many studies that use the return periods as the grading methods [56,57], this study uses 24-h rainfall for the division. According to the meteorological department's standard classification, rainfall between 0.1 mm to 9.9 mm in 24 h is considered light rain, between 10.0 mm to 24.9 mm is moderate rain, between 25.0 mm to 49.9 mm is heavy rain, and greater than 49.9 mm is a rainstorm. Thus, of the 631 rainfall events, 408 were light rain events, 154 were moderate rain events, 57 are heavy rain events and 12 were rainstorm events, and the probability of occurrence of each level of rainfall in Beijing each year is shown in the bottom right of Figure 3. Some studies also used 24-h rainfall [58,59], but their application differed due to the study area and the content of the study.

3.2.2. Rainfall Monitoring in the Field

In this study, field monitoring was conducted for eight rainfall events, focusing on the process of hydrological and water quality changes at the outfalls of the drainage pipes on campus during rainfall.

The sampling points were divided into surface subsurface sampling and underground stormwater network outfall sampling, taking into account the land use types that have a significant impact on surface water pollution. In addition, in combination with human and material resources, rainfall runoff from five underground types, namely, residential areas, high-density traffic areas, medium-density traffic areas, high-rise building rooftops and low-rise building rooftops, was monitored underground. The outfall of the underground storm water network was selected as outfall 3 (outfall to the municipal drainage network), which has the largest catchment area in the study area, for flow and concentration monitoring.

The HOBO weather station was set up on the roof of the low-rise building and recorded detailed data including atmospheric pressure, temperature, relative humidity, solar radiation values, wind speed and rainfall at 5-min intervals. A flow meter (HACH, FL900) was set up at drainage outfall 3 for real-time monitoring of network flow at the same time interval as above; this was subsequently collated as time series data and entered into the model. The samples collected in 500 mL polyethylene bottles were immediately bottled and taken back to the laboratory for water quality determination [60]. Transient runoff samples were mainly collected by hand sampling. This method is more flexible as it allows the sampling interval to be adjusted at any time depending on the prevailing rainfall conditions. Road surface rainwater was collected at road rainwater grates, roof rainwater was collected at down pipes and underground pipe network samples were collected in rainwater wells. In addition, the water quality data were obtained through the analysis of runoff samples collected in the rainfall process. The rainfall information and hydrological and water quality data are shown in Table 2.

Table 2. Water quality conditions of eight representative rainfall events in the study area in 2014.

Rainfall Events	Duration of Rainfall/min	Precipitation /mm	Volume Capture Ratio of Annual Rainfall/%	Average Concentration of COD /(mg·L^{-1})	Average Concentration of SS /(mg·L^{-1})	Average Concentration of TN /(mg·L^{-1})	Average Concentration of TP /(mg·L^{-1})
0804 Light rain	266.00	5.66	0.79	5.06	2.67	0.28	0.01
0809 Light rain	125.00	5.64	0.76	7.96	4.47	0.36	0.01
0823 Moderate rain	50.00	10.40	0.69	39.63	23.47	1.36	0.11
0926 Moderate rain	25.00	7.80	0.71	35.08	21.14	1.10	0.09
0729 Heavy rain	640.00	35.74	0.65	64.10	33.41	3.73	0.22
0830 Heavy rain	115.00	29.00	0.67	62.06	29.97	3.37	0.25
0901 Heavy rain	1885.00	33.60	0.70	76.06	36.67	3.54	0.25
0831 Rainstorm	170.00	70.56	0.68	63.05	31.05	5.76	0.30

All the data were used for subsequent modelling, including 2 light rain events, 2 moderate rain events, 3 heavy rain events and 1 rainstorm event, which helped to analyze the differences in the benefits under the different rainfall levels. The rainfall classification is based on the amount of rainfall received in a 24-h period as previously described, with the exception of rainfall event 0926 which was 7.8 mm, but it only lasted 25 min. Given the intensity of the rainfall, this rainfall was classified as moderate rain.

3.3. Model Construction

In this study, the urban storm sewer and watershed stormwater management modelling software PCSWMM was used to simulate runoff and pollutants in the study area.

The data required for the SWMM simulations include catchment data (land use types, pipe network data, digital elevation data), meteorological data and hydrological observations for the rate and validation of the model [61]. The land use data and pipe network data are taken from the school platform; the digital elevation data are taken from all rainwater nodes (2048 points) and elevation data provided by the sounding company (622 points); and the other data are taken from field monitoring results. Of the parameters to be entered, the main physical characteristics of the catchment characterization data such as area and imperviousness were obtained using Arc GIS analysis of spatial data. The average slope of the sub-catchment area is 1.7%, which is calculated from the DEM data of all nodes; the pipe diameter, pipe length and slope are obtained from the pipe network data.

According to the actual situation, the Horton model was selected for the infiltration simulation model [62], the saturation function was selected for the accumulation model, and exponential flushing was selected for the flushing model.

Model calibration and validation were performed using the data for the eight actual monitored rainfall sites listed in Table 2, where the initial calibration (sensitivity-based radio tuning calibration) was performed using the SRTC tool in PCSWMM [63], followed by more accurate calibration via the availability-aware scheduling algorithm by using the nondominated sorting genetic algorithm II (NSGAII) [64].

The percentage of imperviousness is one of the most sensitive parameters affecting model simulations [65]. However, the data processing calculation process and the regional sub-bedding type decoding process are subject to some errors. In order to reduce the uncertainty, this paper determined the model sub-bedding input data through high resolution land-use maps (5 m × 5 m) and field research, while setting a 10% uncertainty for land use.

The rainfall events used for calibration were 0729, 0804, 0809 and 0823 and for validation were 0830, 0831, 0901 and 0926, taking into account the number of peaks in the rainfall events and the amount of rainfall. Both the calibration and validation NSE values were above 0.8 [66,67], and R^2 exceeded 0.9 [68], which indicates satisfactory simulation results. The simulation of the validation event 0831 was less effective than the other three rainfall events, presumably because the rate period did not include the type of short-duration heavy rainfall like 0831, so the parameter set obtained from the rate was not as effective as the other rainfall events for this type of rainfall simulation. The model parameters are summarized in Table 3, and the results are shown in Figure 4. The SWMM model tends to overestimate peak flows, which can cause peak flow reductions to be somewhat overestimated.

Table 3. Calibration results of main parameters of SWMM in the study area.

Parameter	Calibration Result	Parameter	Calibration Result
N-Imperv	1.20×10^{-2}	Dstore-Asphalt Pavements/mm	1.15
N-Perv	0.80	Dstore-Roofs/mm	1.23
Max.Infil.Rate/(mm·h^{-1})	150.00	Dstore-Concrete Pavements/mm	1.34
Min.Infil.Rate/(mm·h^{-1})	20.00	Dstore-Sports Field 1/mm	1.77
Decay Constant/(h^{-1})	2.00	Dstore-Sports Field 2/mm	1.68
Zero-Imperv/%	25.00	Dstore-Mixed Land/mm	2.22
Pipe Roughness	1.50×10^{-2}	Dstore-Perv/mm	10.20

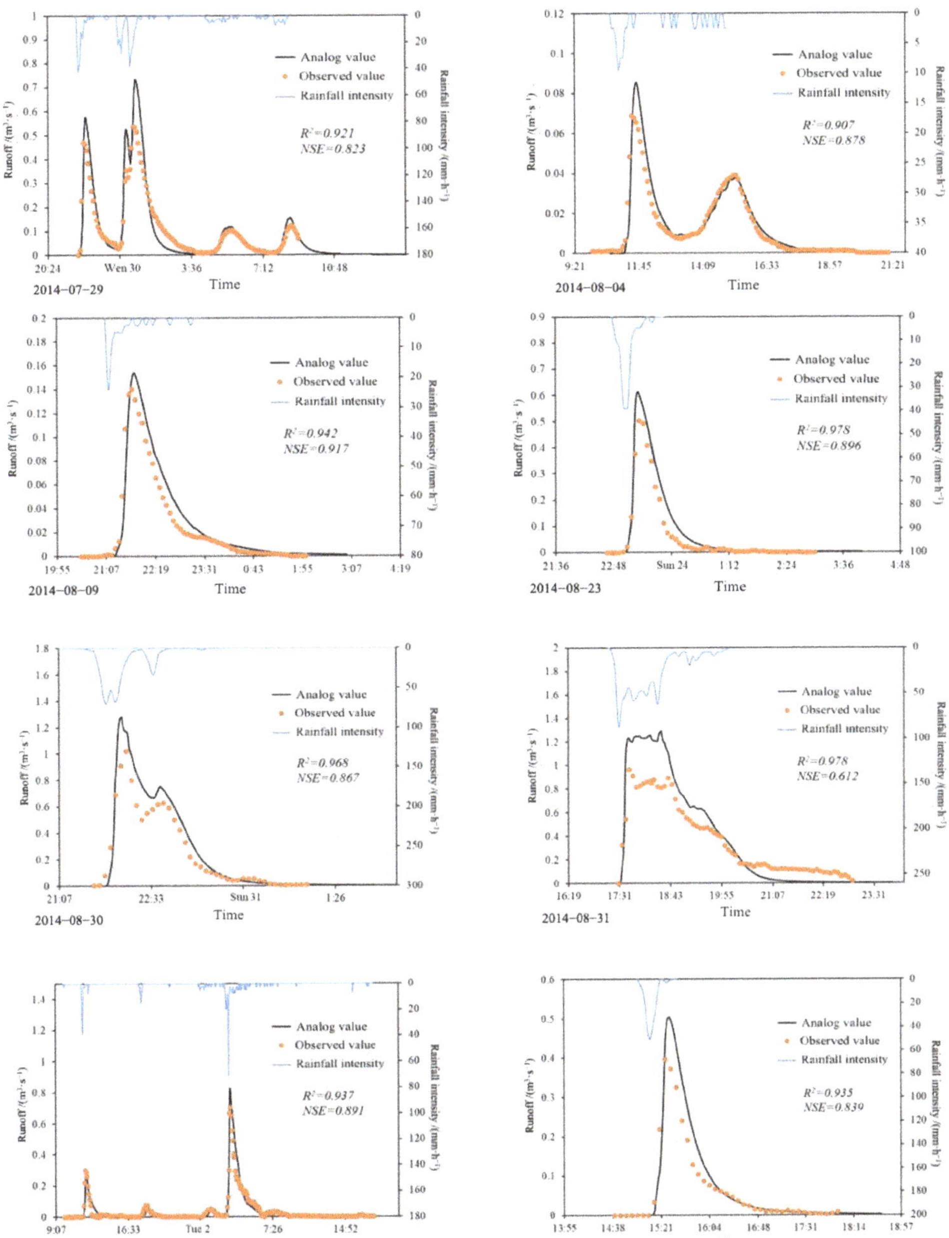

Figure 4. Water quantity simulation calibration and validation results of SWMM in the study area.

Based on the actual monitoring of pollutant characteristics and rainfall pattern characteristics, COD, NH_3-N and TP, which are conventional pollutants, were selected for nonpoint source pollution simulation. TP and SS have a high positive correlation as the majority of the TP is in the particulate state [69–71], and the results of the calibration and validation of TP can to some extent reflect the results of SS. Due to the lack of local data for SS, calibration and validation results from TP were used as a supplement. Therefore, the above three indicators were mainly selected for the determination.

As the contaminant accumulation and flushing processes in the subsurface are influenced by regional characteristics and there is no corresponding reference range of values, AANSGA-II was used directly to calibrate the contaminant accumulation and flushing parameters. The rainfall events collected were 0729, 0804, 0830, 0831 and 0926, and the results are shown in Table 4.

Table 4. Water quality simulation calibration and validation NSE results of SWMM in the study area.

Determination of Model Parameters	Rainfall Events	COD	NH_3-N	TP
Calibration	0729	0.613	0.625	0.594
	0830	0.483	−5.420	0.542
	0926	0.507	0.523	0.341
Validation	0804	0.546	0.487	−0.344
	0831	0.669	0.373	0.477

The validated SWMM was imported into InfoWorks ICM for debugging to ensure the accuracy of the simulation results. Before 2D flooding simulation, digital elevation model (DEM) data for flooding simulation was obtained by interpolation (using the inverse distance-weighted interpolation method) in ArcGIS based on the ground elevation data of 374 inspection wells to create a ground irregular triangular network (TIN) model in InfoWorks ICM, based on the aforementioned total of 2670 elevation data (2048 rainwater nodes points and 622 points provided by the sounding company). A 2D simulation polygon was created within the TIN model, and the building area was defined as a blank area to preclude runoff flow in the simulation process. The polygon was gridded, a 2D flooding calculation was performed, and the data were read to obtain a time–flooded area curve [39].

As the subsequent evaluation of the flood control effect of grey and green facilities mainly considered event 0831, it could not be used to test the validity of the model. Among the remaining seven rainfall events, the one with the highest intensity of rainfall, which is the most prone to flooding, was selected for simulation. The precipitation of 0729, 0830 and 0901 were close to each other, but the rainfall duration was significantly different, so the event 0830 with the shortest rainfall duration was selected for SWMM simulation. The amount of water overflowing from the check wells was selected as the input data for the flooding simulations. The simulated waterlogging points were basically consistent with the actual observed waterlogging locations, which verified the validity of the model simulation process.

3.4. Cost-Effectiveness Evaluation

The calculation parameters are shown in Table 5. As many countries charge plot owners for stormwater fees on the basis of impervious surface area [72,73], it is difficult to reflect the difference in fees for different rainfall levels. The average value of the levy per unit volume of rainwater discharge in Godyń's study [74] was chosen, where the value in his study was used directly because the exchange rate of the EUR against the CNY is close to purchase power parity [75]. Reservoir construction units and reservoir capacity investment reference the national forestry department uniform standard value. The reduced flood control cost per unit area of runoff reduction was based on the annual emergency fund for flood control in Beijing. The amount of carbon dioxide fixed and oxygen released from the green space was based on the field observation results of Li et al. [76]. China launched a national carbon emissions trading system in 2021, but the foundation of the market-based mechanism for carbon price formation is still weak, and the trading structure needs to be improved, so we referred to the median carbon tax rate values of other countries in State and Trends of Carbon Pricing 2021. The price per unit mass of oxygen refers to the China Price Statistical Yearbook. The COD reduction cost per unit was based on the study results of Li et al. [77]. The cost of treating other pollutants and the economic benefits of reducing the negative impacts on water per unit of pollutant treated is with reference to the study by Zhu et al. [42]. She [78] found that the construction cost of pipe networks with pipe diameters of 600–1000 mm was 247–990 CHY/m in China. The construction costs of permeable pavements, green roofs, and pipe network systems in this study area were chosen as average values, i.e., 130 CHY/m^2, 200 CHY/m^2, and 618.5 CHY/m, respectively.

Table 5. Cost-effectiveness evaluation method parameters.

Parameter	Value	Parameter	Value
n	30a	i	5%
P_{ro}	1.23 CNY/(m^3)	P_f	6.11 CNY/(m^3)
F_{CO_2}	0.013 kg/(m^2·d)	F_{O_2}	0.018 kg/(m^2·d)
P_{CO_2}	141 CNY/t	P_{O_2}	1108 CNY/t
P_{1-COD}	4.14 CNY/kg	P_{1-TP}	52.4 CNY/kg
P_{1-TN}	23 CNY/kg	P_{2-SS}	4.96 CNY/kg
P_{2-COD}	1.24 CNY/kg	P_{2-TP}	0.176 CNY/kg
P_{2-TN}	0.996 CNY/kg		

4. Results and Discussion

4.1. Quantitative Analysis of the Benefit Indicators

The runoff control results for the eight rainfall events are shown in Table 6. The retrofitting measures yielded favorable runoff control effects at the different levels of rainfall, but the effect decreased with increasing rainfall level. This is consistent with the findings of Guo et al. [68] that LID facilities are more effective in controlling runoff during smaller and more frequent rainfall events. After retrofitting, runoff was almost not discharged under the light rain and moderate rain scenarios, which suggests that the grey and green infrastructure can absorb the runoff generated under low-level rainfall completely. The runoff control effect of the retrofitting measures under the rainstorm scenario was significantly lower than that under the other scenarios. But the total runoff reduction rate still reached more than 80% relative to before retrofitting, and the peak flow rate was reduced by nearly 70%, which indicates that grey and green infrastructure retrofitting still provided suitable rainwater absorption under heavy rainfall. The effect was still satisfactory.

Table 6. Simulation results of runoff control in the study area.

Rainfall Level	Total Runoff Reduction/m^3	Total Runoff Reduction Rate/%	Peak Flow Reduction/(m^3·s^{-1})	Peak Flow Reduction Rate/%
Light rain	7.53×10^{-7}	99.69	0.12	99.61
Moderate rain	1.59×10^{-6}	97.96	0.55	99.17
Heavy rain	5.58×10^{-6}	90.52	0.84	89.80
Rainstorm	1.07×10^{-5}	80.88	0.89	68.51

At the same time, simulation of event 0831 before and after the renovation revealed that the number of severe overflow nodes in the study area was reduced from 46 to 21, and the total overflow volume was reduced from 5030 to 3915 m^3 after the renovation. This indicates that the renovation of grey and green infrastructure improved the ability to discharge water from the road surface in the study area. It is conducive to reducing the economic loss caused by flooding and the impact on the activities of residents.

The nonpoint source reduction results for the eight rainfall events are shown in Figure 5. Facility retrofitting produced satisfactory reduction and purification effects on COD, SS, TN and TP. The control effect also decreased with increasing rainfall level because the pollutant content is related to the rainfall level, and the pollutant reduction capacity of the retrofitted facilities reached saturation after a certain rainfall level was attained, resulting in a decrease in the pollutant reduction rate. This is also in line with the study by Li et al. [79] where the resilience of LID facilities in sponge cities decreases with the increase in the rainfall return period.

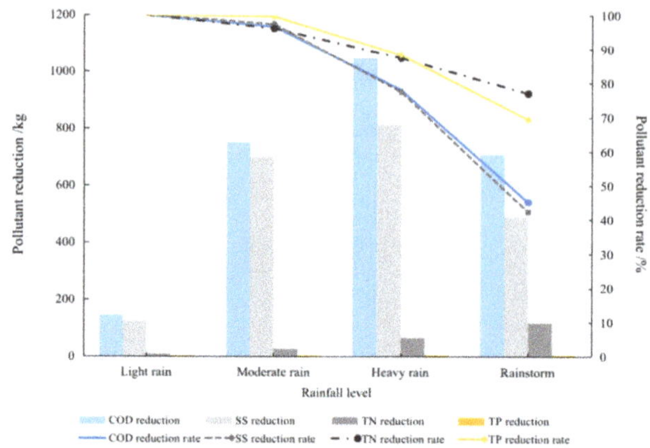

Figure 5. Simulation results of nonpoint source pollution reduction in the study area.

The total amount of each pollutant was significantly reduced in outfall 3 under the light and moderate rainfall scenarios after the renovation, indicating that green infrastructure imposes a strong control effect on the nonpoint source pollution problems generated by low-level rainfall, with the best control effect on the TP level. Under the rainstorm scenario, the water quality effect of the facilities was significantly smaller than that under the other scenarios, among which the COD and SS control effects decreased more significantly than the TN and TP control effects (approximately 30% difference), which is due to the high content and proportion of COD and SSs in runoff and the high runoff accumulation under the rainstorm scenario, resulting in the inability to reduce pollutants promptly. This is in line with She et al. [80] who conducted a study of residential and commercial areas where COD and TSS emissions were significantly higher than TN and TP by tens to thousands of times, with TP reduction rates higher than TSS and COD by about 15–18%.

Due to the small study area, some benefits were not obvious, and the green roof accounted for more than 80% of the total green infrastructure transformation area, which is the main green infrastructure aspect, so the ecological benefits mainly included the carbon sequestration and oxygen release benefits of the green area, which is considered the ecological benefit index in the study area. The calculated carbon dioxide absorption of green roofs in the study area is 356.66 kg/d, and the oxygen release amount is 493.83 kg/d. The green roof retrofitting area accounts for 15.53% of the total roof area in the study area.

According to the Carbon Emission Accounts & Datasets (CEADs) [81], the apparent carbon dioxide emissions in Beijing in 2019 reached 70.61 Mt. Due to the lack of data specific to the study area scale, assuming that the apparent CO_2 emissions per unit area per unit time in Beijing remain the same, i.e., CO_2 emissions of 0.012 kg/(m^2·d), and based on the area of the study area, it can be concluded that the study area emissions are 6872.77 kg/d, so the new green space can reduce the daily CO_2 emissions of the study area by 5.19%, which could contribute to urban carbon emission reduction.

The actual amount of carbon dioxide absorbed and oxygen released may vary due to various factors, such as the specific plant types and ages of lawns in different regions [82], but these differences were not described in this paper due to the difficulty of obtaining statistical information on the variability of lawn plant types, numbers, and distribution patterns.

4.2. Cost Monetization Analysis

The total cost of grey and green infrastructure renovation in the study area is shown in Table 7. The cost of green infrastructure was higher than that of grey infrastructure, namely, 11.30 times higher, which occurred because the renovation mainly involved green renovation, and the total cost of green infrastructure, especially green roof renovation, was high due to the large area.

Table 7. Benefits and costs of grey and green infrastructure in the study area.

Benefits and Costs	Indexes	Light Rain	Moderate Rain	Heavy Rain	Rainstorm
-	Average number of fields/(a^{-1})	40.80	15.40	5.70	1.20
Runoff control benefits	Runoff control benefit/(CNY·field^{-1})	2492.00	3944.24	13,128.96	25,251.08
	Runoff control benefit/(CNY·a^{-1})	101,673.57	60,741.30	74,835.10	30,301.30
Flood control benefits	Flood control benefit/(CNY·field^{-1})	NA	NA	NA	6812.65
	Flood control benefit/(CNY·a^{-1})	NA	NA	NA	8175.18
Water quality benefits	COD control benefit/(CNY·field^{-1})	796.13	4055.82	5657.12	3828.30
	SS control benefit/(CNY·field^{-1})	621.29	3472.64	4045.43	2432.43
	TN control benefit/(CNY·field^{-1})	190.76	634.80	1601.59	2830.35
	TP control benefit/(CNY·field^{-1})	14.20	119.35	251.31	312.30
	Water quality benefit/(CNY·field^{-1})	1622.37	8282.62	11,555.45	9403.39
	Water quality benefit/(CNY·a^{-1})	66,192.85	127,552.27	65,866.07	11,284.07
Hydrological regulation and water quality benefits	Hydrological regulation and water quality benefits/(CNY·a^{-1})	167,866.42	188,293.57	140,701.17	49,760.54
Ecological benefits	Carbon sequestration benefit/(CNY·d^{-1})	50.29			
	Oxygen release benefit/(CNY·d^{-1})	547.16			
	Ecological benefits/(CNY·a^{-1})	218,070.24			
Total benefits	Total benefits/(CNY)	11,755,189.30			
Construction costs of green infrastructure	Construction cost of permeable pavement/(CNY)	723,450.00			
	Construction cost of green roof/(CNY)	5,487,000.00			
	Total costs of green infrastructure/(CNY)	9,074,545.15			
Construction costs of grey infrastructure	Construction cost of pipe network/(CNY)	549,368.40			
	Total costs of grey infrastructure/(CNY)	802,722.56			
Total costs	Total costs/(CNY)	9,877,267.72			

Notes: Through modelling and field visits, no flooding occurred in the study area under rainfall classes other than rainstorm for the eight rainfall events. Therefore, it is considered that only the rainstorm scenario produces a more significant flood control benefit for the annual rainfall events. Therefore, the flood control benefits for the light, moderate and heavy rainfall scenarios were not calculated and are expressed as NA.

4.3. Benefit Monetization Analysis

The runoff control benefits at the different rainfall levels are shown in Table 7, and the benefits were positively correlated with the rainfall level. The runoff control benefits of the retrofitting measures under the heavy rainfall and rainstorm scenarios were higher, namely, 3 to 10 times, than those under the light and moderate rainfall scenarios, which occurred because the rainfall level under the rainstorm scenario was 7 and 14 times that under the moderate and light rainfall scenarios, respectively, and the rainfall level under the heavy rainfall scenario was approximately 3 and 6 times that under the moderate and light rainfall scenarios, respectively.

The flood control benefit under rainstorm conditions was 1.73–2.73 times greater than the runoff control benefits under light and moderate rainfall conditions, and 27–52% of runoff control benefits under heavy rain and rainstorm conditions, with a higher flood control benefit due to the comprehensive use of various grey and green facilities for the renovation of the overflow nodes of the pipe network in this study, with a significant reduction in overflow nodes and a degree of reduction in overflow volume that can be equated to the construction price of a larger volume flood control reservoir.

The monetized results of the single rain nonpoint source reduction benefits for different rainfall levels, calculated on the basis of eight actual observed rainfall events, are listed in Table 7. The TN and TP control benefits are positively correlated with the rainfall level. The COD and SS control benefits are positively correlated with the rainfall level under the nonextreme rainfall scenarios, and the benefits were reduced by 30–40% under the extreme rainfall scenarios. As the COD and SS control benefits accounted for 66–91% of the total benefits, the water quality benefits follow the same trend as the changes in the benefits of both. The main reason for the trend of increasing and then decreasing COD and SS control benefits is that the rate of pollutant transport under extreme rainfall scenarios exceeded the capacity and rate of pollutant absorption at the retrofit facility.

The total annual runoff control benefits, total flood control benefits, and total water quality benefits at each rainfall level in Beijing were calculated, as summarized in Table 7.

The runoff control benefit of the retrofitted facilities in the study area was the highest under the light rainfall scenario, which was 1.67–3.36 times higher than that under the other scenarios. This occurred because the rainfall in Beijing is mainly light rainfall, accounting for more than 60% of the annual rainfall, and the proportion of the other rainfall scenarios is low, so a single high-value rainfall event slightly impacts the total annual benefit.

The annual effectiveness of facility retrofitting in controlling nonpoint pollution in the study area was highest under the moderate rainfall scenario and lowest under the rainstorm scenario, with the retrofitted facilities under the moderate rainfall scenario 11.30 times more effective than those under the rainstorm scenario. The most significant annual economic benefits of retrofitting the facility can therefore be achieved under a medium rainfall scenario. In this regard, the reduction in pollutants relies mainly on green infrastructure rather than grey infrastructure. The vegetation in the facility slows down the flow of runoff and traps and deposits pollutants there, using the biochemical reaction of the plants and the absorption and infiltration of the soil to avoid pollutants from flowing into natural water bodies and causing water pollution.

It is important to note that as the biochemical reaction of vegetation consumes a limited variety and quantity of pollutants and the soil has a certain carrying capacity, there is a risk that when the concentration of pollutants is too high, the sustainable and stable functioning of the green infrastructure is threatened, resulting in a situation where the total effectiveness of nonpoint source reduction decreases with the duration of use of the facility.

Since the flooding control benefits were only examined under the rainstorm scenario with a very low frequency, the runoff control and water quality benefits of the retrofitted facilities were first discussed under the full suite of rainfall scenarios. The total annual runoff control and water quality benefits of the retrofitted facilities were highest under the moderate rainfall scenario and lowest under the rainstorm scenario, while the benefits under each rainfall scenario were 3.38–4.53 times higher than those under the rainstorm scenario, which occurred because the probability of rainstorms in Beijing is considerably lower than that of the other classes and because the retrofitted facilities provide a limited pollutant reduction capacity under the heavy rainfall scenario. Hence, the benefits under the rainstorm scenario were low. The retrofitting of grey and green infrastructure in the study area should focus more on the control of light, moderate, and heavy rainfall.

The flood control benefits are further included in the discussion. The annual flood control benefit accounted for 16.4% of the hydrological regulation and water quality benefits, which brought very significant economic benefits. At the same time, rainstorms are more harmful to the study area, and waterlogging seriously affects the production and life of people. Therefore, specific control measures for rainstorms should be moderately implemented if the budget allows.

The annual nonpoint source reduction benefit and the annual hydrological regulation benefit of the retrofit facility are basically the same, with the hydrological benefit slightly higher than the water quality benefits, exceeding the water quality benefits by 1.8%. The high water-quality benefits were due to the high unit pollutant reduction costs and the fact that the retrofits mainly involved green facilities with large green roofs providing a high ability to absorb and dissipate pollutants. The high hydrological benefit is due to the fact that both grey and green retrofitted facilities generate hydrological control effects, of which green infrastructure focuses on runoff control under lower rainfall scenarios, and grey infrastructure focuses on higher rainfall scenarios. Green infrastructure can retain most of the rainfall when rainfall is low, resulting in significant runoff control benefits; although the degree of grey infrastructure retrofitting is low, the study area has a temperate monsoon climate with limited rainfall, and statistics show that the study area has a low probability of extreme rainfall, so grey infrastructure retrofitting can meet most of the needs of the study area, resulting in significant runoff and flood control benefits.

The annual ecological benefits of the renovated facilities are approximately 80% and 79% of the annual hydrological and water quality benefits, respectively, accounting for 28.5% of the total annual benefits, which is slightly lower than the hydrological and water quality benefits. But it is still about 1/3 of the overall, indicating that the potential for ecological benefits of the renovated facilities is huge. The fact that only green facilities have ecological improvement benefits leads to a significantly higher monetary value of green facilities than grey facilities. The total benefits of green infrastructure are approximately ten times greater than those of grey facilities.

4.4. Cost–Benefit Ratio Analysis

The benefit–cost ratio over the life cycle is 1.19, of which the benefit–cost ratios of green and grey infrastructure are 1.23 and 0.73, respectively, and green infrastructure is slightly more economically effective than grey infrastructure. To completely solve the flooding problem in the study area, a large amount of grey and green infrastructure was renovated and constructed, and the total runoff amount, peak flow, degree of flooding, and runoff pollutants were significantly reduced, resulting in greater mitigation of the water problem in the study area. But at the same time, as a mature community, the construction and maintenance costs of its facilities significantly increased, thus yielding a limited actual economic value and low net benefits. The grey infrastructure in this study exhibited low alteration and maintenance costs, high rainstorm scenario benefits, and notable alteration of the existing pipe network system, which plays a supporting role under the high to heavy rainfall scenarios but provides poor net benefits due to the extremely low frequency of heavy rainfall in Beijing.

In summary, the modification of grey and green infrastructure in the study area can produce certain economic values of rainfall runoff, internal flooding, and water quality benefits, but the benefit–cost ratio is only slightly higher than 1. If the grey and green facilities in the study area are modified, more suitable types of facilities should be used, and their locations should be optimized, considering the costs and benefits of these facilities.

In addition, the low benefit–cost ratio is due to the small study area in this paper, which does not include the external benefits of grey and green infrastructure in the calculation process. In the actual situation, it is still necessary to consider the following: (1) the reduction in external runoff will impose an ameliorating effect on river scouring as well as flooding, thus reducing the occurrence of disasters such as landslides and mudslides caused by excessive scouring as well as the personal and property losses caused by downstream flooding to local residents; (2) the facilities intercept rainwater, increase the amount of rainwater infiltration, replenish groundwater, and raise the groundwater level, and this part of rainwater can be used as urban green space irrigation water, water for road cleaning, water for firefighting, etc., which to a certain extent reduces tap water development and utilization and eases the pressure on the urban water supply; and (3) green infrastructure can alleviate the heat island effect and reduce energy use, while the synergy between the study area and other green infrastructure can reduce the growth rate of urban energy consumption. To better assess the cost–effectiveness of grey and green infrastructure, the retrofitting effect of the entire system should be evaluated on a larger scale to reduce the possibility of misestimation.

4.5. Uncertainty and Applicability Analysis

Transient runoff samples are mainly collected by hand sampling. There are also disadvantages to this approach, namely a certain amount of subjectivity and non-reproducibility. In the future, programmable auto-samplers could be considered to compensate for the current shortcomings.

In terms of model simulation, the SWMM tends to overestimate peak flows, and high peak flow simulations also lead to high total runoff, which can lead to higher peak flow reduction rates and higher total runoff reduction rates than in reality. This results in higher calculated runoff control benefits than actual values, higher hydrological benefits and higher total benefits.

Since no calibration or validation of SS was performed, the SS confidence level was derived in this study by calibration and validation of TP, and therefore all results related to SS are subject to some error. Heavy metal concentrations were derived from SS, so there is some uncertainty in the heavy metal content, and further research is needed to reduce this uncertainty when considering the heavy metal content and the risk it poses. In addition, in terms of benefit calculation, due to some uncertainty in the SS reduction rate, the benefits from SS reduction calculated through it also have some error, which has an impact on the calculation of water quality control benefits. This ultimately leads to uncertainty in the calculation of benefits. In future research, it is necessary to further obtain richer and more accurate data on SS to compensate for this part of the shortcomings and reduce the uncertainty of benefits and risks.

In terms of calculating the benefits of runoff control, as the study area has not yet implemented a mature stormwater charging system, the stormwater fees of other countries are used as a proxy in this paper. As Poland is relatively similar to China in terms of price levels, while the Euro to CNY exchange rate is close to purchasing power parity, using the Polish levy for stormwater fees reduces the error. However, the stormwater fee is the total cost of all urban drainage infrastructure, including cisterns, pumping stations, pipes, etc. The campus drainage facilities in this study are relatively simpler and the maintenance and construction cost expenditure is relatively less than the reference value, resulting in a high benefit calculation. In addition, the Polish stormwater fee levy is related to the capacity of water storage facilities in impervious areas, and this paper uses the average value, which will lead to some error.

Flood control benefits are proxied using the cost of a flood control reservoir of the same volume as the overflow reduction. In practice, the benefits may not be linearly related to the volume of overflow reduction. The actual benefits are quantified monetarily based on the impact of flooding on various aspects such as travel, personal safety and property damage to residents, and corrosion and destruction of buildings when no grey and green facilities are built. However, due to differences in many aspects such as population density, average income of residents, building heights, and the way buildings are constructed with materials, as well as the amount of rainfall in each rainstorm, it is difficult to make specific and detailed calculations of their flood control benefits based on each storm. The use of the shadow engineering method can simplify the calculation steps and prevent the lack of some of the measured data in the study area from making monetary calculations difficult, but it can also lead to certain errors.

The cost per unit of pollutant treated and the economic benefits of reducing the negative impacts of treatment on water are also calculated using empirical values. The cost per unit of COD treatment is calculated using the results of COD reduction during China's 11th Five-Year Plan period. The development of science and technology have led to a reduction in pollutant emissions, which combined will result in some error in the unit COD treatment cost. Other pollutant treatment costs are calculated from studies in recent years, and the influence of the time factor is relatively small, but due to the scale of the studies, there is still some error when applying it to the current study area.

The indicators of ecological benefits are also calculated based on reference values. Due to the varying degrees of variability in climate, hydrological characteristics, land use, economic and other regional characteristics of different regions, the calculation is subject to a certain degree of error. Because of the high human and material costs and the specificity of the study area, it is difficult to generalize, so it is easier to use the average values of the region or country to which the study area belongs and the formulaic monetization method.

In this study, due to the small size of the study area and the simplicity of the renovation facilities, the ecological benefits other than the carbon sequestration and oxygen release benefits may be limited, so they are not calculated in detail in this paper, which may result in small total benefits.

We are currently only evaluating based on limited scenarios and have not evaluated and compared the reliability, resilience and sustainability of different types and proportions of mixed grey and green facilities. Casal-Campos et al. [83] have done a good job in this regard and future comparisons of different mixed facilities could also be conducted for the current study area to achieve a higher benefit–cost ratio.

The current benefits are based on the results of current climate conditions, but as the global climate is changing considerably, rainfall patterns in the study area may change somewhat and Beijing may experience more rainy days or an increase in average daily precipitation and an increase in the proportion of extreme rainfall [84], which may lead to an increase in the benefits of grey facilities over that of green facilities; the current results will change somewhat and the conclusions will change somewhat as a result.

4.6. Risk Analysis

Green facilities reduce the velocity of rainwater runoff and use vegetation and soil to trap pollutants so that they do not enter natural water and cause pollution. Vegetation and microorganisms can break down some pollutants or use their own biochemical reactions to convert some pollutants into harmless substances; however, there are some pollutants such as heavy metals that are difficult to be converted by vegetation and microorganisms, and the runoff containing these pollutants seeps down through the soil pores, causing soil pollution and possibly groundwater contamination. As the pollutant reductions and reduction rates in the study were mainly based on observing the pollutant concentrations at the outfalls and focusing on the pollution of surface water, etc., where the stormwater runoff flows directly, the results of the total reduction in each type of pollutant were obtained, and in practice the threat of pollutants to the soil and groundwater also needs to be looked at.

The actual construction of green infrastructure should therefore refer to relevant design guidelines, such as the design guideline issued by the German Association for Water, Wastewater and Waste (DWA). Specific requirements, limiting dimensions, design parameters for each component of the facility, and facility design loads need to be determined to keep the risk of groundwater contamination within manageable limits.

5. Conclusions

This study comprehensively assessed and monetized the hydrological, nonpoint source, and ecological benefits of grey and green infrastructure. The main findings are as follows:

The retrofitting of grey and green infrastructure could effectively reduce the total volume and peak flow of stormwater runoff and had a good effect on the control of flooding in the study area. Under different rainfall scenarios, the reduction rates of total runoff and peak flow in the study area were higher than 65%. Grey and green infrastructure had a good reduction effect on surface pollution from stormwater, with the reduction in all pollutants close to 100% under the light and moderate rain scenarios, 75–90% under the heavy rain scenario and 40–80% under the rainstorm scenario. The combined benefits of grey and green infrastructure are highest in the medium rainfall scenario, and the economic effectiveness of grey and green facilities in relieving flooding and drainage pressure on the network is limited in the heavy rainfall scenario. Both grey and green facilities can improve the sustainability of urbanized areas and bring significant economic benefits, but green infrastructure has more multifaceted benefits and higher monetized values of benefits.

The relative control of runoff and pollution with a positive benefit can be achieved in smaller scale campus areas with relatively simple grey and green facilities consisting of green roofs, permeable pavements, and pipe modifications, which can be a reference for similarly situated areas. As the scale of the area increases, or as the area becomes more urbanized, additional types or numbers of facilities can be considered to achieve better results.

The comprehensive benefit evaluation system for grey and green facilities in this study can be applied to the comprehensive accounting of hydrological, nonpoint source pollution and ecological benefits as well as costs in other regions, providing some reference on the economic aspects of the feasibility of facility construction in other regions. However, the current method still suffers from ambiguity in the data and uncertainty in the alternative calculation of benefits, and further refinement is needed in region-specific studies.

Green infrastructure traps pollutants and mitigates pollution in water such as rivers, but soil contamination and groundwater contamination from pollutant infiltration needs to be further considered subsequently.

Author Contributions: Writing—original draft preparation by J.W.; writing—review and editing by X.Z., S.W., L.C. and Z.S. All authors have read and agreed to the published version of the manuscript.

Funding: This research was funded by The Fund for Innovative Research Group of the National Natural Science Foundation of China (No. 52221003), and the National Natural Science Foundation of China (No. 42277044).

Data Availability Statement: The SWMM can be downloaded from https://www.epa.gov/water-research/storm-water-management-model-swmm/. The data used in this study can be obtained by emailing the first author.

Conflicts of Interest: The authors declare that they have no known competing financial interests or personal relationships that could have appeared to influence the work reported in this paper.

Appendix A

Table A1. Hydrological, nonpoint source and ecological benefits in selected studies.

Aspects		Results	Authors
Runoff reduction rates	Rain gardens	77%	Glick et al. [16]
	Porous pavements	29%	
	Green roofs	15%	
	Cisterns	15%	
	All LIDs	30%	
Peak flow reduction rate	All LIDs	24%	
Runoff reduction rates	Current conditions	80%	Abduljaleel et al. [17]
	Future climate change conditions	29%	
Peak flow reduction rates	Current conditions	62%	
	Future climate change conditions	13%	
Effect of reducing runoff	Best	Infiltration trenches	
	Worst	Rain gardens Bioretention ponds	
Peak flow reduction rates	General rainfall events	22%	Quichimbo-Miguitama et al. [18]
	Extreme rainfall events	15%	
Runoff reduction rates	-	20%	
Flooded nodes reduction rates	Short-term events	27%	
	Extreme events	4%	
Runoff reduction rates	Conventional medium density cities	29%	Seo et al. [19]
	Conservation medium density cities	25%	
Nitrate loads reduction rates	Conventional medium density cities	31%	
	Conservation medium density cities	30%	
TP reduction rates	Conventional medium density cities	25%	
	Conservation medium density cities	22%	
Runoff reduction rate	-	35.08%	Deng et al. [20]
Peak flow reduction rate	-	26.82%	
Nonpoint source pollution reduction rate	-	45.18%	
Peak flow reduction rate	-	80%	Saadatpour et al. [28]
SS reduction rate	-	81.86%	
Leachate temperature reduction	Permeable pavement	2 degrees Celsius	LeBleu et al. [21]
Temperature reduction	Green roofs before 8 a.m.	1 degrees Celsius	Shen [22]
	Green roofs at 2 p.m.	18 degrees Celsius	
Carbon sequestration/(kg carbon dioxide equivalent·a^{-1})	Green land	5450	Lin et al. [23]
	Rainwater utilization	15,379	
	Runoff pollutant removal	19,552	

References

1. Qi, B.; Xu, P.; Wu, C. Analysis of the Infiltration and Water Storage Performance of Recycled Brick Mix Aggregates in Sponge City Construction. *Water* **2023**, *15*, 363. [CrossRef]
2. Lyu, H.-M.; Xu, Y.-S.; Cheng, W.-C.; Arulrajah, A. Flooding Hazards across Southern China and Prospective Sustainability Measures. *Sustainability* **2018**, *10*, 1682. [CrossRef]
3. Fletcher, T.D.; Andrieu, H.; Hamel, P. Understanding, Management and Modelling of Urban Hydrology and Its Consequences for Receiving Waters: A State of the Art. *Adv. Water Resour.* **2013**, *51*, 261–279. [CrossRef]
4. Vinck, E.; De Bock, B.; Wambecq, T.; Liekens, E.; Delgado, R. A New Decision Support Tool for Evaluating the Impact of Stormwater Management Systems on Urban Runoff Pollution. *Water* **2023**, *15*, 931. [CrossRef]
5. Godyń, I.; Bodziony, M.; Grela, A.; Muszyński, K.; Pamuła, J. Determination of Pollution and Environmental Risk Assessment of Stormwater and the Receiving River, Case Study of the Sudół River Catchment, Poland. *Int. J. Environ. Res. Public Health* **2023**, *20*, 504. [CrossRef]
6. Kshirsagar, M.P.; Khare, K.C. Support Vector Regression Models of Stormwater Quality for a Mixed Urban Land Use. *Hydrology* **2023**, *10*, 66. [CrossRef]
7. Qi, W.; Ma, C.; Xu, H.; Chen, Z.; Zhao, K.; Han, H. A Review on Applications of Urban Flood Models in Flood Mitigation Strategies. *Nat. Hazards* **2021**, *108*, 31–62. [CrossRef]
8. Hou, J.; Mao, H.; Li, J.; Sun, S. Spatial Simulation of the Ecological Processes of Stormwater for Sponge Cities. *J. Environ. Manag.* **2019**, *232*, 574–583. [CrossRef]
9. Nguyen, T.T.; Ngo, H.H.; Guo, W.; Wang, X.C.; Ren, N.; Li, G.; Ding, J.; Liang, H. Implementation of a Specific Urban Water Management—Sponge City. *Sci. Total Environ.* **2019**, *652*, 147–162. [CrossRef]
10. Wang, Z.; Qi, F.; Liu, L.; Chen, M.; Sun, D.; Nan, J. How Do Urban Rainfall-Runoff Pollution Control Technologies Develop in China? A Systematic Review Based on Bibliometric Analysis and Literature Summary. *Sci. Total Environ.* **2021**, *789*, 148045. [CrossRef]
11. Wang, K.; Zhang, L.; Zhang, L.; Cheng, S. Coupling Coordination Assessment on Sponge City Construction and Its Spatial Pattern in Henan Province, China. *Water* **2020**, *12*, 3482. [CrossRef]
12. Qiao, X.-J.; Liao, K.-H.; Randrup, T.B. Sustainable Stormwater Management: A Qualitative Case Study of the Sponge Cities Initiative in China. *Sustain. Cities Soc.* **2020**, *53*, 101963. [CrossRef]
13. Zhang, J.; Han, Y.; Qiao, X.-J.; Randrup, T.B. Citizen Willingness to Pay for the Implementation of Urban Green Infrastructure in the Pilot Sponge Cities in China. *Forests* **2023**, *14*, 474. [CrossRef]
14. Sample, D.J.; Heaney, J.P.; Wright, L.T.; Fan, C.-Y.; Lai, F.-H.; Field, R. Costs of Best Management Practices and Associated Land for Urban Stormwater Control. *J. Water Resour. Plann. Manag.* **2003**, *129*, 59–68. [CrossRef]
15. Cheng, T.; Huang, B.; Yang, Z.; Qiu, J.; Zhao, B.; Xu, Z. On the Effects of Flood Reduction for Green and Grey Sponge City Measures and Their Synergistic Relationship—Case Study in Jinan Sponge City Pilot Area. *Urban Clim.* **2022**, *42*, 101058. [CrossRef]
16. Glick, R.; Jeong, J.; Srinivasan, R.; Arnold, J.G.; Her, Y. Adaptation of SWAT Watershed Model for Stormwater Management in Urban Catchments: Case Study in Austin, Texas. *Water* **2023**, *15*, 1770. [CrossRef]
17. Abduljaleel, Y.; Demissie, Y. Identifying Cost-Effective Low-Impact Development (LID) under Climate Change: A Multi-Objective Optimization Approach. *Water* **2022**, *14*, 3017. [CrossRef]
18. Quichimbo-Miguitama, F.; Matamoros, D.; Jiménez, L.; Quichimbo-Miguitama, P. Influence of Low-Impact Development in Flood Control: A Case Study of the Febres Cordero Stormwater System of Guayaquil (Ecuador). *Sustainability* **2022**, *14*, 7109. [CrossRef]
19. Seo, M.; Jaber, F.; Srinivasan, R.; Jeong, J. Evaluating the Impact of Low Impact Development (LID) Practices on Water Quantity and Quality under Different Development Designs Using SWAT. *Water* **2017**, *9*, 193. [CrossRef]
20. Deng, Y.; Deng, J.; Zhang, C. Sponge City and Water Environment Planning and Construction in Jibu District in Changde City. *Sustainability* **2022**, *15*, 444. [CrossRef]
21. LeBleu, C.; Dougherty, M.; Rahn, K.; Wright, A.; Bowen, R.; Wang, R.; Orjuela, J.A.; Britton, K. Quantifying Thermal Characteristics of Stormwater through Low Impact Development Systems. *Hydrology* **2019**, *6*, 16. [CrossRef]
22. Shen, Z. Green Roof Design of Residential Area Based on Sponge City Theory. *Wirel. Commun. Mob. Comput.* **2022**, *2022*, 861962. [CrossRef]
23. Lin, X.; Ren, J.; Xu, J.; Zheng, T.; Cheng, W.; Qiao, J.; Huang, J.; Li, G. Prediction of Life Cycle Carbon Emissions of Sponge City Projects: A Case Study in Shanghai, China. *Sustainability* **2018**, *10*, 3978. [CrossRef]
24. Wilbers, G.-J.; de Bruin, K.; Seifert-Dähnn, I.; Lekkerkerk, W.; Li, H.; Budding-Polo Ballinas, M. Investing in Urban Blue–Green Infrastructure—Assessing the Costs and Benefits of Stormwater Management in a Peri-Urban Catchment in Oslo, Norway. *Sustainability* **2022**, *14*, 1934. [CrossRef]
25. Wei, Z.; Yumiao, L.; Chengdan, W. A Discussion on the Construction Ideas of Sponge City in A Green Eco-District. *IOP Conf. Ser. Earth Environ. Sci.* **2021**, *657*, 012058. [CrossRef]
26. Li, X.-J.; Deng, J.-X.; Xie, W.-J.; Jim, C.-Y.; Wei, T.-B.; Lai, J.-Y.; Liu, C.-C. Comprehensive Benefit Evaluation of Pervious Pavement Based on China's Sponge City Concept. *Water* **2022**, *14*, 1500. [CrossRef]
27. Raei, E.; Reza Alizadeh, M.; Reza Nikoo, M.; Adamowski, J. Multi-Objective Decision-Making for Green Infrastructure Planning (LID-BMPs) in Urban Storm Water Management under Uncertainty. *J. Hydrol.* **2019**, *579*, 124091. [CrossRef]

28. Saadatpour, M.; Delkhosh, F.; Afshar, A.; Solis, S.S. Developing a Simulation-Optimization Approach to Allocate Low Impact Development Practices for Managing Hydrological Alterations in Urban Watershed. *Sustain. Cities Soc.* **2020**, *61*, 102334. [CrossRef]
29. Song, J.; Wang, J.; Xi, G.; Lin, H. Evaluation of Stormwater Runoff Quantity Integral Management via Sponge City Construction: A Pilot Case Study of Jinan. *Urban Water J.* **2021**, *18*, 151–162. [CrossRef]
30. Grimaldi, S.; Serinaldi, F. Asymmetric Copula in Multivariate Flood Frequency Analysis. *Adv. Water Resour.* **2006**, *29*, 1155–1167. [CrossRef]
31. Dong, X.; Guo, H.; Zeng, S. Enhancing Future Resilience in Urban Drainage System: Green versus Grey Infrastructure. *Water Res.* **2017**, *124*, 280–289. [CrossRef] [PubMed]
32. Su, J.; Li, J.; Gao, X.; Yao, Y.; Jiang, C. Comprehensive Analysis of Waterlogging Control and Carbon Emission Reduction for Optimal LID Layout: A Case Study in Campus. *Env. Sci. Pollut. Res.* **2022**, *29*, 87802–87816. [CrossRef]
33. Zhang, Z.; Gu, J.; Zhang, G.; Ma, W.; Zhao, L.; Ning, P.; Shen, J. Design of Urban Runoff Pollution Control Based on the Sponge City Concept in a Large-Scale High-Plateau Mountainous Watershed: A Case Study in Yunnan, China. *J. Water Clim. Change* **2021**, *12*, 201–222. [CrossRef]
34. Tang, S.; Jiang, J.; Zheng, Y.; Hong, Y.; Chung, E.-S.; Shamseldin, A.Y.; Wei, Y.; Wang, X. Robustness Analysis of Storm Water Quality Modelling with LID Infrastructures from Natural Event-Based Field Monitoring. *Sci. Total Environ.* **2021**, *753*, 142007. [CrossRef] [PubMed]
35. Han, Y.H.; Lau, S.L.; Kayhanian, M.; Stenstrom, M.K. Correlation Analysis among Highway Stormwater Pollutants and Characteristics. *Water Sci. Technol.* **2006**, *53*, 235–243. [CrossRef]
36. Xie, Z.; Xie, L.; Li, J. Direct Subsidies or Tax Credits The Effects of Different R&D Policy Tools. *Int. J. Technol. Manag.* **2021**, *86*, 25. [CrossRef]
37. Xu, Y.-S.; Shen, S.-L.; Lai, Y.; Zhou, A.-N. Design of Sponge City: Lessons Learnt from an Ancient Drainage System in Ganzhou, China. *J. Hydrol.* **2018**, *563*, 900–908. [CrossRef]
38. Samouei, S.; Özger, M. Evaluating the Performance of Low Impact Development Practices in Urban Runoff Mitigation through Distributed and Combined Implementation. *J. Hydroinformatics* **2020**, *22*, 1506–1520. [CrossRef]
39. Cheng, T.; Xu, Z.; Hong, S.; Song, S. Flood Risk Zoning by Using 2D Hydrodynamic Modeling: A Case Study in Jinan City. *Math. Probl. Eng.* **2017**, *2017*, 5659197. [CrossRef]
40. Ma, J.; Liu, D.; Wang, Z. Sponge City Construction and Urban Economic Sustainable Development: An Ecological Philosophical Perspective. *Int. J. Environ. Res. Public Health* **2023**, *20*, 1694. [CrossRef]
41. Zhao, J.; Fonseca, C.; Zeerak, R. Stormwater Utility Fees and Credits: A Funding Strategy for Sustainability. *Sustainability* **2019**, *11*, 1913. [CrossRef]
42. Zhu, Y.; Xu, C.; Yin, D.; Xu, J.; Wu, Y.; Jia, H. Environmental and Economic Cost-Benefit Comparison of Sponge City Construction in Different Urban Functional Regions. *J. Environ. Manag.* **2022**, *304*, 114230. [CrossRef] [PubMed]
43. Hotchkiss, J.L.; Rupasingha, A. Individual Social Capital and Migration. *Growth Change* **2021**, *52*, 808–837. [CrossRef]
44. Gao, Z.; Zhang, Q.H.; Xie, Y.D.; Wang, Q.; Dzakpasu, M.; Xiong, J.Q.; Wang, X.C. A Novel Multi-Objective Optimization Framework for Urban Green-Gray Infrastructure Implementation under Impacts of Climate Change. *Sci. Total Environ.* **2022**, *825*, 153954. [CrossRef] [PubMed]
45. Zhu, Y.; Xu, C.; Liu, Z.; Yin, D.; Jia, H.; Guan, Y. Spatial Layout Optimization of Green Infrastructure Based on Life-Cycle Multi-Objective Optimization Algorithm and SWMM Model. *Resour. Conserv. Recycl.* **2023**, *191*, 106906. [CrossRef]
46. Souza, C.R.B.; de Souza, S.N.M.; Secco, D.; Lenz, A.M.; Cruz Siqueira, J.A. Green Roofs and Their Contribution for the Reduction of Room Temperature in Buildings in Cascavel-State Paraná/Green Roofs and Energy Efficiency. *Acta Sci. Technol.* **2018**, *40*, 35267. [CrossRef]
47. Cascone, S.; Leuzzo, A. Thermal Comfort in the Built Environment: A Digital Workflow for the Comparison of Different Green Infrastructure Strategies. *Atmosphere* **2023**, *14*, 685. [CrossRef]
48. Ouyang, Z.; Zhu, C.; Yang, G.; Xu, W.; Zheng, H.; Zhang, Y.; Xiao, Y. Accounting for Gross Ecosystem Product: Concepts, Accounting Methods and Case Studies. *J. Ecol.* **2013**, *33*, 6747–6761. (In Chinese)
49. Ouyang, Z.; Zheng, H.; Xiao, Y.; Polasky, S.; Liu, J.; Xu, W.; Wang, Q.; Zhang, L.; Xiao, Y.; Rao, E.; et al. Improvements in Ecosystem Services from Investments in Natural Capital. *Science* **2016**, *352*, 1455–1459. [CrossRef]
50. Toledo-Gallegos, V.M.; My, N.H.D.; Tuan, T.H.; Börger, T. Valuing Ecosystem Services and Disservices of Blue/Green Infrastructure. Evidence from a Choice Experiment in Vietnam. *Econ. Anal. Policy* **2022**, *75*, 114–128. [CrossRef]
51. Xie, G.; Zhang, C.; Zhang, L.; Chen, W.; Li, S. Gross ecosystem product: Concept, accounting framework and case study. *J. Nat. Resour.* **2015**, *30*, 1243–1254. (In Chinese)
52. Wang, M.; Zhang, D.; Adhityan, A.; Ng, W.J.; Dong, J.; Tan, S.K. Assessing Cost-Effectiveness of Bioretention on Stormwater in Response to Climate Change and Urbanization for Future Scenarios. *J. Hydrol.* **2016**, *543*, 423–432. [CrossRef]
53. Xie, J.; Chen, H.; Liao, Z.; Gu, X.; Zhu, D.; Zhang, J. An Integrated Assessment of Urban Flooding Mitigation Strategies for Robust Decision Making. *Environ. Model. Softw.* **2017**, *95*, 143–155. [CrossRef]
54. Jiayu, H.; Jiewei, X.; Qiaohui, T.; Yang, S. Application of the Technology in Sponge Campus in China. *Int. J. Archit. Arts Appl.* **2019**, *5*, 105. [CrossRef]

55. Hung, F.; Harman, C.J.; Hobbs, B.F.; Sivapalan, M. Assessment of Climate, Sizing, and Location Controls on Green Infrastructure Efficacy: A Timescale Framework. *Water Resour. Res.* **2020**, *56*, e2019WR026141. [CrossRef]
56. Zhao, Y.; Xia, J.; Xu, Z.; Qiao, Y.; Shen, J.; Ye, C. Impact of Urbanization on Regional Rainfall-Runoff Processes: Case Study in Jinan City, China. *Remote Sens.* **2023**, *15*, 2383. [CrossRef]
57. Merchán-Sanmartín, B.; Carrión-Mero, P.; Suárez-Zamora, S.; Aguilar-Aguilar, M.; Cruz-Cabrera, O.; Hidalgo-Calva, K.; Morante-Carballo, F. Stormwater Sewerage Masterplan for Flood Control Applied to a University Campus. *Smart Cities* **2023**, *6*, 1279–1302. [CrossRef]
58. Awasthi, N.; Tripathi, J.N.; Petropoulos, G.P.; Gupta, D.K.; Singh, A.K.; Kathwas, A.K. Performance Assessment of Global-EO-Based Precipitation Products against Gridded Rainfall from the Indian Meteorological Department. *Remote Sens.* **2023**, *15*, 3443. [CrossRef]
59. Wang, X.; Zhao, Z.; Zhang, Z.; Ren, M.; Sagris, T. Design Rainfall Change of Rainwater Source Control Facility to Meet Future Scenarios in Beijing. *Int. J. Environ. Res. Public Health* **2023**, *20*, 4355. [CrossRef]
60. Cao, Q.; Cao, J.; Xu, R. Optimizing Low Impact Development for Stormwater Runoff Treatment: A Case Study in Yixing, China. *Water* **2023**, *15*, 989. [CrossRef]
61. Sidek, L.M.; Chua, L.H.C.; Azizi, A.S.M.; Basri, H.; Jaafar, A.S.; Moon, W.C. Application of PCSWMM for the 1-D and 1-D–2-D Modeling of Urban Flooding in Damansara Catchment, Malaysia. *Appl. Sci.* **2021**, *11*, 9300. [CrossRef]
62. Li, W.; Wang, H.; Zhou, J.; Yan, L.; Liu, Z.; Pang, Y.; Zhang, H.; Huang, T. Simulation and Evaluation of Rainwater Runoff Control, Collection, and Utilization for Sponge City Reconstruction in an Urban Residential Community. *Sustainability* **2022**, *14*, 12372. [CrossRef]
63. Munir, B.A.; Ahmad, S.R.; Hafeez, S. Integrated Hazard Modeling for Simulating Torrential Stream Response to Flash Flood Events. *ISPRS Int. J. Geo-Inf.* **2020**, *9*, 1. [CrossRef]
64. Krebs, G.; Kokkonen, T.; Valtanen, M.; Setälä, H.; Koivusalo, H. Spatial Resolution Considerations for Urban Hydrological Modelling. *J. Hydrol.* **2014**, *512*, 482–497. [CrossRef]
65. Birkinshaw, S.J.; O'Donnell, G.; Glenis, V.; Kilsby, C. Improved Hydrological Modelling of Urban Catchments Using Runoff Coefficients. *J. Hydrol.* **2021**, *594*, 125884. [CrossRef]
66. Zhang, D.; Mei, C.; Ding, X.; Liu, J.; Fu, X.; Wang, J.; Wang, D. Impacts of Rainstorm Characteristics on Runoff Quantity and Quality Control Performance Considering Integrated Green Infrastructures. *Sustainability* **2022**, *14*, 11284. [CrossRef]
67. Yang, Y.; Zhang, W.; Liu, Z.; Liu, D.; Huang, Q.; Xia, J. Coupling a Distributed Time Variant Gain Model into a Storm Water Management Model to Simulate Runoffs in a Sponge City. *Sustainability* **2023**, *15*, 3804. [CrossRef]
68. Guo, X.; Guo, Q.; Zhou, Z.; Du, P.; Zhao, D. Degrees of Hydrologic Restoration by Low Impact Development Practices under Different Runoff Volume Capture Goals. *J. Hydrol.* **2019**, *578*, 124069. [CrossRef]
69. Wakida, F.T.; Martinez-Huato, S.; Garcia-Flores, E.; Piñon-Colin, T.D.J.; Espinoza-Gomez, H.; Ames-López, A. Pollutant Association with Suspended Solids in Stormwater in Tijuana, Mexico. *Int. J. Environ. Sci. Technol.* **2014**, *11*, 319–326. [CrossRef]
70. Kamali, M.; Alamdari, N.; Esfandarani, M.S.; Esfandarani, M.S. Effects of Rainfall Characteristics on Runoff Quality Parameters within an Industrial Sector in Tennessee, USA. *J. Contam. Hydrol.* **2023**, *256*, 104179. [CrossRef]
71. Vaze, J.; Chiew, F.H.S. Experimental Study of Pollutant Accumulation on an Urban Road Surface. *Urban Water* **2002**, *4*, 379–389. [CrossRef]
72. Wilkerson, B.; Romanenko, E.; Barton, D.N. Modeling Reverse Auction-Based Subsidies and Stormwater Fee Policies for Low Impact Development (LID) Adoption: A System Dynamics Analysis. *Sustain. Cities Soc.* **2022**, *79*, 103602. [CrossRef]
73. Yoo, J.; Park, K. Stormwater Utility Fee Estimation Method for Individual Land Use Areas. *Sustainability* **2022**, *14*, 10211. [CrossRef]
74. Godyń, I. Economic Incentives in Stormwater Management: A Study of Practice Gaps in Poland. *Water* **2022**, *14*, 3817. [CrossRef]
75. Qiu, D.; Deng, X. Does RMB purchasing power parity hold?—A multi-country cointegration analysis based on the RMB-EUR. *Financ. Theory Pract.* **2020**, *4*, 41–51. (In Chinese)
76. Li, H.; Zhao, W.; Gu, R.; Li, Y.; Chen, Z.; Zhang, X. Effects of three different green-lands in plantation structure on the O_2-emitting, CO_2-fixing, heat-absorbing and temperature-decreasing in residential quarters. *Environ. Sci.* **1999**, *6*, 41–44. (In Chinese) [CrossRef]
77. Li, H.; Wang, J.; Ge, C. A cost-benefit analysis of the pollution reduction during the eleventh five-year period in China. *J. Environ. Sci.* **2013**, *33*, 2270–2276. (In Chinese) [CrossRef]
78. She, J. Cost analysis of grooving construction of reinforced concrete drainage pipeline in loess area. *Constr. Econ.* **2019**, *40*, 70–72. (In Chinese) [CrossRef]
79. Li, J.; Jiang, Y.; Zhai, M.; Gao, J.; Yao, Y.; Li, Y. Construction and Application of Sponge City Resilience Evaluation System: A Case Study in Xi'an, China. *Env. Sci Pollut Res* **2023**, *30*, 62051–62066. [CrossRef]
80. She, L.; Wei, M.; You, X. Multi-Objective Layout Optimization for Sponge City by Annealing Algorithm and Its Environmental Benefits Analysis. *Sustain. Cities Soc.* **2021**, *66*, 102706. [CrossRef]
81. Guan, Y.; Shan, Y.; Huang, Q.; Chen, H.; Wang, D.; Hubacek, K. Assessment to China's Recent Emission Pattern Shifts. *Earth's Future* **2021**, *9*, e2021EF002241. [CrossRef]
82. Escobedo, F.; Varela, S.; Zhao, M.; Wagner, J.E.; Zipperer, W. Analyzing the Efficacy of Subtropical Urban Forests in Offsetting Carbon Emissions from Cities. *Environ. Sci. Policy* **2010**, *13*, 362–372. [CrossRef]

83. Casal-Campos, A.; Sadr, S.M.K.; Fu, G.; Butler, D. Reliable, Resilient and Sustainable Urban Drainage Systems: An Analysis of Robustness under Deep Uncertainty. *Environ. Sci. Technol.* **2018**, *52*, 9008–9021. [CrossRef] [PubMed]
84. Wang, Y.; Zhang, Z.; Zhao, Z.; Sagris, T.; Wang, Y. Prediction of Future Urban Rainfall and Waterlogging Scenarios Based on CMIP6: A Case Study of Beijing Urban Area. *Water* **2023**, *15*, 2045. [CrossRef]

water

MDPI

Article

Optimizing Low Impact Development for Stormwater Runoff Treatment: A Case Study in Yixing, China

Qian Cao [1], Jiashun Cao [1,2,*] and Runze Xu [1,2,*]

1 College of Environment, Hohai University, Nanjing 210098, China
2 Key Laboratory of Integrated Regulation and Resource Development on Shallow Lakes, Ministry of Education, Hohai University, Nanjing 210098, China
* Correspondence: caojiashun@163.com (J.C.); runzexu@hhu.edu.cn (R.X.)

Abstract: Low-impact development (LID) practices have been recognized as a promising strategy to control urban stormwater runoff and non-point source pollution in urban ecosystems. However, many experimental and modeling efforts are required to tailor an effective LID practice based on the hydraulic and environmental characteristics of a given region. In this study, the InfoWorks ICM was applied to simulate the runoff properties and determine the optimal LID design in a residential site at Yixing, China, based on four practical rainfall events. Additionally, the software was redeveloped using Ruby object-oriented programming to improve its efficiency in uncertainty analysis using the Generalized Likelihood Uncertainty Estimation method. The simulated runoff was in good agreement with the observed discharge (Nash–Sutcliffe model efficiency coefficients >0.86). The results of the response surface method indicated that when the sunken green belt, permeable pavement, and green roof covered 8.6%, 15%, and 10%, respectively, of the 11.3 ha study area, the designed system showed the best performance with relatively low cost. This study would provide new insights into designing urban rainfall-runoff pollution control systems.

Keywords: first flush effect; InfoWorks ICM; LID optimization; generalized likelihood uncertainty estimation

Citation: Cao, Q.; Cao, J.; Xu, R. Optimizing Low Impact Development for Stormwater Runoff Treatment: A Case Study in Yixing, China. *Water* **2023**, *15*, 989. https://doi.org/10.3390/w15050989

Academic Editor: Enedir Ghisi

Received: 5 February 2023
Revised: 27 February 2023
Accepted: 3 March 2023
Published: 4 March 2023

1. Introduction

Urbanization, characterized by continuous growth in population and land development, has altered the urban water cycle. Increasing urban impervious areas disrupted the infiltration process and resulted in a significant increase in the amount of surface water runoff, intensifying the frequency and severity of floods [1–3]. Besides, the growing population and industry largely augment the pollution load, such as through emissions from vehicles, the use of pharmaceuticals and personal care products, and the release of micro-/nano-plastics [4–6]. These pollutants exhibit stronger interactions with each other and may enrich adverse substances such as antibiotic resistance genes in surface water and floods [7,8]. One of the greatest issues in urban runoff is the first flush effect (FFE), which implies a greater discharge rate of pollutant mass or concentration in the early part of the runoff as compared with later in the storm [9–11]. Chow and Yusop (2014) [9] examined the water quality of 52 rainfall events and concluded that the first 10 mm of rainfall carried about 50% of the total pollutants. Wang et al. (2016) [10] proposed that intercepting the first 30–40% of the surface runoff was the most effective in stormwater quality management. As a result, controlling the first flush is critical for stormwater management.

Low impact development (LID) has been regarded as a promising strategy to compensate for the influence of urbanization on hydrology and water quality by simulating the pre-development site hydrology with site design techniques [12,13]. LID, as an effective and environmentally friendly practice for urban runoff management, is capable of significantly reducing urban runoff pollution loads [14]. This strategy was first introduced by the U.S. Environmental Protection Agency in the 1990s and has been widely used in many cities [12,15,16]. Green infrastructures such as bioretention cells, green roofs, and permeable pavements are implemented for LID purposes [17,18]. However, the design and implementation of these green infrastructures require optimization to achieve better performance [14,19].

Due to the random and uneven distribution of surface runoff, optimizing LID practices based on model simulation has become an ideal strategy. SWMM, MIKE, and InfoWorks ICM were often used by researchers to simulate the quality and quantity of surface runoff [20–22]. SWMM is a commonly used software that is easy to operate and commonly applied for secondary development. However, its input was complicated, and the results were difficult to visualize [20,22]. MIKE was feasible to simulate the hydraulics and quality, but some models required to be coupled manually [21]. InfoWorks ICM, developed by Wallingford, facilitated the operation and visualization of the urban water cycle simulation, making it a preferable choice for this study [23,24]. For example, Fan et al. (2022) [25] applied InfoWorks ICM to analyze the hydrological and pollution reduction in outfall and storage under different hydrological patterns, vertical parameter settings, and green infrastructure installation locations. However, most of the stormwater management practices in China only focused on reducing the volume rather than the FFE, which is key to a more effective LID practice design and stormwater runoff management [23,26].

Therefore, the aims of this study are: (1) to establish a model using InfoWorks ICM for stormwater runoff quality monitoring and estimation; (2) to conduct sensitivity analysis, calibration, validation, and uncertainty analysis on the established model; and (3) to optimize various LID facilities to maximize their performance while minimizing the cost. This study presents a promising method for urban runoff management, and the results are available for decision-makers to use in future planning.

2. Materials and Methods

2.1. Site Description

This study took place in Yixing (31°07′~31°37′ N, 119°31′~120°03′ E), a city located in the Southern part of Jiangsu Province, China (Figure 1). Yixing is hilly in the south, flat in the north, and has Taihu Lake to the east. The city has a humid subtropical climate and is influenced by the East Asian monsoon, which results in four distinct seasons and dense river networks. The average annual rainfall is 1177 mm and mostly occurs during the spring and summer. Rapid urbanization results in a growing amount of waste and poses a threat to the environment, especially the Taihu Lake. Therefore, the control and management of non-point pollution are of great significance. The study was conducted in a 0.1113 km^2 residential area with 8.6% green area, 48.4% construction area, and the remaining 43% roads.

2.2. Stormwater Sampling and Data Acquisition

Flowrates were monitored at 5–10 min intervals, and the samples for water quality analysis were manually collected in 500 mL polyethylene bottles. The data for four rainfall events (7 November 2015, 22 August 2015, 5 April 2018, and 23 April 2018) were acquired from the automatic rain gages at the 104 freeway, which recorded every 0.2 mm.

Then, the collected samples were sent to the laboratory for water quality analysis. The samples were kept in the fridge before analysis, and all the experiments were performed within 24 h. The concentrations of suspended solids (SS), NH_4^+-N, chemical oxygen demand (COD), and total phosphorus (TP) were measured by the weighing method, Nessler's reagent spectrophotometry method, rapid digestion spectrophotometry method, and Mo-Sb anti-spectrophotometric method, respectively.

2.3. Data from Stormwater Monitoring

Four different storm events were used to calibrate and validate the rainfall-runoff model (Table 1): storm events on 22 August 2015 and 11 July 2015 for the calibration process and storm events on 4 May 2018 and 23 April 2018 for validation.

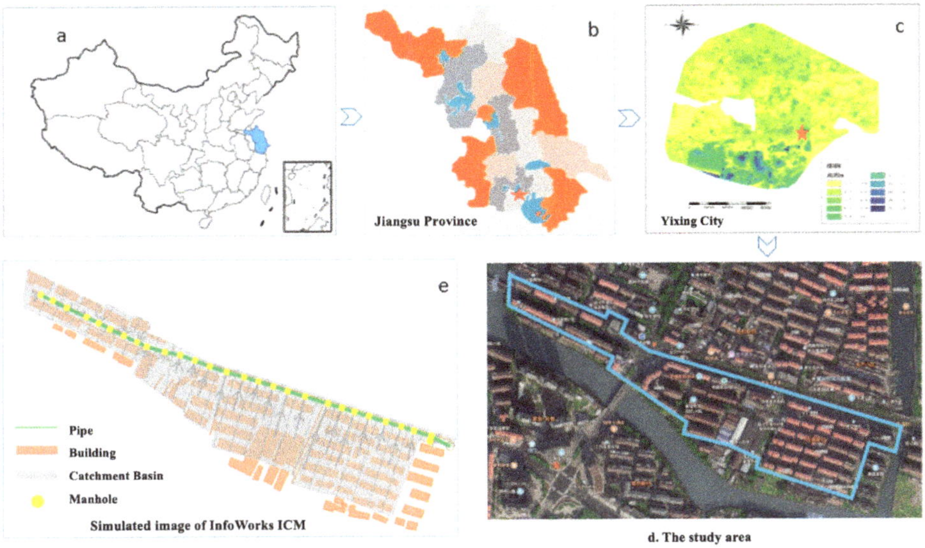

Figure 1. Map of the study area: (**a**) location of Jiangsu Province in China; (**b**) location of Yixing City in Jiangsu Province; (**c**) the DEM (Digital Elevation Model) of Yixing City; (**d**) study area; (**e**) simulated image of InfoWorks ICM.

Table 1. Data from stormwater monitoring.

	Previous Dry Duration (h)	Depth (mm)	Duration (min)	Average Intensity (mm/h)	Characterization
22 August 2015	39	20.2	420	2.89	Moderate
11 July 2015	18	13.0	150	5.20	Heavy
4 May 2018	22	9.0	160	3.38	Moderate
23 April 2018	3	55.0	576	5.73	Heavy

2.4. Rainfall-Runoff Pollution Model Setup

InfoWorks ICM was applied to set up an urban rainfall-runoff pollution model, including a hydrologic module and a water quality module (Figure 2). In the hydrologic module, the surface runoff in impervious areas, including roads and buildings, was calculated by the rational method, while the infiltration in the green areas was estimated using the Horton equation [26,27]. The nonlinear reservoir method was applied in the routing model. Since the study area has a separate sewer system, only the storm drain was simulated using the Saint-Venant equations. In the model, the pipelines were generalized into connecting lines between nodes, and the boundary conditions were the water outlet or head loss.

Based on the hydrologic module, the water quality module simulates the accumulation, erosion, and transport processes of the pollutants [28]. In InfoWorks ICM, the accumulation process was assumed to be linear, and the accumulation rate decreased exponentially as the mass of surface sediments increased. The buildup of pollutants is calculated by the Euler approximation equation as shown below:

$$\frac{dM}{dt} = Ps - K_1 gM \quad (1)$$

where M is the mass of accumulations (per unit area) (kg/ha), Ps is the pollutant accumulation coefficient (kg/(ha×day)), K_1 is the decay factor (day^{-1}).

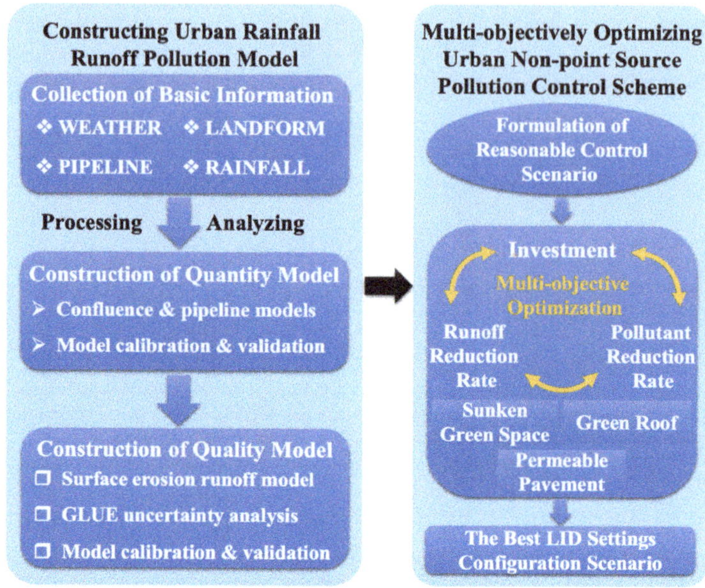

Figure 2. Flowchart for optimizing LID size.

The wash-off process is modeled as the function of accumulated pollutants and rainfall intensity:

$$\frac{dM}{dt} = -KagM(t) \tag{2}$$

$$Ka(t) = C_1 I(t)^{C_2} - C_3 I(t) \tag{3}$$

where *Ka(t)* is the wash-off rate; *I(t)* is the rainfall intensity; and C_1, C_2, and C_3 are wash-off coefficients.

The runoff is calculated by the single linear reservoir confluence equation. The model also assumes that the quantity of pollutants in surface runoff equals the product of surface sediments and the efficacy coefficient, which remains unchanged during a rainfall event.

2.5. Model Calibration and Uncertainty Analysis

For the hydrologic model, the sensitive parameters as well as their range were referred to previous studies [12,29]. Then, the runoff model was calibrated by two rainfall events in 2015 (22 August 2015 and 11 July 2015), and the remaining two (4 May 2018 and 23 April 2018) were applied for verification. The accuracy of the model was evaluated by Nash–Sutcliffe Efficiency (NSE) graphically and statistically.

For the water quality model, the sensitivity, uncertainty analysis, and calibration were analyzed by the Generalized Likelihood Uncertainty Estimation (GLUE) method, which is a global analysis method that concludes several optimal parameter sets to avoid interactions between parameters.

The first step of the GLUE method is to determine the likelihood function (Equation (4)).

$$L(\alpha|y) = \left(1 - \frac{\sigma_\varepsilon^2}{\sigma_0^2}\right)^N \tag{4}$$

where $L(\alpha|y)$ is the likelihood of parameter set α, given the observed data (y). The quantities σ_ε^2 and σ_0^2 refer to the error variance between model simulations and observed data and the variance of the observed data, respectively.

Then, the data of two rainfall events (22 August 2015 and 11 July 2015) are applied for the rough calibration to narrow the range of parameters. By assuming that the distribution of the parameters was uniform, 2000 sets of parameters were randomly chosen. The batch input of model parameters and the automatic output of model results were realized by redeveloping the InfoWorks ICM via RUBY. Then, the values of the likelihood function were calculated using the VBA function in Excel.

2.6. LID Module

In InfoWorks ICM, the LID facilities are attached to the model as discrete elements, and their performance is simulated by a unit-based process (Table S1). The model generalizes each LID facility into a space composed of multiple vertical layers, including surface layers, pavement layers, soil layers, storage layers, an underdrain, and a drainage mat. Then, the simulation is achieved by estimating the water quantity and quality in different layers [12].

A sensitivity analysis was conducted to investigate how the parameters of the LID facilities impact the volume and pollutant reduction in surface runoff. The sensitivity analysis was performed by the Morris screening method using a random One-factor-At-a-Time (OAT) design, in which only one input parameter e_i is modified between two successive runs of the model. The change induced on the model can then be unambiguously attributed to such a modification using an elementary effect defined by

$$e_i = (y(x_i) - y)/\Delta x \tag{5}$$

where $y(x_i)$ is the new outcome, y the previous one, and Δx is the variation in the parameter x.

The rainfall event used in this sensitivity analysis has a return period of three years, a duration of 2 h, a peak coefficient of 0.4, and a 48-h dry period.

2.7. Optimizing LID Configuration

Since the size of LID facilities is the key to LID design, the *EMC* (Event Mean Concentration) equation (Equation (6)) was applied to analyze the influence of different LID facility sizes on the volume and pollutants reduction in surface runoff. *EMC* is often used in water quality evaluation, and when the reduction in pollutants is larger than the volume, the *EMC* value is larger than 0.

$$EMC = \frac{\sum C_t Q_t \Delta t}{\sum Q_t \Delta t} = \frac{\sum C_t V_t}{\sum V_t} \tag{6}$$

where Δt is the calculation interval, Q_t is the flux during the time interval, V_t is the volume of runoff, and C_t is the concentration of the pollutants in Δt.

The performance of LID facilities of different sizes was tested by rainfall events with a peak coefficient of 0.4, a duration of 2 h, a dry period of 48 h and return periods of 1, 3, 5, and 10 years. By changing the portion of the bioretention cell (from 2.5% to 20%), permeable pavement (from 10% to 90%), and green roof (5–50%) in the study area, the results of water quality, volume, and EMC reduction versus the size were plotted as figures for further analysis.

The response surface method (RSM) based on the Box–Behnken design (BBD) was applied for the multi-purpose optimization calculated by Design-expert. The portion of the bio-retention cell, permeable pavement, and green roof were set as factors, and the water volume reduction rate ($f_1(x)$), pollution removal rate ($f_2(x)$), and cost ($f_3(x)$) were set as responses. The results were analyzed by the least-squares method, and individual linear, quadratic, and interaction terms were determined by the analysis of variance (ANOVA). The optimal design parameters were the values of the factors with the largest desirability in the numerical optimization process (Equation (7)) (Table 2).

$$D = \left(d_1^{w_1} \times d_2^{w_2} \times \cdots\cdots \times d_n^{w_n}\right)^{1/(w_1+w_2+\cdots\cdots+w_n)} \tag{7}$$

where d_n is the dimensionless response and w_n is the weights of each response.

Table 2. Objective, weights, and range of responses.

	Objective	Weights	Minimum	Maximum
Quantity reduction	Maximize	4	0	100
Pollutant removal	Maximize	2	0	100
Cost	Minimize	4	0	15

As the study area has 8.6% of the green area, 43.0% of the road, and 48.4% of the construction area, the portions of the sunken green belt (x_1), permeable pavement (x_2), and green roof (x_3) were set to be smaller than 8.6%, 43.0%, and 48.4%, respectively. These three green infrastructures were chosen due to their wide applications as LID [17,30,31]. The cost ($f_3(x)$) is the sum of the area times the unit price of each LID facility, and the price was collected from the market and relevant papers: the sunken green belt is \$14.3/m^2, the permeable pavement is \$28.6/m^2, and the green roof is \$25.3/m^2. The design and solutions are shown in Table 3.

Table 3. Experimental design and results for response surface analysis.

	Design Matrix			Solutions			Solutions		
Run	X_1	X_2	X_3	$f_1(x)$	$f_2(x)$	$f_3(x)$	$f_1(x)$	$f_2(x)$	$f_3(x)$
1	7.6	10	30	44.95	84.43	6.74	35.15	83.12	6.74
2	7.6	30	10	49.86	92.50	6.97	38.57	91.30	6.97
3	7.6	20	20	66.49	95.58	11.19	56.54	95.00	11.19
4	7.6	10	10	71.35	97.62	11.42	59.94	97.68	11.42
5	6.6	20	30	47.43	84.43	7.08	39.74	83.12	7.08
6	8.6	20	30	52.34	92.50	7.30	43.16	91.30	7.30
7	8.6	10	20	63.99	85.66	10.86	51.95	95.00	10.86
8	8.6	20	10	68.85	97.62	11.08	55.35	97.68	11.08
9	7.6	20	20	39.14	82.77	4.96	30.64	81.04	4.96
10	7.6	30	30	60.69	94.10	9.41	52.16	93.21	9.41
11	7.6	20	20	55.72	94.10	8.75	42.97	93.21	8.75
12	8.6	30	20	77.11	98.52	13.20	64.37	98.51	13.20
13	6.6	30	20	58.21	94.10	9.08	47.56	93.21	9.08
14	6.6	10	20	58.21	94.10	9.08	47.56	93.21	9.08
15	7.6	20	20	58.21	94.10	9.08	47.56	93.21	9.08
16	6.6	20	10	58.21	94.10	9.08	47.56	93.21	9.08
17	7.6	20	20	58.21	94.10	9.08	47.56	93.21	9.08

3. Results and Discussion

3.1. Calibration, Validation, and Sensitivity Analysis

3.1.1. Hydrologic Model

Before calibration, the sensitive parameters of the hydrologic model were selected from the InfoWorks ICM manuals, most of which suggested that the percent of impervious area and the depth of depression storage on the impervious portion of the study area were the most sensitive parameters (Table 4). The difference between the observed and simulated flow is shown graphically in Figure 3. The Nash–Sutcliffe model efficiency coefficients (NSEs) were 0.86, 0.89, 0.89, and 0.88 for the rainfall events on 7 November 2015, 22 August 2015, 5 April 2018, and 23 April 2018, respectively. As for the peak flow, the differences in volume and time between the observed and simulated data were less than 20%. The results indicated that the simulated runoff was in good agreement with the observed discharge and was acceptable for further analysis.

Table 4. Parameters for the rainfall-runoff model.

Models	Parameters	Value
Runoff model	Runoff coefficients for impervious pavements	0.93
	F_0 (Horton) initial infiltration (mm/h)	76.20
	F_c (Horton) Permeability rate (mm/h)	3.81
	K (Horton) reduction rate (L/h)	0.01
	Horton recover rate (L/h)	0.014
Routing model	Slope of the impermeable surface (m/m)	0.003
	Slope of the permeable surface (m/m)	0.00
	Manning roughness of the impermeable pavement	0.013
	Manning roughness of the permeable surface	0.15
	Initial loss the impermeable surface (mm)	0.7
	Initial loss the permeable surface (mm)	1.0

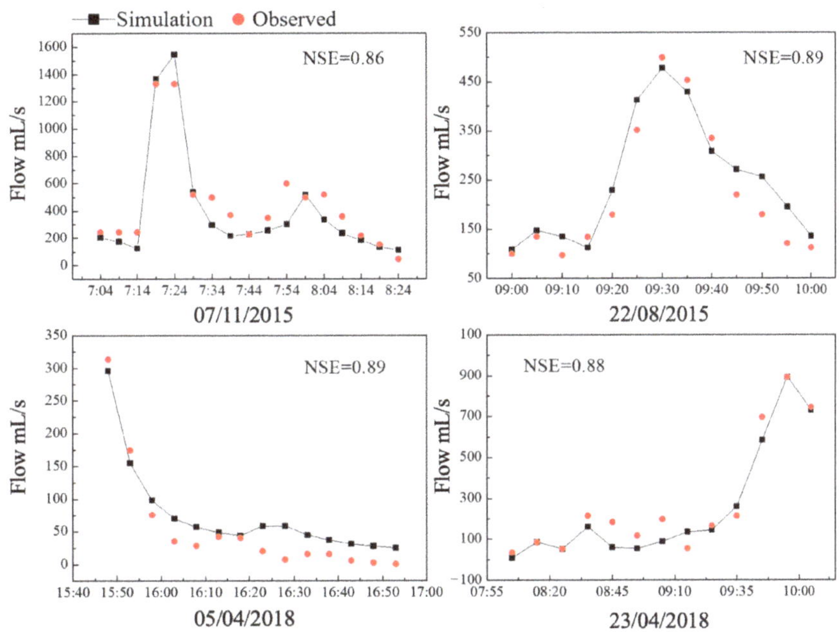

Figure 3. Observed and simulated flow rates in calibration.

3.1.2. Water Quality Model

For the water quality model, the uncertainty analysis was conducted by the GLUE method, and the calibrated parameters are shown in Table S2.

The simulations of SS, COD, and TP are acceptable; however, the modeled NH_4^+-N concentration is not accurate enough (Figure 4). The inaccuracy in NH_4^+-N modeling indicated that the concentration of NH_4^+-N in runoff might not be linearly related to the concentration of SS [32]. According to the likelihood distribution of different parameters, it can be concluded that the SS simulation is most sensitive to C_3, and the NSE value plateaus when C3 falls in the range of −8 to −6. The model is not sensitive to the value of C_1, since a high NSE value always occurs whatever C1 is in the range. The COD modeling is sensitive to both γ_{1COD} and γ_{3COD}, and the NSE value reaches the maximum when γ_{1COD} is around 1 and γ_{3COD} is around 0.25. For the NH_4^+-N simulation, $\gamma_{1NH_4^+-N}$ and $\gamma_{3NH_4^+-N}$ are the sensitive parameters, while for TP, the sensitive parameters are γ_{1TP} and γ_{3TP}.

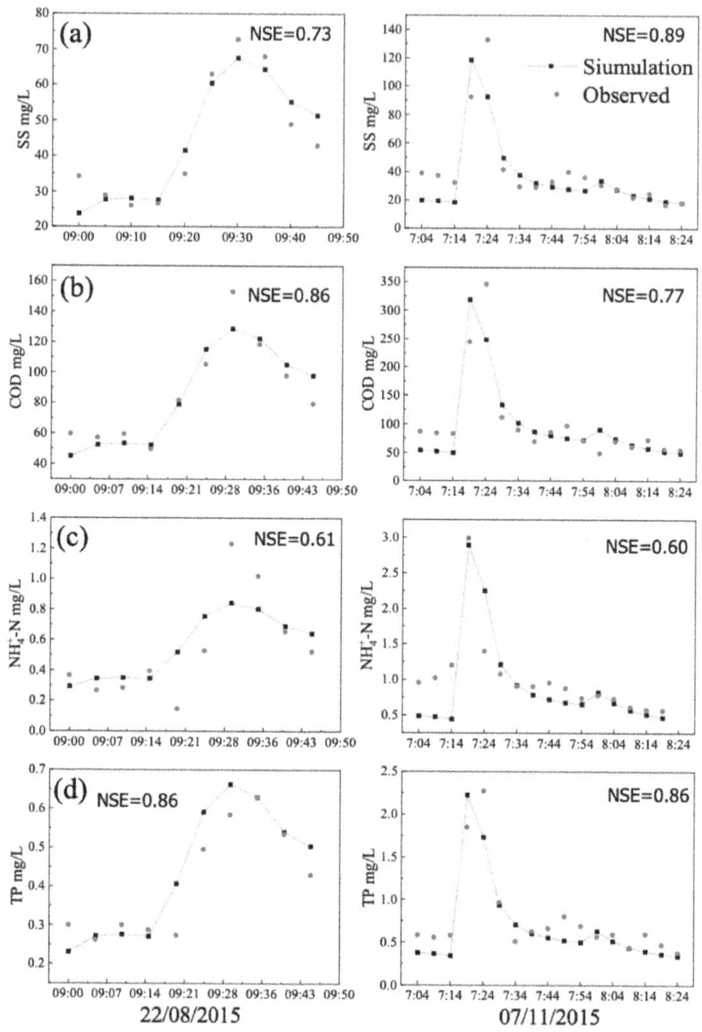

Figure 4. Observed and simulated (**a**) SS, (**b**) COD, (**c**) NH_4^+-N, (**d**) TP concentrations in calibration.

Then, the upper and lower ranges of the uncertainty analysis falling within the 90% confidence interval were plotted with the observed data. The observed data for SS and COD falls in the uncertainty range, while some of the observed NH_4^+-N and TP concentrations were not in the model's uncertainty range. This exclusion can be explained by the assumptions of the water quality simulation in InfoWorks ICM that the concentrations of pollutants are linearly related to the concentration of SS and the coefficient is consistent in a storm event [23]. Therefore, it neglected the relationship between pollutants and rainfall characteristics and might result in errors. In addition, Deletic et al. (2012) [33] pointed out that such integrated models containing several interdependent sub-models might cause over-parametrization and enlarge the uncertainty of the models.

The model was then calibrated with rainfall events on 5 April 2018 and 23 April 2018. The NSE values of the SS and COD modeling of both rainfall events imply that the model is accurate enough; however, for NH_4^+-N and TP modeling, with NSE values larger than 0.6 and a similar peaking time, the results can be regarded as acceptable.

3.2. Impact of LID Sizes on the Volume and Pollutant Reduction of Surface Runoff

Size is the most important parameter in LID design because it directly influences the volume and quality of the runoff in a catchment. As shown in Figure 5, when the area of the sunken green belt is enlarged, the volume reduction increases linearly; however, the pollutant reduction rate increased at first and then declined. The curve of pollutant reduction is due to the FFE, for the initial rainfall carries most of the pollutants, and the concentration of pollutants declines as the rainfall goes on. Likewise, according to the figures, the FFE increased with the increase in rainfall intensity, the runoff interception decreased, and the change in pollutant removal was negligible.

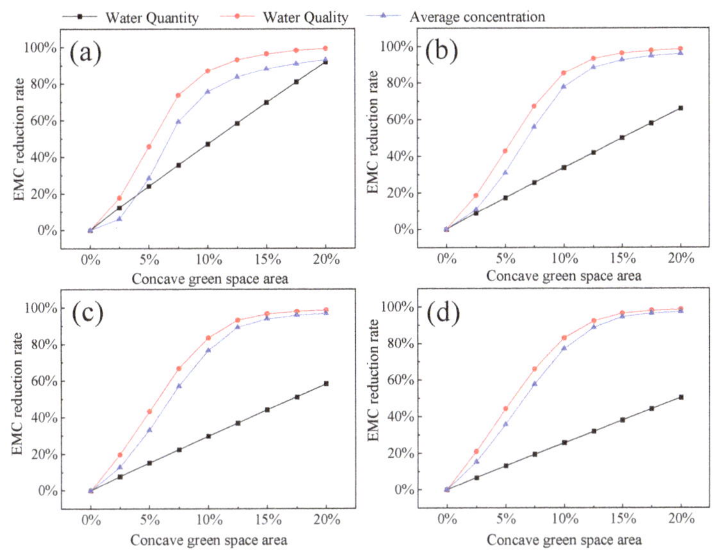

Figure 5. The impact of the sunken green belt on runoff water quality and quantity. (**a**) rainfall intensity once a year. (**b**) rainfall intensity once every three years. (**c**) rainfall intensity once every five years. (**d**) rainfall intensity once every ten years.

Comparing the figures in Figure 5, it can be concluded that when the peak and duration were consistent, the water quantity reduction rate decreased significantly with the increase in rainfall intensity. However, the change in water quality reduction rate was negligible. Besides, the FFE were enhanced with the increase in rainfall intensity. Therefore, even though less water was retained by the sunken green belt when the precipitation intensified, the change in intercepted pollutants was negligible.

Since permeable pavements only intercept rainfall on the surface, the reduction in water quantity is directly proportional to the area of the LID facility with a slope of around one (Figure 6). The pollutants accumulated on the surface were also intercepted with the runoff, and thus the pollutant removal curve is almost the same as the quantity reduction curve.

In the model, the green roof only received rainwater that fell directly on it, so the water reduction rate was proportional to the area. When the rainfall intensity increased, there would be an overflow on the green roof [34]. The water reduction rate would thereby decrease, but the changes in the pollution removal rate were negligible (Figure 7). In this model, the accumulated pollutants on the surface of green roofs were not considered. As a result, the total amount of accumulated pollutants decreases when green roofs take up more space in the study area, and therefore the pollutants in the stormwater are diminished.

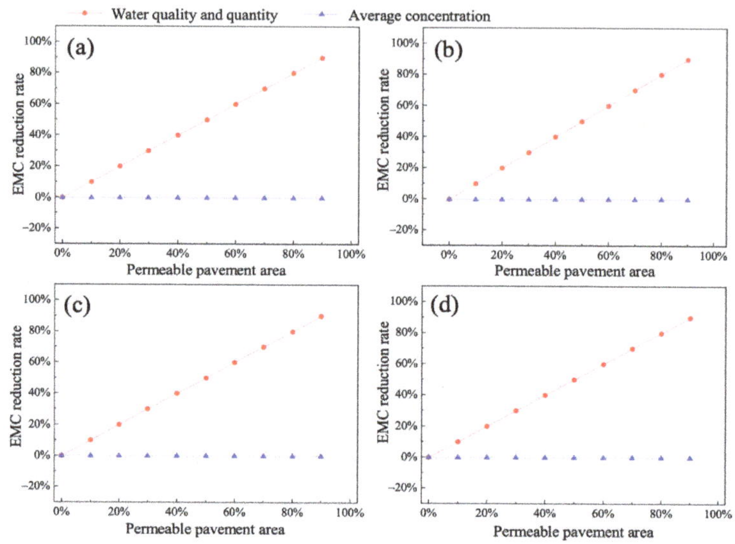

Figure 6. The impact of permeable pavement area on runoff water quality and quantity. (**a**) rainfall intensity once a year. (**b**) rainfall intensity once every three years. (**c**) rainfall intensity once every five years. (**d**) rainfall intensity once every ten years.

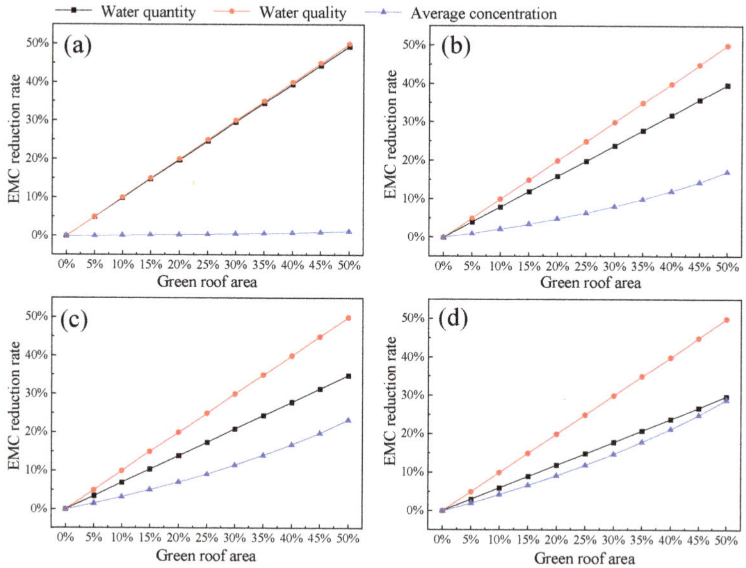

Figure 7. The impact of green roof area on runoff water quality and quantity. (**a**) rainfall intensity once a year. (**b**) rainfall intensity once every three years. (**c**) rainfall intensity once every five years. (**d**) rainfall intensity once every ten years.

3.3. Optimization of LID Facilities

For water quantity reduction, the results were fitted with a first-order polynomial equation (Equations (8) and (9)), and the value of the regression coefficients was calculated for rainfall events with a three-year and ten-year return period. The p-Value for either rainfall intensity is less than 0.0001, which indicates that the model is significant enough.

For each rainfall event, since the determination coefficients R^2, Adj R^2,and Pred R^2 were all 100%, the model fits perfectly.

$$f_{1three-years}(x) = 1.59606 + 2.44357X_1 + 1.07443X_2 + 0.82593X_3 \tag{8}$$

$$f_{1ten-years}(x) = 0.93942 + 1.70496X_1 + 1.07295X_2 + 0.61173X_3 \tag{9}$$

The response surfaces for runoff quantity reduction are shown in Figure S1, in which the results indicate that there were no interactions between the portion of the sunken green belt, permeable pavements, and green roof. The water reduction rate increased with any one of the above variables when other variables remained unchanged.

For pollutant removal, the regression coefficients were calculated, and the response variable was fitted with the following second-order polynomial equations:

$$\begin{aligned} f_{2\,three-years}(x) &= -41.52042 + 19.28233X_1 + 2.23699X_2 + 2.23699X_3 - 0.15076X_1X_2 \\ &-0.15076X_1X_3 - 0.01728X_2X_3 - 0.70556X_1^2 - 8.63834E^{-0.03}X_2^2 \\ &-8.63834E^{-0.03}X_3^2 \end{aligned} \tag{10}$$

$$\begin{aligned} f_{2\,ten-years}(x) &= -34.15688 + 16.9891X_1 + 2.17877X_2 + 2.17877X_3 - 0.13749X_1X_2 \\ &-0.13749X_1X_3 - 0.017156X_2X_3 - 0.5773X_1^2 - 8.57779E^{-0.03}X_2^2 \\ &-8.57779E^{-0.03}X_3^2 \end{aligned} \tag{11}$$

The ANOVA analysis indicated that the model was significant because either p-Value was smaller than 0.0001 and the Adj R^2 for rainfall events with different intensities was 0.9995, 0.9991, respectively. The determination coefficients (R^2) for two simulated rainfall events were 99.98% and 99.96%, respectively, which indicated that the model was adequate for prediction and that only 0.02% and 0.04% of the total variation could not be explained by the model within the range of experimental variables.

The response surface analysis of runoff pollution removal rate was also considered accurate because the p values were smaller than 0.0001. The contour lines in Figure S2 were almost straight and parallel, indicating that the portion of the sunken green belt, the permeable pavement area, and the green roof had little impact on the runoff pollutant removal rate. Additionally, when other variables remained the same, the runoff pollutant removal rate increased with any one of the above variables.

The combinatorial optimization function in Design-expert was employed to calculate the optimal LID configuration under two rainfall intensities. The optimal solutions were the overlapped areas of the contour figures (Figure 8). The maximal desirability for a 3-year return period rainfall event was 0.585. The optimal configuration of LID facilities in the investigated area was 8.6% of the sunken green belt, 15% of the permeable pavement, and 10% of the green roof, with an estimated cost of about $900,000. Under this LID design, the runoff volume reduction rate and pollutant removal rate were 47.3% and 90.4%, respectively. The outcomes were similar to the prediction values, which proved the effectiveness of the response surface method in LID design optimization.

For precipitation with a 10-year return period, the maximal expectation was 0.538. The optimal portion of the sunken green belt, permeable pavement, and green roof in the study area was 8.6%, 19%, and 10%, respectively, with a total cost of about $1,000,000. The runoff volume reduction rate and pollutant removal rate of this LID design were 42.1% and 90.9%, respectively. These results indicated that the sunken green belt was a better choice compared with permeable pavement and a green roof in terms of water quality improvement and price. Therefore, in the LID design, the sunken green belt should take priority over permeable pavement and green roof, and the optimal area of these two LID facilities needs to be calculated [35,36]. However, Shen and Xu (2021) [30] found that the runoff generated by impermeable roads drained directly into the rainwater wells and not through the green belts. Therefore, the confluence relationship between impermeable roads and green belt areas should be changed from parallel to series to improve the volume control effect on rainfall-runoff [30].

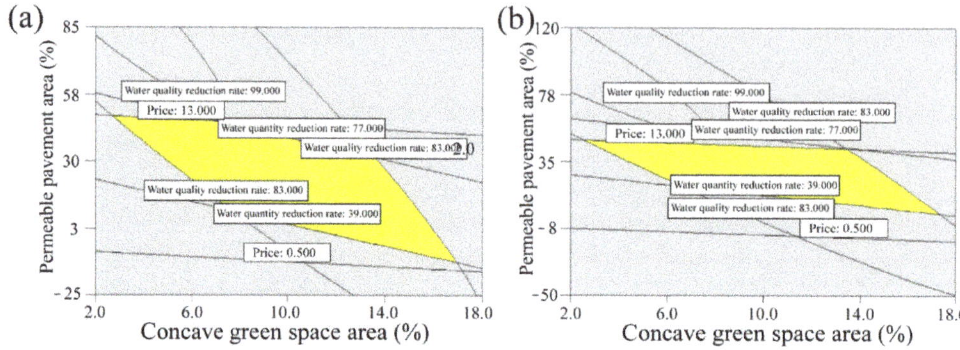

Figure 8. Contour overlay diagram. (**a**) rainfall intensity once every three years. (**b**) rainfall intensity once every ten years.

3.4. Perspectives and Limitations

This study established a rainfall-runoff model and calibrated and validated the data of real-case rainfall events in InfoWorks ICM. The impact of the sizes of these three different types of LID facilities on the water quantity as well as the water quality was examined by different rainfall events with different intensities, and the optimal design in a certain study area was determined by the response surface methodology using the Design Expert. Ho et al. (2022) [31] reported that the green roofs and permeable pavements had a higher unit cost reduction rate than the rain barrels. However, through our methods, we found that the sunken green belt was a better choice compared with permeable pavement and a green roof in terms of water quality improvement and price. Our study provided a strategy for optimizing the design of LID facilities for stormwater runoff treatment, which could provide insight into the future planning of LID facilities in urban ecosystems.

The limitations of this study lie in the water quality estimation of both the model and the LID facilities. In InfoWorks ICM, the LID facilities are only modeled to remove the pollutant from the intercepted runoff and neglect the physical removal processes, such as filtration and sedimentation. Moreover, in this research, only runoff quantity reduction, pollutant removal, and total investment were taken for optimization purposes; however, other objectives, such as social and human benefits, and difficulty in construction should be included.

4. Conclusions

This paper provided a strategy for optimizing the design of LID facilities for stormwater runoff treatment through the rainfall-runoff model and the response surface methodology. Using the GLUE method in the calibration and uncertainty analysis of the water quality model avoided equifinality and improved the accuracy of the parameters as well as the efficiency of model calibration. Results showed that LID facilities only removed pollutants in the intercepted runoff, so the initial flush effect cannot be significantly alleviated. Meanwhile, the sunken green belt was more effective and economical in reducing the runoff volume and improving runoff quality. However, for areas with limited green space, the optimal ratio of various LID facilities can be obtained from the surface response method and the overall expectation function method. The optimization process used the response surface methodology, which incorporates hydrological responses, water quality dynamics, and the investment of different LID designs. However, other objectives, such as social and human benefits and difficulty in construction, should be included in the future. This proposed methodology may be helpful in stormwater management facility planning.

Supplementary Materials: The following supporting information can be downloaded at: https://www.mdpi.com/article/10.3390/w15050989/s1. Table S1: Parameters for LID facilities. Table S2: The values of water quality model parameters. Figure S1. Response surfaces for water quantity reduction rate. (a) rainfall intensity once every three years. (b) rainfall intensity once every ten years. Figure S2. Response surface diagram for water quality reduction rate. (a) rainfall intensity once every three years. (b) rainfall intensity once every ten years.

Author Contributions: Conceptualization, Q.C. and R.X.; Formal analysis, Q.C. and J.C.; Investigation, Q.C. and R.X.; Methodology, Q.C. and R.X.; Supervision, J.C.; Writing—original draft, Q.C. and R.X.; Writing—review and editing, Q.C., R.X. and J.C. All authors have read and agreed to the published version of the manuscript.

Funding: This research received no external funding.

Data Availability Statement: Supporting data can be found by emailing runzexu@hhu.edu.cn.

Conflicts of Interest: The authors declare no conflict of interest.

References

1. Saraswat, C.; Kumar, P.; Mishra, B.K. Assessment of stormwater runoff management practices and governance under climate change and urbanization: An analysis of Bangkok, Hanoi and Tokyo. *Environ. Sci. Policy* **2016**, *64*, 101–117. [CrossRef]
2. Kaykhosravi, S.; Khan, U.T.; Jadidi, A. A Comprehensive Review of Low Impact Development Models for Research, Conceptual, Preliminary and Detailed Design Applications. *Water* **2018**, *10*, 1541. [CrossRef]
3. Krisnayanti, D.S.; Rozari, P.d.; Garu, V.C.; Damayanti, A.C.; Legono, D.; Nurdin, H. Analysis of Flood Discharge due to Impact of Tropical Cyclone. *Civ. Eng. J.* **2022**, *8*, 1752–1763. [CrossRef]
4. Shi, H.; Yin, D.; Li, X.; Gong, Y.; Li, J. Urban stormwater runoff thermal characteristics and mitigation effect of low impact development measures. *J. Water Clim. Chang.* **2019**, *10*, 53–62. [CrossRef]
5. Chen, Z.; Shi, X.; Zhang, J.; Wu, L.; Wei, W.; Ni, B.-J. Nanoplastics are significantly different from microplastics in urban waters. *Water Res. X* **2023**, *19*, 100169. [CrossRef] [PubMed]
6. Hong, J.; Lee, B.; Park, C.; Kim, Y. A colorimetric detection of polystyrene nanoplastics with gold nanoparticles in the aqueous phase. *Sci. Total Environ.* **2022**, *850*, 158058. [CrossRef] [PubMed]
7. Luo, T.; Dai, X.; Chen, Z.; Wu, L.; Wei, W.; Xu, Q.; Ni, B.-J. Different microplastics distinctively enriched the antibiotic resistance genes in anaerobic sludge digestion through shifting specific hosts and promoting horizontal gene flow. *Water Res.* **2023**, *228*, 119356. [CrossRef]
8. Chen, Z.; Wei, W.; Liu, X.; Ni, B.J. Emerging electrochemical techniques for identifying and removing micro/nanoplastics in urban waters. *Water Res.* **2022**, *221*, 118846. [CrossRef] [PubMed]
9. Chow, M.F.; Yusop, Z. Sizing first flush pollutant loading of stormwater runoff in tropical urban catchments. *Environ. Earth Sci.* **2014**, *72*, 4047–4058. [CrossRef]
10. Wang, M.; Zhang, D.; Adhityan, A.; Ng, W.J.; Dong, J.; Tan, S.K. Assessing cost-effectiveness of bioretention on stormwater in response to climate change and urbanization for future scenarios. *J. Hydrol.* **2016**, *543*, 423–432. [CrossRef]
11. Suwarno, I.; Ma'arif, A.; Raharja, N.M.; Nurjanah, A.; Ikhsan, J.; Mutiarin, D. IoT-based lava flood early warning system with rainfall intensity monitoring and disaster communication technology. *Emerg. Sci. J.* **2021**, *4*, 154–166. [CrossRef]
12. Baek, S.S.; Choi, D.H.; Jung, J.W.; Lee, H.J.; Lee, H.; Yoon, K.S.; Cho, K.H. Optimizing low impact development (LID) for stormwater runoff treatment in urban area, Korea: Experimental and modeling approach. *Water Res.* **2015**, *86*, 122–131. [CrossRef]
13. Abduljaleel, Y.; Demissie, Y. Identifying Cost-Effective Low-Impact Development (LID) under Climate Change: A Multi-Objective Optimization Approach. *Water* **2022**, *14*, 3017. [CrossRef]
14. Rong, Q.; Liu, Q.; Xu, C.; Yue, W.; Su, M. Optimal configuration of low impact development practices for the management of urban runoff pollution under uncertainty. *J. Environ. Manag.* **2022**, *320*, 115821. [CrossRef] [PubMed]
15. Jeon, M.; Guerra, H.B.; Choi, H.; Kim, L.-H. Long-Term Monitoring of an Urban Stormwater Infiltration Trench in South Korea with Assessment Using the Analytic Hierarchy Process. *Water* **2022**, *14*, 3529. [CrossRef]
16. Kaykhosravi, S.; Khan, U.T.; Jadidi, M.A. The Effect of Climate Change and Urbanization on the Demand for Low Impact Development for Three Canadian Cities. *Water* **2020**, *12*, 1280. [CrossRef]
17. Chuang, W.-K.; Lin, Z.-E.; Lin, T.-C.; Lo, S.-L.; Chang, C.-L.; Chiueh, P.-T. Spatial allocation of LID practices with a water footprint approach. *Sci. Total Environ.* **2023**, *859*, 160201. [CrossRef]
18. Lee, J.M.; Park, M.; Min, J.-H.; Kim, J.; Lee, J.; Jang, H.; Na, E.H. Evaluation of SWMM-LID Modeling Applicability Considering Regional Characteristics for Optimal Management of Non-Point Pollutant Sources. *Sustainability* **2022**, *14*, 14662. [CrossRef]
19. Xiong, L.; Xu, Z.; Xu, J. Combined Optimization of LID Patches and the Gray Drainage System to Control Wet Weather Discharge Pollution. *ACS ES&T Water* **2022**, *2*, 1734–1746. [CrossRef]
20. Gironás, J.; Roesner, L.A.; Rossman, L.A.; Davis, J. A new applications manual for the Storm Water Management Model (SWMM). *Environ. Modell. Softw.* **2010**, *25*, 813–814. [CrossRef]

21. Li, J.; Zhang, B.; Mu, C.; Chen, L. Simulation of the hydrological and environmental effects of a sponge city based on MIKE FLOOD. *Environ. Earth Sci.* **2018**, *77*, 32. [CrossRef]
22. Rosa, D.J.; Clausen, J.C.; Dietz, M.E. Calibration and Verification of SWMM for Low Impact Development. *JAWRA J. Am. Water Resour. Assoc.* **2015**, *51*, 746–757. [CrossRef]
23. He, Q.; Chai, H.; Yan, W.; Shao, Z.; Zhang, X.; Deng, S. An integrated urban stormwater model system supporting the whole life cycle of sponge city construction programs in China. *J. Water Clim. Chang.* **2019**, *10*, 298–312. [CrossRef]
24. Li, J.; Deng, C.; Li, Y.; Li, Y.; Song, J. Comprehensive Benefit Evaluation System for Low-Impact Development of Urban Stormwater Management Measures. *Water Resour. Manag.* **2017**, *31*, 4745–4758. [CrossRef]
25. Fan, G.; Lin, R.; Wei, Z.; Xiao, Y.; Shangguan, H.; Song, Y. Effects of low impact development on the stormwater runoff and pollution control. *Sci. Total Environ.* **2022**, *805*, 150404. [CrossRef] [PubMed]
26. Zhang, Z.; Gu, J.; Zhang, G.; Ma, W.; Zhao, L.; Ning, P.; Shen, J. Design of urban runoff pollution control based on the Sponge City concept in a large-scale high-plateau mountainous watershed: A case study in Yunnan, China. *J. Water Clim. Chang.* **2021**, *12*, 201–222. [CrossRef]
27. Kong, F.; Ban, Y.; Yin, H.; James, P.; Dronova, I. Modeling stormwater management at the city district level in response to changes in land use and low impact development. *Environ. Modell. Softw.* **2017**, *95*, 132–142. [CrossRef]
28. Urich, C.; Rauch, W. Modelling the urban water cycle as an integrated part of the city: A review. *Water Sci. Technol.* **2014**, *70*, 1857–1872. [CrossRef] [PubMed]
29. Li, M.; Yang, X. Global Sensitivity Analysis of SWMM Parameters Based on Sobol Method. *China Water Wastewater* **2020**, *36*, 95–102.
30. Shen, H.B.; Xu, Z.X. Monitoring and Evaluating Rainfall-Runoff Control Effects of a Low Impact Development System in Future Science Park of Beijing. *J. Am. Water Resour. Assoc.* **2021**, *57*, 638–651. [CrossRef]
31. Ho, H.C.; Lee, H.Y.; Tsai, Y.J.; Chang, Y.S. Numerical Experiments on Low Impact Development for Urban Resilience Index. *Sustainability* **2022**, *14*, 8696. [CrossRef]
32. de Macedo, M.B.; Pereira de Oliveira, T.R.; Oliveira, T.H.; Gomes Junior, M.N.; Texeira Brasil, J.A.; Ferreira do Lago, C.A.; Mendiondo, E.M. Evaluating low impact development practices potentials for increasing flood resilience and stormwater reuse through lab-controlled bioretention systems. *Water Sci. Technol.* **2021**, *84*, 1103–1124. [CrossRef] [PubMed]
33. Deletic, A.; Dotto, C.B.S.; McCarthy, D.T.; Kleidorfer, M.; Freni, G.; Mannina, G.; Uhl, M.; Henrichs, M.; Fletcher, T.D.; Rauch, W.; et al. Assessing uncertainties in urban drainage models. *Phys. Chem. Earth Parts A/B/C* **2012**, *42-44*, 3–10. [CrossRef]
34. Ouellet, V.; Khamis, K.; Croghan, D.; Gonzalez, L.M.H.; Rivera, V.A.; Phillips, C.B.; Packman, A.I.; Miller, W.M.; Hawke, R.G.; Hannah, D.M.; et al. Green roof vegetation management alters potential for water quality and temperature mitigation. *Ecohydrology* **2021**, *14*, e2321. [CrossRef]
35. Malekinezhad, H.; Sepehri, M.; Hosseini, S.Z.; Santos, C.A.G.; Rodrigo-Comino, J.; Meshram, S.G. Role and Concept of Rooftop Disconnection in Terms of Runoff Volume and Flood Peak Quantity. *Int. J. Environ. Res.* **2021**, *15*, 935–946. [CrossRef]
36. Koc, K.; Ekmekcioglu, O.; Ozger, M. An integrated framework for the comprehensive evaluation of low impact development strategies. *J. Environ. Manag.* **2021**, *294*, 113023. [CrossRef]

water

MDPI

Article

Analysis of Preferential Flow in Artificial Substrates with *Sedum* Roots for Green Roofs: Experiments and Modeling

Xuan Chen [1,2], Ruifen Liu [1,*], Defu Liu [1,3] and Xiaokang Xin [4]

1 Hubei Key Laboratory of Ecological Remediation of River-Lakes and Algal Utilization, School of Civil Engineering, Architecture and Environment at Hubei University of Technology, Wuhan 430068, China
2 Hubei YongYeHang Appraisal Consulting Co., Ltd., Wuhan 430061, China
3 Engineering Research Center of Eco-Environment in Three Gorges Reservoir Region, Ministry of Education, China Three Gorges University, Yichang 443002, China
4 Changjiang Water Resources Protection Institute, Wuhan 430051, China
* Correspondence: ruifen1986@aliyun.com; Tel.: +86-186-9612-6075

Abstract: The occurrence of preferential flow in vegetated artificial substrates can weaken the stormwater management performance of green roofs. To explore preferential flow, various plant–substrate combinations that involved two *Sedum* species (*Sedum sarmentosum* and *Sedum lineare*) and two artificial substrates for three depths of 6, 10, and 14 cm were established. Artificial substrates without plants were either perlite-based (namely, PAS) or vermiculite-based (namely, VAS), and they were also set as controls. Thereafter, solute breakthrough experiments were conducted, followed by inverse and forward modeling in Hydrus-1D. Skewness coefficients of all solute breakthrough curves were non-zero, suggesting a prevalence of preferential flow. The Nash–Sutcliffe efficiency coefficients during calibration and validation were greater than 0.7. The obtained hydraulic parameters were different among various vegetated PAS and pure PAS without plants, but appeared the same for the VAS case. Rainfall intensity, plant species, and substrate depth, and the interaction of plant species and substrate depth all had significant effects on PAS preferential flow outflow and index (PFI). Substrate depth had a significant effect on VAS preferential flow and PFI. Since a 10 cm-PAS with *S. lineare* had the smallest PFI of 43.16% in simulation scenarios, its use may better control preferential flow in green roofs.

Keywords: green roofs; preferential flow; artificial substrate; *Sedum* roots; solute breakthrough experiments; HYDRUS-1D

Citation: Chen, X.; Liu, R.; Liu, D.; Xin, X. Analysis of Preferential Flow in Artificial Substrates with *Sedum* Roots for Green Roofs: Experiments and Modeling. *Water* **2023**, *15*, 914. https://doi.org/10.3390/w15050914

Academic Editor: Aldo Fiori

Received: 3 February 2023
Revised: 22 February 2023
Accepted: 24 February 2023
Published: 27 February 2023

1. Introduction

Urban stormwater issues, such as inland flooding and water pollution [1], remain challenging in many Chinese cities, and "Sponge City" is a Chinese concept to tackle these issues [2]. Green roof, especially that which is categorized as "extensive green roof" [3] with flat or gentle slope [4], has become one of the important stormwater control measures for sponge city construction. A green roof usually consists of multiple function layers, among which the vegetation layer and substrate layers play important roles in retaining rainwater and detaining runoff [5,6]. When rainwater falls on a green roof, the plant leaves, stems, and branches intercept rainwater, and the substrate layer stores rainwater in its pore structure. Due to the limited depth of the substrate layer, there is a maximum amount of water that substrate can hold within its structure against the pull of gravity [7] (that is, water-holding capacity (WHC) [6] or maximum water capacity [8]). Normally, it is believed that green roof runoff will not occur until the rainwater stored in the substrate pores exceeds the WHC of the substrate [6]. The retained rainwater refers to the difference between rainwater and runoff, and green roofs can be effective in reducing rainwater volume [3]. Detention refers to the temporal delay that occurs between rainwater that is not retained and green roof runoff, and this process can determine the timing and magnitude of peak runoff [9].

Sedum species, which are extremely resistant to the harsh conditions of summer drought and winter cold on roofs with stable coverage [10], have been widely used in green roof projects [11–13]. According to the Guidelines for the Planning, Execution, and Upkeep of Green-Roof Sites (hereinafter referred to as FLL standards) [7], an artificial substrate consisting of 80% to 90% *v/v* lightweight aggregate and 10% to 20% *v/v* organic matter favors plant growth, with a quality of nutrient-rich, lightweight, and good permeability. Artificial substrate has now become the preferred choice for green roofs, compared to garden soil and improved soil [14].

Previous green roof studies have showed that plant selection, as well as substrate type and depth, influenced stormwater management in green roofs [15–18]. It should be noted that there is a link between plant root traits and the stormwater management performance of green roofs. MacIvor and Lundholm [19] monitored the hydrological performance of 15 green roofs, each with a monoculture of different plants, but the same substrate consisting of potting soil, brick, perlite, sand, peat, and compost, in the Atlantic Canada coastal region. The monitoring results indicated that the greater the plant root density, the less the rainwater retained. Hu et al. [20] conducted continuous hydrological monitoring on four green roofs with a monoculture of different plants (*Callisia repens*, *Portulaca grandiflora*, *Plectranthus prostratus*, and *Sedum lineare*), but the same substrate consisting of peat soil, perlite, and vermiculite, in Shenzhen, China. It was noted that the larger the diameter of individual roots, the less the rainfall retained. The above facts clearly show that plant roots influence the green roof hydrological performance, and runoff differences among substrates with different plants may link to root-induced changes in pore structures and hydraulic properties of substrates [21]. However, quantitative studies on plant root traits and hydraulic properties of vegetated substrates for green roofs are relatively rare.

Quite often, the rainwater retention effect of green roofs decreases with the increasing amount of rainfall [22]. This can be explained by the limited WHCs of green roofs and may also be associated with the preferential flow in the root-induced macropore channels during large rainfall events [22]. Preferential flow is a non-uniform, non-equilibrium flow [23], a common form of water movement and solute transport [24,25]. With large pores as the preferred paths, the occurrence of preferential flow can cause a rapid transport of water and solute and insufficient contact between substrate and water; as a result, substrate can generate runoff before it reaches its WHC [21,26]. The generation of preferential flow will make green roofs less capable of retaining rainwater and detaining runoff [27], especially for large rainfall events that are critical for urban drainage and flood control [28]. Moreover, in consideration of the interaction between water, heat, and solute [29,30], preferential flow will also influence green roofs' other performances, such as cooling effect and runoff quality improvement.

The generation of preferential flow in green roofs can be the result of a combination of water conditions and internal factors [31]. Water conditions referring to initial water content, rainfall intensity, etc. may affect the time of runoff occurrence and volume [32,33]. Internal factors are mainly characteristics related to the vegetation layer and substrate layer. An artificial substrate compliant with the FLL standards will contain a considerable amount of large particles (>2 mm) and have a limited portion of fine particles (<63 μm). This composition would create numerous large pores (e.g., 0.03–3.00 mm) that are likely to cause preferential flow to occur [34]. Liu and Fassman-Beck [35] detected preferential flow in non-vegetated substrate by indoor experiments and simulation methods, indicating the occurrence of preferential flow in substrate with porous structures at low water content. However, this study did not consider the role of plants. Plant roots account for a large proportion of the green roof substrate layer [36], and the pore channels formed by plant root are also one of the important mechanisms for preferential flow generation [37]. Zhang et al. [38] showed that in both the mixture and as a monoculture, an herbaceous plant (*Stypandra glauca*) created preferential flow pathways in green roofs. However, very few studies have provided quantitative data about root traits (e.g., dimeter and volume density) of commonly used *Sedum* species for the stormwater management purpose, and detailed investigation about preferential flow in green roof substrates with *Sedum* species remains lacking.

The purposes of this paper are: (1) to detect the occurrence of preferential flow in various plant–substrate combinations by indoor solute breakthrough experiments, (2) to characterize the substrate hydraulic properties of each combination, and (3) to analyze the effects of plant species, substrate depth, rainfall intensity, and initial water content on the preferential flow development in plant–substrate combinations. Two artificial substrates with various plant species and substrate depths were subjected to solute breakthrough experiments to detect the occurrence of preferential flow. The Hydrus-1D model, validated by experimental data, was used to obtain the hydraulic parameters of each combination and to investigate the influence of different factors on preferential flow.

2. Materials and Methods

2.1. Plants and Substrates

Two *Sedum* species indigenous to China, *Sedum sarmentosum* (*SS*) and *Sedum lineare* (*SL*), and two artificial substrates in accordance with the FLL standards, perlite-based artificial substrate (PAS) and vermiculite-based artificial substrate (VAS), were selected for experiments. The basic physical (e.g., saturated hydraulic conductivity (K_s)) and chemical properties of PAS and VAS are shown in Table 1 [39]. For each artificial substrate, both plant species were propagated by stem cuttings in monoculture at three depths (6 cm, 10 cm, and 14 cm) in 10 cm-diameter acrylic cylinders and grown in an artificial climate chest for 103 days to ensure an excellent plant coverage [39]. Thereafter, plant root characteristics, such as root volume density, and K_s of these 12 vegetated substrates of varying depths (Table 2) were measured [39].

Table 1. Characteristics of the artificial substrates [mean (SE)].

Characteristics	Substrate Type	
	PAS	VAS
Components (% by volume)	90% perlite (<6 mm) 10% chicken manure	90% vermiculite (<5 mm) 10% chicken manure
Bulk density (g/cm^3)	0.21 (0.01)	0.34 (0.01)
Total porosity (%)	91.40 (0.01)	78.80 (0.02)
WHC (%)	36.65 (1.33)	64.05 (1.55)
Organic matter content (g/kg)	31.15 (2.72)	38.64 (2.60)
K_s (cm/min)	54.45 (0.19)	18.48 (1.39)

Table 2. Characteristics of plant roots and vegetated artificial substrates [mean (SE)].

Substrate Depth-Substrate Type-Plant Species	Root Volume Density /(mm^3/cm^3)	Root Volume Density of 0.2–0.4 mm Roots/(mm^3/cm^3)	K_s /(cm/min)
6 cm-PAS-*SS*	0.63 (0.00)	0.33 (0.00)	2.12 (0.17)
10 cm-PAS-*SS*	0.42 (0.01)	0.33 (0.01)	1.97 (0.19)
14 cm-PAS-*SS*	1.37 (0.02)	0.20 (0.01)	0.56 (0.00)
6 cm-PAS-*SL*	0.71 (0.03)	0.31 (0.01)	2.64 (0.09)
10 cm-PAS-*SL*	0.02 (0.00)	0.01 (0.00)	0.68 (0.06)
14 cm-PAS-*SL*	1.13 (0.02)	0.15 (0.00)	0.74 (0.05)
6 cm-VAS-*SS*	2.46 (0.04)	0.37 (0.02)	19.19 (0.54)
10 cm-VAS-*SS*	0.16 (0.00)	0.09 (0.00)	14.14 (1.31)
14 cm-VAS-*SS*	5.66 (0.06)	0.14 (0.01)	12.77 (0.57)
6 cm-VAS-*SL*	1.59 (0.02)	0.59 (0.01)	16.94 (1.28)
10 cm-VAS-*SL*	1.31 (0.02)	0.44 (0.01)	15.44 (0.67)
14 cm-VAS-*SL*	8.98 (0.24)	0.15 (0.01)	13.60 (0.67)

2.2. Solute Breakthrough Experiments

For preferential flow detection in various plant–substrate settings, solute breakthrough experiments were conducted. For each substrate type (PAS or VAS), possible influential factors, such as plant species, substrate depth, rainfall event, and initial water content, were all considered (Figure 1). The level settings of each factor are shown in Figure 1, while plant species, substrate depth, and rainfall event were set as 3-level and initial water content was set as a virtual level, producing a 4-factor, 3-level, orthogonal experimental design for each substrate type. The applied rainfalls of 2 a, 5 a, and 10 a corresponded to 1 h-duration storms for return periods of 2, 5, and 10 years in Wuhan City in central China, with rainfall intensities of 3.8 cm/h, 5.4 cm/h, and 6.6 cm/h, respectively [40]. The initial water content was either water-holding capacity (WHC) or mild drought conditions (MDC) to represent a typical wet/dry moisture condition in green roofs in Wuhan's climate [40]. For each substrate type, nine sets of experiments were conducted, as referring to nine lines in Figure 1. Therefore, there were 18 sets of experiments for the two substrates, and with each set repeating for three times, 54 sets of solute breakthrough experiments were conducted in total.

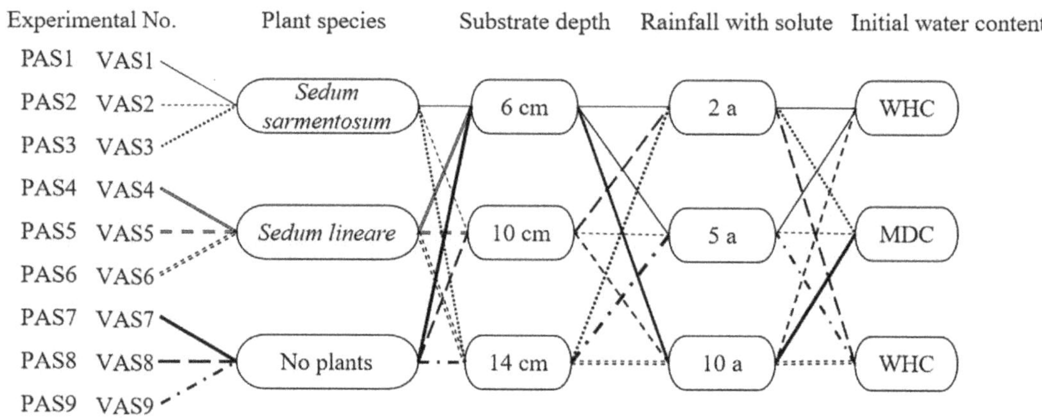

Figure 1. Orthogonal design of solute breakthrough experiments.

Figure 2 shows the apparatus for solute breakthrough experiments, which consisted of a Mariotte bottle (volume 4 L), a rainfall device (37 needles of 0.45 mm diameter for inflow), an acrylic column (diameter 10 cm, loading plant–substrate), an outflow collection container, and an automatic weighing scale (capacity 30 kg, accuracy ±0.1 g). NaCl solution of concentration 1 g/L was used as the tracer, which was dosed through the Mariotte bottle. After 60 min of dosing, the NaCl solution was replaced with deionized water without changing intensity [40]. The outflows from the acrylic column were measured automatically by the weighing instrument at 3 min intervals, and outflow water samples were also collected manually for NaCl concentration measurements. Since the K_s values (Tables 1 and 2) of different plant–substrate combinations were notably larger than the applied rainfall intensities, neither ponding nor overflow occurred in all experiments, and the measured outflow was equal to runoff. The cumulative outflow mass [g] was first converted into outflow volume [L] by assuming a water density of 1 g/cm^3, and was then converted into water depth [cm] by dividing the column surface area of 10 cm diameter. Taking PAS1 as an example (Figure 1), the cumulative outflow process is shown in Figure 3a. The negative values represent the outflow direction vertical downward. The result of the related solute breakthrough curve is also shown in Figure 3b. The C and C_0 [g/L] are the outflow and inflow solute concentration, respectively. V [L] is the cumulative outflow volume with time. V_0 [L] is the infiltrated water volume within the substrate pores, equal to the volume of rainfall minus the volume of outflow in the same time.

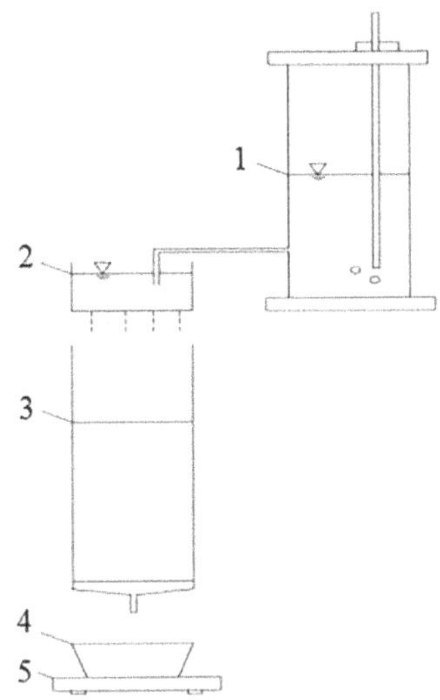

Figure 2. Apparatus for solute breakthrough experiments. 1—Mariotte bottle, 2—rainfall device, 3—acrylic column, 4—outflow collection container, 5—automatic weighing scale.

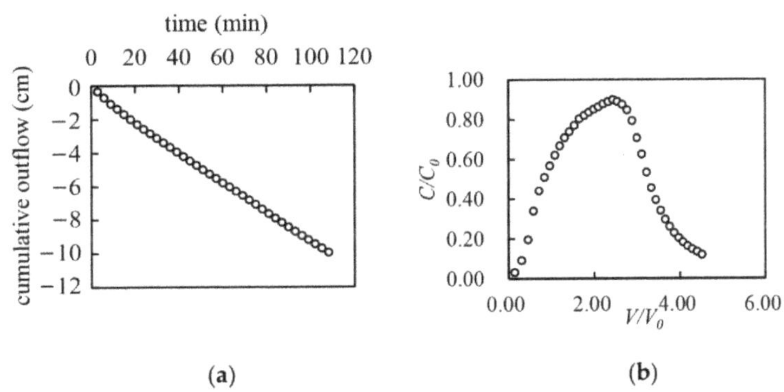

Figure 3. An example of measured data. (**a**) Cumulative outflow curve; (**b**) Solute breakthrough curve.

2.3. Preferential Flow Detection

The shape of the solute breakthrough curve can be quantified by the temporal moments method to determine whether preferential flow is occurring [41]. The temporal moments (M) are described as [42]:

$$M_P = \int_0^\infty T^P c(Z,T)/c_0 \, \mathrm{d}T \quad p = 0,1,2,\ldots \tag{1}$$

where p is the order of the moments; T is equal to V/V_0; $c(Z,T)$ and c_0 are the time-dependent outflow and the initial solute concentrations [g/L], respectively; and Z is the dimensionless spatial coordinate.

The temporal moments method also uses the standard moments (μ'_p) and central moments (μ_P), as defined by the following equations:

$$\mu'_p = M_P/M_0 \tag{2}$$

$$\mu_P = \frac{1}{M_0}\int_0^\infty \left(T - \mu'_1\right)^P c(Z,T)/c_0 \mathrm{d}T \quad p = 0,1,2,\ldots \tag{3}$$

The first-order standard moments (μ_1') describe the breakthrough time of the tracer in the solute breakthrough experiment, and the second-order central moments (μ_2) describe the degree of dispersion of the solute breakthrough curve. The third-order central moments (μ_3) are used to quantitatively describe the asymmetry of the solute breakthrough curve. The dimensionless skewness coefficient (S) then can be obtained as below:

$$S = \mu_3/{\mu_2}^{3/2} \tag{4}$$

If $S < 0$, the solute breakthrough curve is a rightward biased curve, and if $S > 0$, the solute breakthrough curve is a leftward biased curve, and if $S = 0$, the solute breakthrough curve is symmetric [43,44]. When preferential flow occurs, $S \neq 0$.

2.4. Determination of Substrate Hydraulic Parameters

2.4.1. Calculation of Substrate Hydraulic Parameters

Substrate hydraulic parameters were calculated using inverse modeling by Hydrus-1D [45]. In Hydrus-1D, the water flow module and the inversion module were activated, in which the cumulative outflow data (Figure 3a) were set as the objective function, and the nonlinear Levenberg–Marquardt algorithm was used to minimize the objective function [46]. Once preferential flow was detected (Section 2.3), a dual permeability model [47,48] was selected to describe water flow movement. The dual-permeability model (Equations (5)–(9)) assumes that the porous substrate consists of two interacting, overlapping pore domains. The micropores with relatively low permeability are the matrix domain (subscript m), and the high-permeability preferential flow paths, such as large pores and fractures between the matrix, are the preferential flow domain (subscript f). Both domains are quantified separately using the two coupled Darcy–Richard equations (Equations (5) and (6)). In these equations, substrate hydraulic parameters (θ_r, θ_s, a, n, K_s, and l) defining water retention curves and hydraulic conductivity functions were needed, and the van Genuchten–Mualem formula [46] was used to fit these parameters.

$$\frac{\partial \theta_f}{\partial t} = \frac{\partial}{\partial z}\left[K_f\left(\frac{\partial h_f}{\partial z}+1\right)\right] - S_f - \frac{\Gamma_w}{\omega} \tag{5}$$

$$\frac{\partial \theta_m}{\partial t} = \frac{\partial}{\partial z}\left[K_m\left(\frac{\partial h_m}{\partial z}+1\right)\right] - S_m + \frac{\Gamma_w}{1-\omega} \tag{6}$$

$$\theta = \omega\theta_f + (1-\omega)\theta_m \tag{7}$$

$$\Gamma_w = \alpha_w\left(h_f - h_m\right) \tag{8}$$

$$\alpha_w = \frac{\beta}{a^2}\Gamma K_a \tag{9}$$

where θ_f, θ_m, θ are the water content of the preferential flow domain, matrix domain, and the entire domain, respectively, [cm^3·cm^{-3}]; t is the simulation time [min]; z is the vertical coordinate positive upward [cm]; S_f and S_m are the plant water uptake rates of the preferential flow domain and matrix domain, respectively [min^{-1}]; K_f and K_m are the unsaturated hydraulic conductivity of the preferential flow domain and the matrix domain, respectively [cm·min^{-1}]; h_f and h_m are the matric potential of the preferential flow domain and the matrix domain, respectively [kPa]; ω is the proportion of the preferential flow domain to the whole domain [dimensionless]; Γ_w is the water exchange rate between the two domains [min^{-1}]; α_w is the first-order mass transfer coefficient for water [cm^{-1}·min^{-1}]; β is a dimensionless geometry-dependent shape factor; Γ is a dimensionless scaling factor; a is the distance between the center of the matrix domain and the boundary of the preferential flow domain [cm]; and K_a is the effective hydraulic conductivity of the fracture-matrix interface [cm·min^{-1}].

For inverse modeling, substrate geometry values and initial water contents were set according to the orthogonal design of solute breakthrough experiments (Figure 1). The corresponding spatial-temporal discretization settings were given in Chen's previous study [39]. Since constant rainfall intensities were applied in the experiments, the upper boundary was set as the constant flux boundary. According to the apparatus setting (Figure 2), the lower boundary was set as the seepage face. Due to the short duration of rainfall, evaporation and plant uptake were not considered [49], and therefore, S_f,S_m in Equation (5) and Equation (6) were 0. Some parameters, such as a, β, and Γ in Equation (9), were determined as 0.1 [33,47], 15 (spherical shape assumption), and 0.4 (empirical value) [50], respectively. Still, there were other parameters needed for Equations (5)–(9), including hydraulic parameters of the matrix domain (θ_{rm} (taking the value of 0), θ_{sm}, a_m, n_m, K_{sm}, l_m (pore curvature, generally taking the value of 0.5)), hydraulic parameters of the preferential flow domain (θ_{rf}(taking the value of 0.5), θ_{sf}, α_f, n_f, K_{sf}, l_f(taking the value of 0.5)), the parameter of the interface (K_a), and the dimensionless factor (ω). Constraints on those unspecified parameters were given to ensure an overall unique solution and convergence in the parameter optimization [51]. Based on substrate physical properties (Table 2), the constraint of saturated water content (that is, the sum of θ_{sm} and θ_{sf}) of PAS was set as <0.90, and the constraint of saturated water content of VAS was set as <0.78. Since $\alpha_m, \alpha_f, n_m, n_f$ were related to the physical properties of the particles, and the empirical parameter range was set as $\alpha \in (0.001, 0.01), n \in (2,5)$ [52]. The constraint of hydraulic conductivities of the two domains was set as $K_{sm} + K_{sf} \leq K_s$. The empirical range of K_a was 10^{-7}–10^{-4} when preferential flow occurred [47]. Based on the measured and modeled values of the objective function, the coefficient of determination R^2 (Equation (10), [52]) and the Nash–Sutcliffe efficiency coefficient NSE (Equation (11), [53]) were calculated to determine the optimal parameters.

$$R^2 = \left[\frac{\sum_{i=1}^{N}(O_i - \bar{O})(P_i - \bar{P})}{\left[\sum_{i=1}^{N}(O_i - \bar{O})^2\right]^{0.5}\left[\sum_{i=1}^{N}(P_i - \bar{P})^2\right]^{0.5}} \right]^2 \tag{10}$$

$$\text{NSE} = 1 - \frac{\sum_{i=1}^{N}(P_i - O_i)^2}{\sum_{i=1}^{N}(O_i - \bar{O})^2} \tag{11}$$

where N is the total number of observations; P_i and O_i are, respectively, the ith modeled and observed values (i = 1, 2, ..., N); and $\bar{P}$ and $\bar{O}$ are the mean modeled and observed values, respectively. The R^2 values close to 1 indicate that variations of the observed values can be captured well in the modeling. NSE can range from $-\infty$ to 1, with a closer value of 1 representing a more perfect match [52,53].

2.4.2. Validation of Substrate Hydraulic Parameters

Substrate hydraulic parameters obtained from the inverse modeling were validated by the forward modeling for the solute transport process in Hydrus-1D, and the dual permeability model (Equations (5)–(9)) was used for the associated water flow process. The classical convection-dispersion equation to describe the solute transport process based on water transport is as follows [46]:

$$\frac{\partial \theta_f c_f}{\partial t} + \rho \frac{\partial s_f}{\partial t} = \frac{\partial}{\partial z}\left(\theta_f D_f \frac{\partial c_f}{\partial z}\right) - \frac{\partial q_f c_f}{\partial z} - \phi_f - \frac{\Gamma_s}{w} \tag{12}$$

$$\frac{\partial \theta_m c_m}{\partial t} + \rho \frac{\partial s_m}{\partial t} = \frac{\partial}{\partial z}\left(\theta_m D_m \frac{\partial c_m}{\partial z}\right) - \frac{\partial q_m c_m}{\partial z} - \phi_m + \frac{\Gamma_s}{1-w} \tag{13}$$

$$\Gamma_s = \omega_{dp}(1-w)\theta_m\left(c_f - c_m\right) + \Gamma_w c^* \tag{14}$$

where C_f,C_m are the concentrations of the preferential flow domain and the matrix domain [g·cm^{-3}]; ρ is the bulk density of the substrate [g·cm^{-3}]; D_f, D_m are the sorbed concentrations of the preferential flow domain and the matrix domain [g·g^{-1}]; q_f, q_m are the volumetric fluid flux densities of the preferential flow domain and the matrix domain [cm·s^{-1}]; ϕ_f, ϕ_m are sink-source terms that account for various zero- and first-order or other reactions in both domains [g·cm^{-3}·s^{-1}]; Γ_s is the solute mass transfer term [g·cm^{-3}·min^{-1}]; ω_{dp} is the first-order solute mass transfer coefficient [min^{-1}]; and $c^* = c_f$ for $\Gamma_w > 0$ and $c^* = c_m$ for $\Gamma_w < 0$.

Most settings of the water flow process for the forward modeling in Hydrus 1D were the same as those for the inverse modeling, such as spatial-temporal discretization, initial values, and boundary conditions. The additional inputs as required by the solute transport process were set according to the solute breakthrough experiments (Section 2.2). The molecular diffusion coefficient in free water D_w for Cl^- was 1.7 cm^2/day, and the dispersion coefficient D_L was 1/10 of the corresponding substrate depth (Figure 1). Substrate bulk densities are given in Table 1. The incoming solute concentration was 1 g/L, and the solute dosing time was 60 min (Section 2.2). The forward modeling predicted solute concentrations at different moments, and based on modeled and observed concentrations, R^2 and NSE were calculated to assess the rationality of the substrate hydraulic parameters.

2.5. *Preferential Flow and Influential Factors*

Based on the constructed Hydrus-1D model with validated parameters, four influential factors, including plant species, substrate depth, rainfall intensity, and initial water content, can be varied, according to the control variable method [49] to explore the law of preferential outflow for different plant–substrates. A total of 54 simulated conditions were established [39]. This study focuses on conceptual understanding and describing the flow process rather than performing parameter optimization or stochastic model analysis.

For each simulated condition, the solute breakthrough curve was obtained, and its skewness coefficient was calculated for the preferential flow detection (Section 2.3). In addition, the preferential outflow and the preferential flow index (PFI, the percentage of the preferential outflow to the total water flow [33]) based on simulation results were also obtained. Multi-factor ANOVA was used to test whether the main effects and interaction effects of different influential factors on preferential outflow and PFI were significant. The coefficient of variation, C_v [54], was used to describe the variance of preferential outflow and PFI among simulation conditions [55]. According to Nielsen's classification criteria [56], $C_v \leq 10\%$ indicates a weak coefficient of variation, $10\% < C_v < 100\%$ indicates a medium coefficient of variation, and $C_v \geq 100\%$ indicates a strong coefficient of variation.

3. Results and Discussion

3.1. *Preferential Flow Detection*

The characteristics of the solute breakthrough curves corresponding to the 18 experimental sets of the 2 artificial substrates are shown in Table 3. It can be seen that the S values of all curves are not 0. Among them, the S values of the vegetated and non-vegetated PAS are −0.06–0.37 (PAS1–PAS6) and 0.01–0.21 (PAS7–PAS9), respectively. The S values of the vegetated and non-vegetated VAS are 0.01–0.30 (VAS1–VAS6) and 0.17–0.61 (VAS7–VAS9), respectively. The results indicate that preferential flow commonly occurs in the green roof plant–substrate combinations. According to the existing literature [57,58], the occurrence of preferential flow is related to the non-homogeneity of the substrate, plant roots, and moisture conditions, which will be discussed later in Section 3.3.

Table 3. Summary of solute breakthrough curve characteristics.

Experimental No.	M_0	M_1	μ_1'	μ_2	μ_3	S	Experimental No.	M_0	M_1	μ_1'	μ_2	μ_3	S
PAS1	2.34	0.53	0.23	1.61	−0.12	−0.06	VAS1	1.28	0.58	0.5	0.41	0.01	0.05
PAS2	1.18	0.48	0.41	0.47	0.09	0.27	VAS2	0.79	0.49	0.62	0.22	0.01	0.06
PAS3	0.81	0.35	0.43	0.29	0.05	0.34	VAS3	0.45	0.47	1.05	0.07	0.01	0.30
PAS4	1.13	0.42	0.37	0.57	0.16	0.37	VAS4	0.87	0.51	0.58	0.24	0.01	0.05
PAS5	1.26	0.48	0.38	0.57	0.16	0.37	VAS5	0.94	0.52	0.55	0.26	0.03	0.23
PAS6	1.25	0.50	0.40	0.49	0.07	0.22	VAS6	0.87	0.50	0.58	0.25	0.00	0.01
PAS7	3.39	0.62	0.18	2.46	0.44	0.11	VAS7	1.90	0.48	0.25	1.37	0.98	0.61
PAS8	0.77	0.40	0.52	0.31	0.00	0.01	VAS8	0.62	0.39	0.63	0.18	0.01	0.17
PAS9	0.76	0.37	0.49	0.26	0.03	0.21	VAS9	0.57	0.37	0.64	0.17	0.02	0.26

3.2. Substrate Hydraulic Parameters

3.2.1. Results

Figure 4 show the modeled cumulative outflows from the inverse modeling and the corresponding observed outflows. It can be seen that the calculated R^2 (Equation (10)) are in the range of 0.998–0.999, and NSE (Equation (11)) are in the range of 0.741–0.997. Figure 5 shows the predicted outflow concentrations from the forward modeling and the corresponding measured concentrations. It shows that R^2 are in the range of 0.937–0.993, and NSE are in the range of 0.741–0.973. These data indicate that the substrate hydraulic parameters (Table 4) obtained from the inverse modeling are validated for the forward modeling and can be further used for the preferential flow simulation of different plant–substrates combinations (Section 3.3). In Table 4, the hydraulic parameters of the matrix domain remain constant for each substrate, irrespective of plant–substrate combinations, as the dual permeability model assumes that the root system only make changes to the preferential flow domain [59]. Considering the significant effect of plant root traits on K_s of PAS [60], the hydraulic parameters of the preferential flow domain of PAS are varied. However, since there was no insignificant difference in K_s of VAS due to the root system [60], hydraulic properties of the preferential flow domain of VAS can be viewed as the same.

Table 4. Summary of substrate hydraulic parameters.

Name	Matrix Domain				Preferential Flow Domain				ω	K_a/ (cm·min^{-1})
	θ_{sm}/ (cm^3·cm^{-3})	α_m / (cm^{-1})	n_m	K_{sm} / (cm·min^{-1})	θ_{sf} / (cm^3·cm^{-3})	α_f / (cm^{-1})	n_f	K_{sf} / (cm·min^{-1})		
6 cm-PAS-*SS*					0.33	0.050	2.1	2.0	0.12	
10 cm-PAS-*SS*					0.35	0.050	2.1	1.8	0.11	
14 cm-PAS-*SS*					0.27	0.002	1.5	0.4	0.05	
6 cm-PAS-*SL*	0.150	0.008	2.50	0.100	0.36	0.054	2.0	2.5	0.14	0.75×10^{-6}
10 cm-PAS-*SL*					0.30	0.005	1.8	0.5	0.07	
14 cm-PAS-*SL*					0.30	0.005	1.8	0.6	0.07	
pure PAS					0.75	0.009	3.8	54.4	0.60	0.16×10^{-6}
6 cm-VAS-*SS*								19.1		
10 cm-VAS-*SS*								14.1		
14 cm-VAS-*SS*								12.7		
6 cm-VAS-*SL*	0.131	0.011	2.41	0.105	0.60	0.008	2.618	16.9	0.026	0.75×10^{-6}
10 cm-VAS-*SL*								15.4		
14 cm-VAS-*SL*								13.6		
pure VAS								18.8		

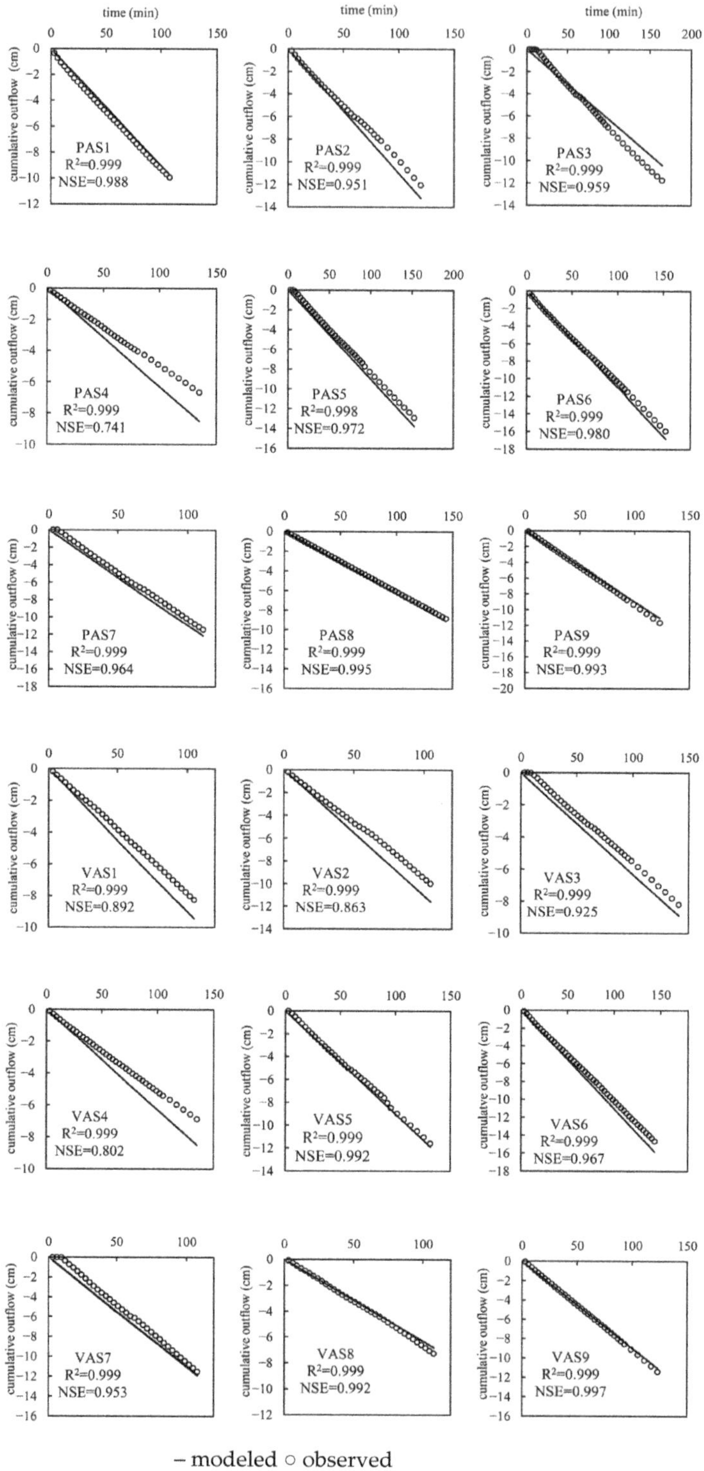

Figure 4. Comparison of modeled and observed values of cumulative outflow in inverse modeling.

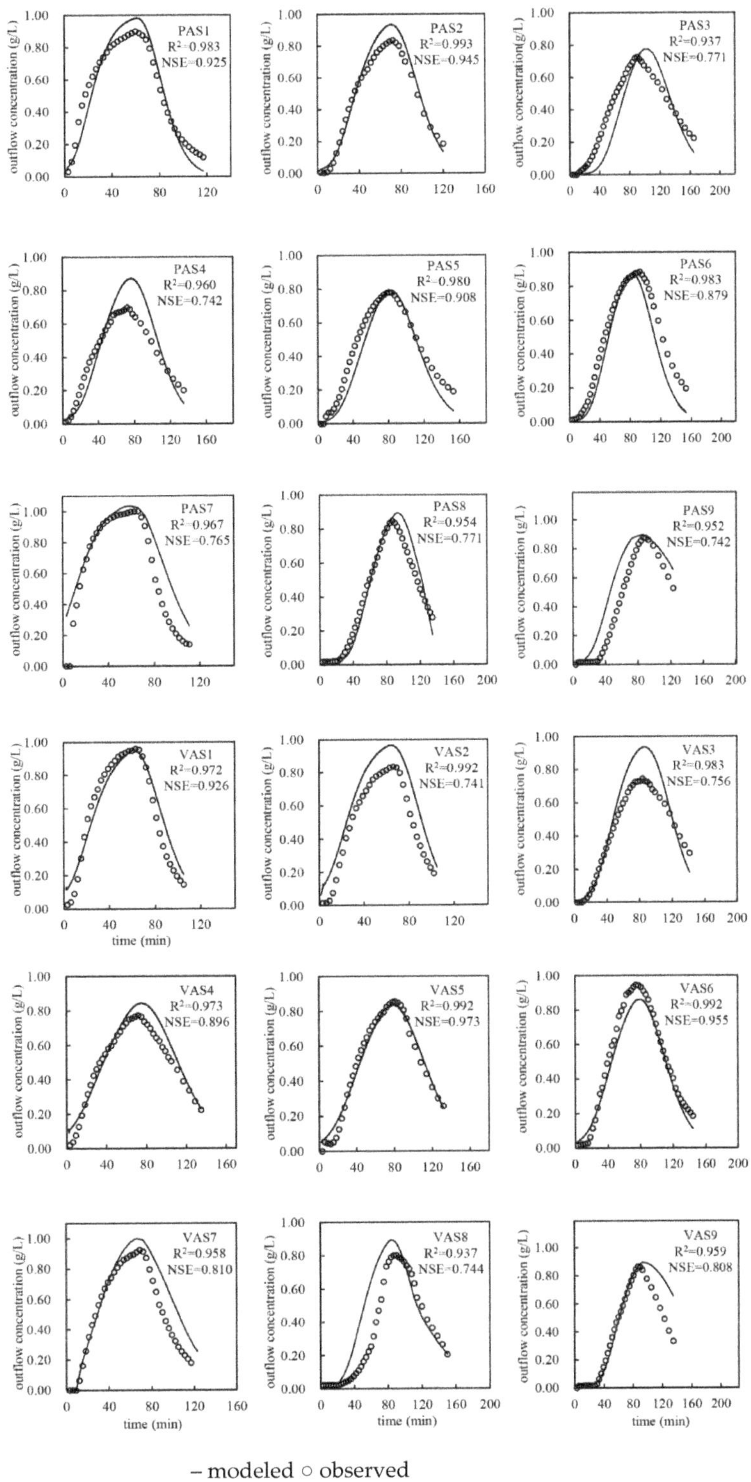

Figure 5. Comparison of predicted and measured outflow concentrations in forward modeling.

3.2.2. Implications

Based on the obtained hydraulic parameters (Table 4), the corresponding water retention curves of PAS in preferential flow domain can be plotted. A water retention curve reflects the variation of pore water in the substrate with the matric potential, and also indirectly reflects the distribution of pore size in the substrate [61,62]. It can be seen from Figure 6 that curves from vegetated PAS are significantly different from that from pure PAS. The initial stable water content in high matrix potentials (i.e., around 0 kPa) from pure PAS is noticeably greater than those from the other curves, and afterwards, the decrease in water content, along with lower matrix potentials, is much steeper than those from the other curves. Among these vegetated PAS, curves also show various differences in terms of the initial stable water content, the decreasing slope, and the specific matric potential that the stable water content starts to decrease. Those differences in the water retention curves of PAS indicate different pore structures are present due to different root characteristics, and further exploration, therefore, is made below.

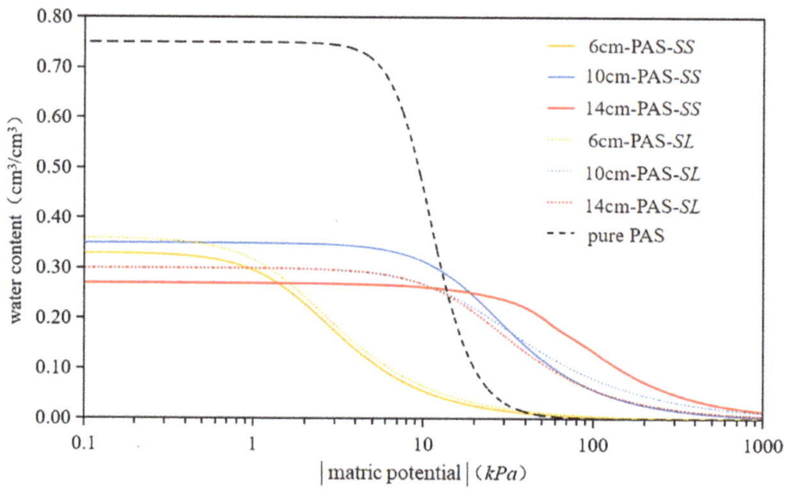

Figure 6. Water retention curves of PAS (preferential flow domain).

To further analyze the root-induced changes of hydraulic properties, the capillary model [63] was combined with a water retention curve for analysis. The capillary model considers that the matric potential S is mainly the result of capillary forces acting on circular capillaries of a certain range of pore sizes. In the model, σ is used to denote the water surface tension coefficient (7.5×10^{-4} N/cm at room temperature), r_0 denotes the capillary radius, and D denotes the capillary diameter (i.e., equivalent pore diameter, $D = 2r_0$). The relationship between the equivalent pore diameter and the matric potential is $D = 4\sigma/S$. When the matric potential is S_1, the corresponding equivalent pore diameter is D_1. Only in the pore diameter less than D_1 are capillary pores filled with water, and the water content is θ_1. When the matric potential is S_2 ($S_1 < S_2$), D_2, θ_2 are obtained in the same way. The ratio of the pore volume occupied by pores with an equivalent pore size between D_2 and D_1 to the total volume of substrate pores is called the equivalent pore volume ratio (θ_1–θ_2). Based on the above theory, equivalent pore volume ratios of PAS between 0–−10 kPa, −10–−100 kPa, −100–−1000 kPa were calculated (Table 5). As can be seen from Table 5, compared to vegetated PAS, pure PAS has greater volume ratios of pores, with diameters > 0.03 mm and between 0.003–0.03 mm, but lower ratios of pores with diameters <0.003 mm. According to the agronomic criteria, pores larger than 0.03 mm in diameter tend to act as macropores for water permeable and aeration, and water in pores between 0.003–0.03 mm are most easily accessible to plants [35]. The differences in pore structure between vegetated PAS and pure PAS reflect that the presence of roots in PAS can effectively block its macropores. Further analysis combined with the root characteristics (Table 2) reveal that the volume ratio of pores > 0.03 mm in diameter is linearly correlated (correlation coefficient $r = 0.92$) with the root volume density of 0.2–0.4 mm roots [39]. This indicates that for vegetated PAS, although shallow-rooted *Sedums* with a large portion of fine roots often lead to a reduction of macropores, the presence of 0.2–0.4 mm roots can effectively offset the reduction. The root characteristics, in turn, are associated with interactions between plant species, substrate type, and substrate depth [60]. The relatively high (28.48–30.63%) macropores are present in 6 cm-PAS-*SS*, 10 cm-PAS-*SS*, and 6 cm-PAS-*SL* (Table 5). It is noted that a deeper PAS does not promote the development of 0.2–0.4 mm roots (Table 2); on the contrary, it will foster a root system resulting in macropore blockage and K_s reduction [60].

Table 5. Equivalent pore volume ratio of PAS.

Name	**Range of Equivalent Pore Sizes (Corresponding Matric Potentials)**		
	>0.03 mm (0–−10 kPa)	**0.003–0.03 mm (−10–−100 kPa)**	**0.0003–0.003 mm (−100–−1000 kPa)**
6 cm-PAS-*SS*	28.48%	5.34%	0.42%
10 cm-PAS-*SS*	30.60%	6.34%	0.61%
14 cm-PAS-*SS*	0.99%	12.47%	12.30%
6 cm-PAS-*SL*	30.63%	6.12%	0.60%
10 cm-PAS-*SL*	4.18%	18.80%	6.92%
14 cm-PAS-*SL*	4.30%	21.05%	5.42%
pure PAS	41.63%	48.41%	0.11%

3.3. Preferential Flow and Influential Factors

3.3.1. Perlite-Based Substrate (PAS)

With the substrate hydraulic parameter (Table 4) in the Hydrus-1D model, various simulation conditions (Table 6) were set up to systematically investigate the effects of plant species, substrate depth, rainfall intensity, and initial water content on the preferential outflow in PAS. The results of the skewness coefficient (S), preferential outflow, and PFI under each simulation condition are shown in Table 6. All of the S values are not zero, suggesting that the occurrence of preferential flow is prevalent. The preferential outflow all exceed 2.49 cm, and the PFI ranges from 33.00% to 100.00%.

Based on the simulation results (Table 6), multi-factor ANOVA was performed, and the results are shown in Table 7. It can be seen that for PAS, rainfall intensity, plant species, substrate depth, and the interaction of plant species and substrate depth, all had significant effects on the preferential outflow and PFI, while the initial water content had no significant effect on both. Therefore, simulation conditions with the initial water content of WHC were excluded from the following analysis, which focuses on rainfall intensity, plant species, and substrate depth for 27 simulation conditions only.

Rainfall intensity: It can be seen from F values in Table 7 that rainfall intensity has the greatest effect on preferential outflow ($F = 268.98$), and a correlation analysis for the 27 simulation conditions shows a positive (correlation coefficient $r = 0.83$) linear relationship between the two. When the rainfall intensity varied with fixed other factors (plant species, substrate depth, etc.), among all the simulation conditions (Table 8), preferential outflow produced from 10 a-rainfall was greater than that from 5 a-rainfall, which in turn was greater than that from 2 a-rainfall (Table 6). However, the rainfall intensity influenced PFI to a lesser extent ($F = 8.175$). The positive (correlation coefficient $r = 0.\ 78$) linear correlation between PFI and rainfall intensity also exists for the 27 simulation conditions. It is noted that in Table 8, with rainfall intensity varying, high mean PFI (≥67.19%), but low C_v (≤2.97%) are present in non-vegetated PAS, 6 cm- and 10 cm-PAS with *SS*, and 6 cm-PAS with *SL*. Table 5 reveals that these plant–substrate combinations have high portions of macropores (28.48–41.63%), comprising pore networks favoring preferential flow development [34]. Therefore, preferential flow development in these plant–substrates is mainly influenced by internal pore structure and less correlated with rainfall intensity, resulting in high mean PFI, but low C_v values. In contrast, for 14 cm-PAS with *SS*, and 10 and 14 cm-PAS with *SL*, small portions of macropores (0.99–4.30%, Table 5) provide few preferential paths, and therefore, the related preferential flow development can be influenced by both internal pore structure and rainfall intensity. Correspondingly, the associated mean PFI are 57.51–82.50%, 36.04–51.06%, and 33.00–65.68%, respectively, and the C_v values are 17.77%, 17.47%, and 38.63%, all showing considerable degrees of variability (Table 8).

Plant species: The effect of plant species on preferential outflow (F = 118.54, Table 7) is second only to rainfall intensity (F = 268.98, Table 7). When plant species varied with fixed other factors (substrate depth, rainfall intensity, etc.), among all the simulation conditions (Table 9), preferential outflow C_v values from 6 cm-PAS subject to various rainfalls were less than 10%, while those from 10 cm-PAS and 14 cm-PAS subject to various rainfalls were greater than 20%. Considering the secondary importance of plant species for preferential outflow (Table 7), the change in preferential outflow C_v can be attributed to plant species, and the effect of plant species on preferential outflow becomes more prominent for deeper substrates. In addition, for any three simulations of varying plant species, but fixed other factors (Table 9), non-vegetated PAS had the largest preferential outflow (Table 6). This may be due to a high volume ratio of macropores (>0.03 mm) in the non-vegetated PAS (41.63%, Table 5), which was 1.36–42.05 times larger than that in the vegetated PAS, and since macropores are potential preferential flow paths, eventually the largest preferential outflow occurred in non-vegetated PAS. Table 7 also shows that plant species have the greatest effect on PFI (F = 84.98). Similar to preferential outflow, based on changes in PFI C_v values for simulations of varying plant species, it can be concluded that the effect of plant species on PFI also becomes more prominent for deeper substrates. As plants make changes to the pore structures of PAS (Table 5), further analysis for the 27 simulation conditions shows a positive (correlation coefficient r = 0.92) linear correlation between macropore volume ratio and PFI. Since the macropore volume ratio is also significantly and positively correlated with the root volume density of 0.2–0.4 mm roots (Section 3.2.2), it indicates that *Sedum* roots of 0.2–0.4 mm diameter promote the development of preferential flow.

Substrate depth: The effect of substrate depth on the preferential outflow is the smallest (F = 31.66, Table 7). When the substrate depth varied with fixed other factors (plant species, rainfall intensity, etc.), among all the simulation conditions (Table 10), preferential outflow C_v values from non-vegetated PAS subject to various rainfalls were less than 1%, followed by less than 20% from PAS with *SS*, and greater than 27% from PAS with *SL* (Table 10). This indicates that PAS with *SL* is more influenced by substrate depth in terms of preferential outflow, compared to non-vegetated PAS and PAS with *SS*. In addition, for any three simulations of varying substrate depth, but fixed other factors (Table 10), 6 cm-vegetated PAS had the largest preferential outflow (Table 6). Similar to preferential outflow, the effect of substrate depth on PFI was the smallest (F = 23.94, Table 7), and based on PFI C_v changes, it is concluded that PAS with *SL* is more influenced by substrate depth in terms of PFI, compared to non-vegetated PAS and PAS with *SS*. Likewise, for any three simulations of varying substrate depth, but fixed other factors (Table 10), 6 cm-vegetated PAS had the largest PFI (Table 6). The 6 cm depth of vegetated PAS is associated with high root volume densities of 0.2–0.4 mm roots (0.33 mm^3/cm^3 for 6 cm-PAS-*SS* and 0.31 mm^3/cm^3 for 6 cm-PAS-*SL*, Table 2) that can play positive roles for preferential flow development.

Table 6. Simulation results of PAS under different simulation conditions.

Simulation No.	Plant Species	Substrate Depth/(cm)	Rainfall Intensity/(a)	Initial Water Content/(%)	S	Preferential Outflow/(cm)	PFI/(%)
1			2	WHC	−0.35	6.38	84.29
2				MDC	−0.35	6.38	84.29
3		6	5	WHC	−0.06	9.01	83.43
4				MDC	−0.06	9.01	83.43
5			10	WHC	−0.31	11.05	83.65
6				MDC	−0.31	11.05	83.65
7			2	WHC	0.64	5.08	67.19
8				MDC	0.64	5.08	67.19
9	*Sedum sarmentosum*	10	5	WHC	0.04	7.59	70.32
10				MDC	0.04	7.59	70.32
11			10	WHC	0.27	9.37	70.77
12				MDC	0.04	9.37	70.77
13			2	WHC	0.34	4.35	57.51
14				MDC	0.34	4.35	57.51
15		14	5	WHC	0.12	7.71	71.43
16				MDC	0.12	7.71	71.43
17			10	WHC	0.56	10.89	82.50
18				MDC	0.56	10.89	82.50
19			2	WHC	0.37	6.73	89.00
20				MDC	0.37	6.73	89.00
21		6	5	WHC	−0.27	9.56	88.55
22				MDC	−0.27	9.56	88.55
23			10	WHC	−0.39	11.58	87.73
24				MDC	0.39	11.58	87.73
25			2	WHC	0.88	2.72	36.04
26				MDC	0.88	2.72	36.04
27			5	WHC	0.37	4.61	42.37
28	*Sedum lineare*	10		MDC	0.37	4.61	42.37
29			10	WHC	−0.52	6.74	51.06
30				MDC	−0.52	6.74	51.06
31			2	WHC	0.07	2.49	33.00
32				MDC	0.07	2.49	33.00
33		14	5	WHC	0.18	4.09	38.01
34				MDC	0.18	4.09	38.01
35			10	WHC	0.22	8.66	65.68
36				MDC	0.22	8.66	65.68
37			2	WHC	0.09	7.59	99.97
38				MDC	0.09	7.59	99.97
39		6	5	WHC	0.10	10.85	100.0
40				MDC	0.10	10.85	100.0
41			10	WHC	0.11	13.21	99.92
42				MDC	0.11	13.21	99.92
43			2	WHC	0.01	7.56	99.89
44				MDC	0.01	7.56	99.89
45	No-plants	10	5	WHC	0.59	10.73	99.91
46				MDC	0.59	10.73	99.91
47			10	WHC	0.63	13.08	99.92
48				MDC	0.63	13.08	99.92
49			2	WHC	0.18	7.59	99.87
50				MDC	0.18	7.59	99.87
51		14	5	WHC	0.21	10.76	99.91
52				MDC	0.21	10.76	99.91
53			10	WHC	0.46	13.19	99.77
54				MDC	0.46	13.19	99.77

Table 7. Multi-factor ANOVA results for preferential outflow and PFI of PAS.

Sources of Variance	*F* Values	
	Preferential Outflow	PFI
Plant species	118.54 **	84.98 **
Substrate depth	31.66 **	23.94 **
Plant species × Substrate depth	13.96 **	10.55 **
Rainfall intensity	268.98 **	8.175 *
Initial water content	0.00	0.00

Note: * ($p < 0.05$) reached a significant level and ** ($p < 0.01$) reached a highly significant level.

Table 8. Mean values and C_v of PAS preferential outflow and PFI for different rainfall intensities.

Simulation No. (Rainfall Intensity Varying)	Fixed Variables: Plant Species, Substrate Depth, Initial Water Content	Preferential Outflow		PFI	
		Mean Values /(cm)	C_v /(%)	Mean Values /(cm)	C_v /(%)
2, 4, 6	*SS*, 6 cm, MDC	8.81	26.56	83.79	0.53
20, 22, 24	*SL*, 6 cm, MDC	9.29	26.22	88.43	0.73
38, 40, 42	No-plants, 6 cm, MDC	10.55	26.75	99.96	0.04
8, 10, 12	*SS*, 10 cm, MDC	7.35	29.30	69.36	2.97
26, 28, 30	*SL*, 10 cm, MDC	4.69	42.88	43.16	17.47
44, 46, 48	No-plants, 10 cm, MDC	10.46	26.49	99.91	0.02
14, 16, 18	*SS*, 14 cm, MDC	7.65	42.75	70.48	17.77
32, 34, 36	*SL*, 14 cm, MDC	3.20	63.03	45.56	38.63
50, 52, 54	No-plants, 14 cm, MDC	10.51	26.71	99.85	0.07

Table 9. Mean values and C_v of PAS preferential outflow and PFI for different plant species.

Simulation No. (Plant Species Varying)	Fixed Variables: Substrate Depth, Rainfall Intensity, Initial Water Content	Preferential Outflow		PFI	
		Mean Values /(cm)	C_v /(%)	Mean Values /(cm)	C_v /(%)
2, 20, 38	6 cm, 2 a, MDC	6.90	9.02	91.09	8.83
4, 22, 40	6 cm, 5 a, MDC	9.81	9.63	90.66	9.36
6, 24, 42	6 cm, 10 a, MDC	11.95	9.42	90.43	9.36
8, 26, 44	10 cm, 2 a, MDC	5.12	47.26	67.71	47.16
10, 28, 46	10 cm, 5 a, MDC	7.64	40.04	70.87	40.60
12, 30, 48	10 cm, 10 a, MDC	9.73	32.74	73.92	33.26
14, 32, 50	14 cm, 2 a, MDC	4.81	53.66	63.46	53.31
16, 34, 52	14 cm, 5 a, MDC	7.52	44.40	69.78	44.40
18, 36, 54	14 cm, 10 a, MDC	10.91	20.76	82.65	20.62

Table 10. Mean values and C_v of PAS preferential outflow and PFI for different substrate depths.

Simulation No. (Substrate Depth Varying)	Fixed Variables: Plant Species, Rainfall Intensity, Initial Water Content	Preferential Outflow		PFI	
		Mean Values /(cm)	C_v /(%)	Mean Values /(cm)	C_v /(%)
2, 8, 14	*SS*, 2 a, MDC	5.08	20.01	67.19	19.98
4, 10, 16	*SS*, 5 a, MDC	8.10	9.72	75.06	9.69
6, 12, 18	*SS*, 10 a, MDC	10.44	8.88	78.97	9.03
20, 26, 32	*SL*, 2 a, MDC	3.98	59.91	52.68	59.78
22, 28, 34	*SL*, 5 a, MDC	6.09	49.60	56.31	49.73
24, 30, 36	*SL*, 10 a, MDC	8.99	27.10	68.16	27.08
38, 44, 50	No-plants, 2 a, MDC	7.58	0.23	99.91	0.05
40, 46, 52	No-plants, 5 a, MDC	10.78	0.58	99.94	0.05
42, 48, 54	No-plants, 10 a, MDC	13.16	0.53	99.87	0.09

3.3.2. Vermiculite-Based Substrate (VAS)

Since the hydraulic parameters of varying VAS remain unchanged or appear comparable (Table 4), minor changes in the pore structures of VAS due to *Sedum* root system are expected. Therefore, the influential factors considered for VAS were substrate depth, rainfall intensity, and initial water content, for which simulation conditions involving no-plants only were set up in Table 11. The resulted skewness coefficient S, preferential outflow, and PFI are also listed in Table 11. None of the S values are zero, which indicates the prevalence of preferential flow occurrence in VAS. The minimum values of preferential outflow and PFI are 7.04 cm and 93.06%, respectively.

Table 11. Simulation results of VAS under different simulation conditions.

Simulation Conditions	Substrate Depth /(cm)	Rainfall Intensity /(a)	Initial Water Content /(%)	S	Preferential Outflow /(cm)	PFI /(%)
1	6	2	WHC	0.05	7.42	98.27
2			MDC	0.05	7.42	98.27
3		5	WHC	0.06	10.62	98.33
4			MDC	0.06	10.62	98.33
5		10	WHC	0.30	13.01	98.26
6			MDC	0.30	13.01	98.26
7	10	2	WHC	0.05	7.24	95.81
8			MDC	0.05	7.24	95.81
9		5	WHC	0.23	10.36	95.84
10			MDC	0.23	10.36	95.84
11		10	WHC	0.01	12.67	95.84
12			MDC	0.01	12.67	95.84
13	14	2	WHC	−0.61	7.04	93.09
14			MDC	−0.61	7.04	93.09
15		5	WHC	0.17	10.05	93.06
16			MDC	0.17	10.05	93.06
17		10	WHC	0.26	12.31	93.12
18			MDC	0.26	12.31	93.12

Multi-factor ANOVA was also performed based on the above simulation results (Table 12). It can be seen that for VAS, both rainfall intensity and substrate depth had significant effects on preferential outflow, and substrate depth had a significant effect on PFI. However, initial water content had no significant effect on preferential outflow and PFI. Therefore, simulation conditions with the initial water content of WHC were excluded from the following analysis, which focuses on rainfall intensity and substrate depth for nine simulation conditions only.

Table 12. Multi-factor ANOVA results for preferential outflow and PFI of VAS.

Sources of Variance	*F* Values	
	Preferential Outflow	PFI
Substrate depth	104.095 **	50,845.585 **
Rainfall intensity	10,207.964 **	1.098
Initial water content	0.000	0.000

Note: ** ($p < 0.01$) reached a highly significant level.

Rainfall intensity: The F value of 10,207.964 (Table 12) clearly shows the dominant role of rainfall intensity on preferential outflow, and a correlation analysis for the nine simulation conditions also shows a positive (correlation coefficient $r = 0.98$) linear relationship between the two. The larger the rainfall intensity, the greater the preferential outflow produced. In contrast, rainfall intensity had no significant effect on PFI (Table 12). When the nine simulation conditions of varying rainfall intensity were grouped by the substrate depth

(Table 13), the resulting mean PFI values are high (≥93.06%), while the C_v values are extremely low (≤0.1%).

Table 13. Mean values and C_v of VAS preferential outflow and PFI for different rainfall intensities.

Simulation No. (Rainfall Intensity Varying)	Fixed Variables: Substrate Depth, Plant Species, Initial Water Content	Preferential Outflow		PFI	
		Mean Values /(cm)	C_v /(%)	Mean Values /(cm)	C_v /(%)
2, 4, 6	6 cm, no-plants, MDC	10.35	27.09	98.29	0.04
8, 10, 12	10 cm, no-plants, MDC	10.09	26.98	95.83	0.02
14, 16, 18	14 cm, no-plants, MDC	9.80	26.97	93.09	0.03

Substrate depth: In addition to the rainfall intensity, VAS depth also had an effect on preferential outflow (Table 12, $F = 104.095$). When the nine simulation conditions of varying substrate depth were grouped by the rainfall intensity (Table 14), it was noted that in a rainfall event, preferential outflow from 6 cm-VAS was the largest and from 14 cm-VAS was the smallest (Table 11). The C_v values of preferential outflow were small (<3%, Table 14). The PFI was only influenced by the VAS depth (Table 12), and its variation pattern, along with different substrate depths, was consistent with that of the preferential outflow (e.g., 6 cm-VAS had the largest PFI, and the C_v values had limited variations (Table 14)).

Table 14. Mean values and C_v of VAS preferential outflow and PFI for different substrate depths.

Simulation No. (Substrate Depth Varying)	Fixed Variables: Rainfall Intensity, Plant Species, Initial Water Content	Preferential Out-flow		PFI	
		Mean Values /(cm)	C_v /(%)	Mean Values /(cm)	C_v /(%)
2, 8, 14	2 a, no-plants, MDC	7.24	2.62	95.72	2.71
4, 10, 16	5 a, no-plants, MDC	10.34	2.76	95.74	2.75
6, 12, 18	10 a, no-plants, MDC	12.66	2.76	95.74	2.69

Considering the negative role of preferential flow on green roof stormwater performance [23,64], the degree of preferential flow should be minimized as much as possible in green roofs. The above analysis shows that preferential flow development in PAS and VAS are controlled by different factors. For PAS, rainfall intensity, plant species, and substrate depth all had significant effects on PFI (Table 7), while for VAS, only substrate depth played a role on PFI. After reviewing Tables 8 and 13, it is known that 10 cm-PAS-*SL* has the lowest mean PFI of 43.16%, regardless of rainfall intensity, and all VAS have large mean PFI ranging from 93.09% to 98.29%. Therefore, for the preferential flow control purpose, a combination of 10 cm-PAS-*SL* may be recommended for the plant–substrate design in green roofs. It should be noted that this recommendation is made based on green roof performance for individual large rainfalls. Future research focusing on improving green roof performance for both large and small rainfalls over a long period may come up with a better plant–substrate design recommendation.

4. Conclusions

In order to investigate the law of preferential outflow in various green roof plant–substrate combinations, two *Sedum* plants (namely *Sedum sarmentosum* and *Sedum lineare*) were planted in two artificial substrates (namely, PAS and VAS) at three different depths, and pure artificial substrates were also set as controls. Thereafter, indoor solute breakthrough experiments and water flow and solute transport simulations in Hydrus-1D were conducted. The in-door experimental results showed that the skewness coefficients of all solute breakthrough curves were non-zero, indicating preferential flow generally occurred in green roof plant–substrate combinations. The hydraulic parameters of different substrates were obtained from the inverse modeling in Hydrus-1D. The correlation coefficients between the modeled and measured values of the cumulative outflow for PAS and VAS were in the range of 0.998–0.999, and the Nash–Sutcliffe efficiency coefficients were in the range of 0.741–0.997. For PAS, the different hydraulic parameters of the vegetated PAS at different depths were due to the differences in root-induced pore structure. In contrast, hydraulic parameters from different VAS can be viewed as the same. According to the forward modeling results in Hydrus-1D, it is concluded that for PAS, rainfall intensity, plant species, substrate depth, and the interaction of plant species and substrate depth all had significant effects on the preferential outflow and PFI, while the initial water content had no significant effect on both. For VAS, both rainfall intensity and substrate depth had significant effects on the preferential outflow, and substrate depth also had a significant effect on PFI. Likewise, initial water content had no significant effect on VAS preferential outflow and PFI. The 10 cm-PAS with *S. lineare* may be recommended for preferential flow control purposes. Further research considering both preferential flow control and stormwater retention improvement for green roof design is needed.

Author Contributions: Investigation, X.C.; Formal analysis, X.C. and R.L.; Methodology, R.L. and D.L.; Writing—Original Draft Preparation, X.C. and R.L.; Writing—Review & Editing, R.L. and X.X. All authors have read and agreed to the published version of the manuscript.

Funding: The research was funded by the National Natural Science Foundation of China, grant number [51909081].

Data Availability Statement: The data presented in this study are available from the first author upon reasonable request.

Conflicts of Interest: The authors declare no conflict of interest.

References

1. Chang, N.B.; Lu, J.W.; Chui, T.F.M.; Hartshorn, N. Global policy analysis of low impact development for stormwater management in urban regions. *Land Use Policy* **2018**, *70*, 368–383. [CrossRef]
2. Fowdar, H.; Payne, E.; Deletic, A.; Zhang, K.F.; McCarthy, D. Advancing the Sponge City Agenda: Evaluation of 22 plant species across a broad range of life forms for stormwater management. *Ecol. Eng.* **2022**, *175*, 106501. [CrossRef]
3. Liu, R.F.; Stanford, R.L.; Deng, Y.; Liu, D.F.; Liu, Y.; Yu, S.L. The influence of extensive green roofs on rainwater runoff quality: A field-scale study in southwest China. *Environ. Sci. Pollut. Res.* **2019**, *27*, 12932–12941. [CrossRef]
4. MoHURD. *Technical Guidelines of the Sponge City Development—Low Impact Development Systems for Storm Water*; Ministry of Housing and Urban Rural Development: Beijing, China, 2014.
5. Chen, Y.; Chen, H. The collective strategies of key stakeholders in Sponge City Construction: A tripartite game analysis of governments, developers, and consumers. *Water* **2020**, *12*, 1087. [CrossRef]
6. Vijayaraghavan, K. Green roofs: A critical review on the role of components, benefits, limitations and trends. *Renew. Sustain. Energy Rev.* **2016**, *57*, 740–752. [CrossRef]
7. FLL. *Guidelines for the Planning, Construction and Execution of Green-Roof Sites*; Forschungsgesellschaft Landschaftsentwicklung Landschaftsbau e.V.: Bonn, Germany, 2008.
8. She, N.; Pang, J. Physically based green roof model. *J. Hydrol. Eng.* **2010**, *15*, 458–464. [CrossRef]
9. Peng, Z.; Smith, C.; Stovin, V. The importance of unsaturated hydraulic conductivity measurements for green roof detention modelling. *J. Hydrol.* **2020**, *590*, 125273. [CrossRef]
10. Farrell, C.; Mitchell, R.E.; Szota, C.; Rayner, J.P.; Williams, N.S.G. Green roofs for hot and dry climates: Interacting effects of plant water use, succulence and substrate. *Ecol. Eng.* **2012**, *49*, 270–276. [CrossRef]
11. Yang, H. Studies on Plants and Substrates Selection for Extensive Green Roofs in Beijing. Master's Thesis, Beijing Forestry University, Beijing, China, 2012.
12. Zhang, F.F. A Study on Technology of Producing *Sedum* spp. Mat for Light Roof-Greening. Master's Thesis, Shanghai Jiao Tong University, Shanghai, China, 2009.
13. Zhou, Y.; Xu, D.Y.; Dong, Y.F.; Chen, F.Z.; Tong, J.; Xie, Y.F.; Chen, Y.X.; Guo, C.X. Study on drought resistance of 9 *Sedums* for light roof greening. *Chin. Agric. Sci. Bull.* **2012**, *28*, 294–301. [CrossRef]
14. Xiao, M.; Lin, Y.; Han, J.; Zhang, G. A review of green roof research and development in China. *Renew. Sustain. Energy Rev.* **2014**, *40*, 633–648. [CrossRef]
15. Sang, M.; Zhang, W.; Zhong, X.; Che, W. Influence from conventional substrate on rainwater retention and outflow water quality characteristics of extensive green roof facility. *Water Resour. Hydropower Eng.* **2018**, *49*, 163–170. [CrossRef]
16. Li, X.X.; Cao, J.J.; Xu, P.X.; Fei, L.; Dong, Q.; Wang, Z.L. Green roofs: Effects of plant species used on runoff. *Land Degrad. Dev.* **2018**, *29*, 3628–3638. [CrossRef]
17. Nagase, A.; Dunnett, N. Amount of water runoff from different vegetation types on extensive green roofs: Effects of plant species, diversity and plant structure. *Landsc. Urban Plan.* **2012**, *104*, 356–363. [CrossRef]

18. Zhang, S.X.; Zhang, S.H.; Zhang, Y.; Wu, S.T. Impacts of vegetation on quantity and quality of runoff from green roofs. *Environ. Sci.* **2019**, *40*, 3618–3625. [CrossRef]
19. MacIvor, J.S.; Lundholm, J. Performance evaluation of native plants suited to extensive green roof conditions in a maritime climate. *Ecol. Eng.* **2011**, *37*, 407–417. [CrossRef]
20. Hu, Y.C.; Qin, H.P.; Lin, Z.X. Variation and influencing factors of rainwater retention of green roofs in Shenzhen. *J. Shenzhen Univ. Sci. Eng.* **2020**, *37*, 347–354. [CrossRef]
21. Stovin, V.; Poë, S.; De-Ville, S.; Berretta, C. The influence of substrate and vegetation configuration on green roof hydrological performance. *Ecol. Eng.* **2015**, *85*, 159–172. [CrossRef]
22. Ge, D.; Zhang, S.H. Impacts of vegetation on hydrological performances of green roofs under different rainfall conditions. *Environ. Sci.* **2018**, *39*, 5015–5023. [CrossRef]
23. Skorobogatov, A.; He, J.X.; Chu, A.; Valeo, C.; van Duin, B. The impact of media, plants and their interactions on bioretention performance: A review. *Sci. Total Environ.* **2020**, *715*, 136918. [CrossRef]
24. Zhang, W.J.; Yuan, S.S. Characterizing preferential flow in landfilled municipal solid waste. *Waste Manag.* **2019**, *84*, 20–28. [CrossRef] [PubMed]
25. Zhang, Y.H.; Zhang, Z.M.; Ma, Z.W.; Chen, J.X.; Akbar, J.; Zhang, S.Y.; Che, C.G.; Zhang, M.X.; Cerdà, A. A review of preferential water flow in soil science. *Can. J. Soil Sci.* **2018**, *98*, 604–618. [CrossRef]
26. Locatelli, L.; Mark, O.; Mikkelsen, P.S.; Arnbjerg-Nielsen, K.; Jensen, M.B.; Binning, P.J. Modelling of green roof hydrological performance for urban drainage applications. *J. Hydrol.* **2014**, *519*, 3237–3248. [CrossRef]
27. Niu, J.Z. *Study on Preferential Flow in Forest Ecosystems*; Science Publishing House: Beijing, China, 2013.
28. Wang, J.B.; Zhao, J.S.; Shen, Z.Y.; Wang, H.; Lei, X.H. Discussion about the two rainfall control approaches in Sponge City Construction. *J. Hydraul. Eng.* **2017**, *48*, 1490–1498. [CrossRef]
29. Hashemi, S.S.G.; Mahmud, H.B.; Ashraf, M.A. Performance of green roofs with respect to water quality and reduction of energy consumption in tropics: A review. *Renew. Sustain. Energy Rev.* **2015**, *52*, 669–679. [CrossRef]
30. Sandoval, V.; Bonilla, C.A.; Gironás, J.; Vera, S.; Victorero, F.; Bustamante, W.; Rojas, V.; Leiva, E.; Pastén, P.; Suárez, F. Porous media characterization to simulate water and heat transport through green roof substrates. *Vadose Zone J.* **2017**, *16*, 1–14. [CrossRef]
31. She, N.; Liu, J. Using Preferential Flow to Model Green Roofs. In Proceedings of the World Environmental and Water Resources Congress, Cincinnati, OH, USA, 19–23 May 2013.
32. Alaoui, A.; Caduff, U.; Gerke, H.H.; Weingartner, R. Preferential flow effects on infiltration and runoff in grassland and forest soils. *Vadose Zone J.* **2011**, *10*, 367–377. [CrossRef]
33. Wu, Q.H.; Zhu, G.S.; Cui, H.D.; Zhang, J.F.; Zhang, F.W. Effect of precipitation intensities on preferential flow and its numerical modeling. *Trans. Chin. Soc. Agric. Eng.* **2014**, *30*, 118–127. [CrossRef]
34. Lu, J.R.; Zhang, Q.; Werner, A.D.; Li, Y.L.; Jiang, S.L.; Tan, Z.Q. Root-induced changes of soil hydraulic properties—A review. *J. Hydrol.* **2020**, *589*, 125203. [CrossRef]
35. Liu, R.F.; Fassman-Beck, E. Hydrologic response of engineered media in living roofs and bioretention to large rainfalls: Experiments and modeling. *Hydrol. Process.* **2017**, *31*, 556–572. [CrossRef]
36. Quinn, R.; Dussaillant, A. The impact of macropores on heavy metal retention in sustainable drainage systems. *Hydrol. Research.* **2018**, *49*, 517–527. [CrossRef]
37. Gao, C.X.; Xu, X.X.; Zhao, J.N.; Zhao, C.P.; Zhang, S.N. Review on macropore flow in soil. *Sheng Tai Xue Bao* **2014**, *34*, 2801–2811. [CrossRef]
38. Zhang, Z.; Szota, C.; Fletcher, T.D.; Williams, N.S.G.; Werdin, J.; Farrell, C. Influence of plant composition and water use strategies on green roof stormwater retention. *Sci. Total Environ.* **2018**, *625*, 775–781. [CrossRef] [PubMed]
39. Chen, X. Effect of *Sedum* Roots on Preferential Flow Transport of Artificial Substrates on Green Roofs. Master's Thesis, Hubei University of Technology, Wuhan, China, 2022.
40. Zhou, M.L. Study on Combination Screening and Application Effect of Plant-Artificial Matrix based on Heavy Rainfall Control. Master's Thesis, Hubei University of Technology, Wuhan, China, 2021.
41. Kamra, S.K.; Lennartz, B. Quantitative indices to characterize the extent of preferential flow in soils. *Environ. Model. Softw.* **2005**, *20*, 903–915. [CrossRef]
42. Young, D.F.; Ball, W.P. Column experimental design requirements for estimating model parameters from temporal moments under nonequilibrium conditions. *Adv. Water Resour.* **2000**, *23*, 449–460. [CrossRef]
43. André, L.; Durante, M.; Pauss, A.; Lespinard, O.; Ribeiro, T.; Lamy, E. Quantifying physical structure changes and non-uniform water flow in cattle manure during dry anaerobic digestion process at lab scale: Implication for biogas production. *Bioresour. Technol.* **2015**, *192*, 660–669. [CrossRef] [PubMed]
44. Lamy, E.; Lassabatere, L.; Bechet, B.; Andrieu, H. Modeling the influence of an artificial macropore in sandy columns on flow and solute transfer. *J. Hydrol.* **2009**, *376*, 392–402. [CrossRef]
45. Schwinger, J.; van Genuchten, M.T. Modeling nonequilibrium flow and transport processes using HYDRUS. *Vadose Zone J.* **2008**, *7*, 782–797. [CrossRef]
46. Rassam, D.; Šimůnek, J.; Mallants, D.; van Genuchten, M.T. *The HYDRUS-1D Software Package for Simulating the One-Dimensional Movement of Water, Heat, and Multiple Solutes in Variably-Saturated Media: Tutorial*; CSIRO Land and Water: Adelaide, Australia, 2018.

47. Gerke, H.H.; van Genuchten, M.T. A dual-porosity model for simulating the preferential movement of water and solutes in structured porous media. *Water Resour. Res.* **2010**, *29*, 305–319. [CrossRef]
48. Arora, B.; Mohanty, B.P.; McGuire, J.T. Inverse estimation of parameters for multidomain flow models in soil columns with different macropore densities. *Water Resour. Res.* **2011**, *47*, W04512. [CrossRef]
49. Zhang, W.; Wang, H.Y.; Zhao, H.Y. Hydrological regulation simulation of rainwater runoff of bioretention based on HYDRUS-1D. *Water Resour. Prot.* **2022**, *38*, 102–108. [CrossRef]
50. Li, F.; Jiao, X.Y.; Li, P.P.; Zhai, D. Parametric inversion of soil water characteristic curves of farmland. *J. Hohai Univ. (Nat. Sci.)* **2009**, *37*, 373–377. [CrossRef]
51. Li, Y.B. Study on Inverse Method of Soil Hydraulic Parameters under Two-dimensional Negative and Positive Pressure Infiltration. Ph.D. Thesis, Northwest A&F University, Xianyang, China, 2018.
52. Ren, D.Y.; Xu, X.; Hao, Y.Y.; Huang, G.H. Modeling and assessing field irrigation water use in a canal system of Hetao, upper Yellow River basin: Application to maize, sunflower and watermelon. *J. Hydrol.* **2016**, *532*, 122–139. [CrossRef]
53. Nash, J.E.; Sutcliffe, J.V. River flow forecasting through conceptual models part I—A discussion of principles. *J. Hydrol.* **1970**, *10*, 282–290. [CrossRef]
54. Jia, J.P. *Statistics*, 5th ed.; China Renmin University Press: Beijing, China, 2012.
55. Chen, X.B.; Zhang, H.J.; Cheng, J.H.; Zhang, F.; Zhang, X.; Ruan, X.Z. Quantitative evaluation of preferential flow development degree based on dyed image variability analysis. *Trans. Chin. Soc. Agric. Mach.* **2015**, *46*, 93–100. [CrossRef]
56. Nielsen, D.R.; Bouma, J. Soil Spatial Variability. In Proceedings of the A Workshop of the ISSS and the SSSA, Las Vegas, NV, USA, 30 November–1 December 1984.
57. Zhang, Y.; Niu, J.Z.; Zhang, M.X.; Xiao, Z.X.; Zhu, W.L. Interaction between plant roots and soil water flow in response to preferential flow paths in Northern China. *Land Degrad. Dev.* **2017**, *28*, 648–663. [CrossRef]
58. Lin, H. Linking principles of soil formation and flow regimes. *J. Hydrol.* **2010**, *393*, 3–19. [CrossRef]
59. Shao, W.; Bogaard, T.A.; Bakker, M.; Greco, R. Quantification of the influence of preferential flow on slope stability using a numerical modelling approach. *Hydrol. Earth Syst. Sci.* **2015**, *19*, 2197–2212. [CrossRef]
60. Chen, X.; Liu, R.F.; Ma, J. The Interaction between *Sedum* Root Traits and Engineered Media in Green Roofs. In Proceedings of the 7th International Conference on Green Materials and Environmental Engineering, Changsha, China, 16–17 January 2022.
61. Lei, Z.D. *Soil Hydro-Dynamics*; Tsinghua University Press: Beijing, China, 1988.
62. Hu, C.W.; Wang, H.; Liu, C.; Yuan, H.; Li, Y.Y. Difference analysis of hydraulic characteristics of typical soils in South China. *J. Soil Water Conserv.* **2017**, *31*, 97–102. [CrossRef]
63. Jiang, W.J.; Wang, S.J.; Li, X.; Li, D. Soil moisture characteristic curve and model analysis of purple soil with residual plastic film. *J. Drain. Irrig. Machinery Eng.* **2021**, *39*, 844–850. [CrossRef]
64. Ge, D.; Zhang, S.H. Influence of types and depths of substrates on hydrological performances of green roofs. *Sci. Soil Water Conserv.* **2019**, *17*, 31–38. [CrossRef]

water

MDPI

Article

Field Performance of Rain Garden in Red Soil Area in Southern China

Chunli Chen [1,2], Yanqi Li [3], Wencai Le [4], Chengyun You [5], Zhenzhong Liu [1,2], Wei Liu [1] and Ru Zhang [1,2,*]

1 School of Resource & Environment, Nanchang University, Nanchang 330031, China
2 Key Laboratory of Poyang Lake Environment and Resource Utilization, Ministry of Education, Nanchang 330029, China
3 Jiangxi Province Ecological and Environmental Montoring Center, Nanchang 330000, China
4 Powerchina Jiangxi Electric Power Construction Co., Ltd., Nanchang 330000, China
5 Jiangxi Huagan Environmental Group Co., Ltd., Nanchang 330000, China
* Correspondence: zru@ncu.edu.cn; Tel.: +86-177-7088-4879

Abstract: Sponge City, as a new concept in urban stormwater management, utilizes on-site or local hydrologic processes for runoff control and therefore is highly dependent on the geographical location (soil type) and site-specific climatic conditions. Field studies are valuable because of the insufficient quantity of field performance data in low-impact development (LID)-related research. Rain gardens are recommended for LID to manage stormwater. A rain garden was designed as a pilot project in Nanchang city, which is one of the typical red soil areas in southern China. Red soil is usually not conducive to runoff infiltration due to its low organic carbon, strong acidity and low permeability rainfall characteristics, but the permeability of the filter media layer is an important parameter in LID design. The construction depth of the rainwater garden was 600 mm, and 30% sand, 10% compost and 60% laterite were used as combined matrix; the permeability coefficient of medium layer was 1.48×10^{-5} m·s^{-1}. Rainfall runoff control and pollutant removal efficiencies were studied based on the on-site conditions. The analysis of almost 2 years of field data showed that volume capture ratio of annual rainfall was 78.9%, the mean load removal of TSS, NH_3-N, TP, TN, COD and NO_3-N were 92.5%, 85.3%, 82.9%, 80.5%, 79.8% and 77.5%, respectively, which could meet the technical guidelines for sponge city construction in Nanchang. The research results could provide a basis for sponge city design in low organic carbon and low permeability areas.

Keywords: rainfall runoff; low impact development; runoff control; pollutant removal; Sponge City

Citation: Chen, C.; Li, Y.; Le, W.; You, C.; Liu, Z.; Liu, W.; Zhang, R. Field Performance of Rain Garden in Red Soil Area in Southern China. *Water* **2023**, *15*, 267. https://doi.org/10.3390/w15020267

Academic Editor: Renato Morbidelli

Received: 13 December 2022
Revised: 3 January 2023
Accepted: 5 January 2023
Published: 8 January 2023

1. Introduction

The rapid urbanization in China over recent decades has led to significant changes in surface hydrological characteristics, such as permeability and detention/storage, etc., and thus resulted in severe non-point source pollution and urban flooding [1–3]. In China, the Sponge City projects initiative has been promoted, and implemented since 2014, as a new approach to urban storm water management, which utilizes on-site or local hydrologic processes for runoff control and therefore is highly dependent on the geographical location (soil type) and the site-specific climatic conditions. In recent years, Sponge City technical guidelines have been issued for major cities such as Beijing, Shanghai, Wuhan and Chongqing, but are not available as yet for many other locations. For some areas, a number of demonstration projects have been completed for selected management practices such as rain gardens that led to good treatment results [4].

The rain garden is one of the most commonly used low-impact development (LID) measures due to its characteristics which reflect the natural water-cycle processes [5,6]. Rain gardens play a vital role in reducing rainwater volume and flow, preventing assets' destruction, removing pollutants from urban runoff, and recharging groundwater [7]. Rain gardens use plants, soil, and their associated microbial communities to reduce or remove pollutants with mechanisms such as filtration, evapotranspiration, adsorption and biotransformation [8–11]. Studies have shown that native soils can effectively be used in the design of rain gardens as long as volume removal goals are achieved [12–14]. Additionally, rain gardens can attenuate runoff peak flow and reduce runoff volumes through the process of detention and retention [15–18]. At present, rain gardens are widely used world-wide because of their flexibility in size and location, ecological values to the landscape and treatment cost-effectiveness compared to conventional runoff treatment methods [14,19].

In China, research efforts in relation to rain gardens have thus far mostly focused on theoretical aspects and laboratory experiments, some on policy standards and construction technology, rather than the individual, event-based or even, to a lesser extent, long-term treatment performance [20–22]. Laboratory tests could be far from reality in terms of replicating real field conditions [23]. Consequently, to date there are still insufficient available field-performance data in the literature [15]. Field tests are very much needed since rain garden performance mainly depends on site-specific infiltration and evapotranspiration, and their effectiveness would be significantly impacted by such parameters as soil type and conditions, types of plants, plant survival, rainfall patterns, pollutant levels, ground use types and other hydrological properties [24].

The precipitation pattern in Southern China is typically characterized by high-intensity, localized and uneven temporally distributed storms. The urban drainage facilities have not generally been correspondingly upgraded [25]. Urban flooding and runoff-induced pollution have thus become the most frequent hazards in many cities in Southern China. Additionally, red soil is the typical soil type in the humid areas of subtropical China, with an area of 56.9 million hm^2, including most of Jiangxi and Hunan Provinces. Red soil is not conducive to the infiltration of runoff and the construction of rain gardens due to its low permeability [26].

Therefore, it is of great interest to study the feasibility of using infiltration-type LID practices such as rain gardens in the red soil region in Southern China. The present study was thus conceived, and a full-scaled rain garden was constructed and tested at a college campus in Nanchang, Jiangxi Province. To achieve this objective, it was decided to use flow and water-quality monitoring to quantify the retention of flow and pollutant and load reductions by the rain garden system. Based on the field experimental data, an in-depth discussion on the design and construction of the facility and its treatment performance on runoff pollutant was presented, which could provide much-needed guidance for the planning and design of rain gardens in red soil regions in the world.This paper was conducted to close the gap in our theoretical research and treatment performance of these LID facilities for Sponge City.

2. Materials and Methods

2.1. Study Area

Nanchang, a typical city in Southern China (Figure 1a), has a subtropical humid monsoon climate characterized by high temperature, rainy summers and mild winters with less rain. The mean annual temperature is 17.7 °C, with an annual precipitation ranged from 941 to 1764 mm, which is unevenly distributed (42.96% of it falls during summer months). The annual runoff volume of Nanchang is 6.153 billion m^3 (831.1mm), the annual average runoff volume in the flood season is 40.5 billion m^3, with 18.9% of it in June. The analysis of rainfall statistics of Nanchang showed that the occurrence probability of acid rain was higher than 90% and the mean pH was lower than 5.6 [27]. This is an important cause of local red soil acidification.

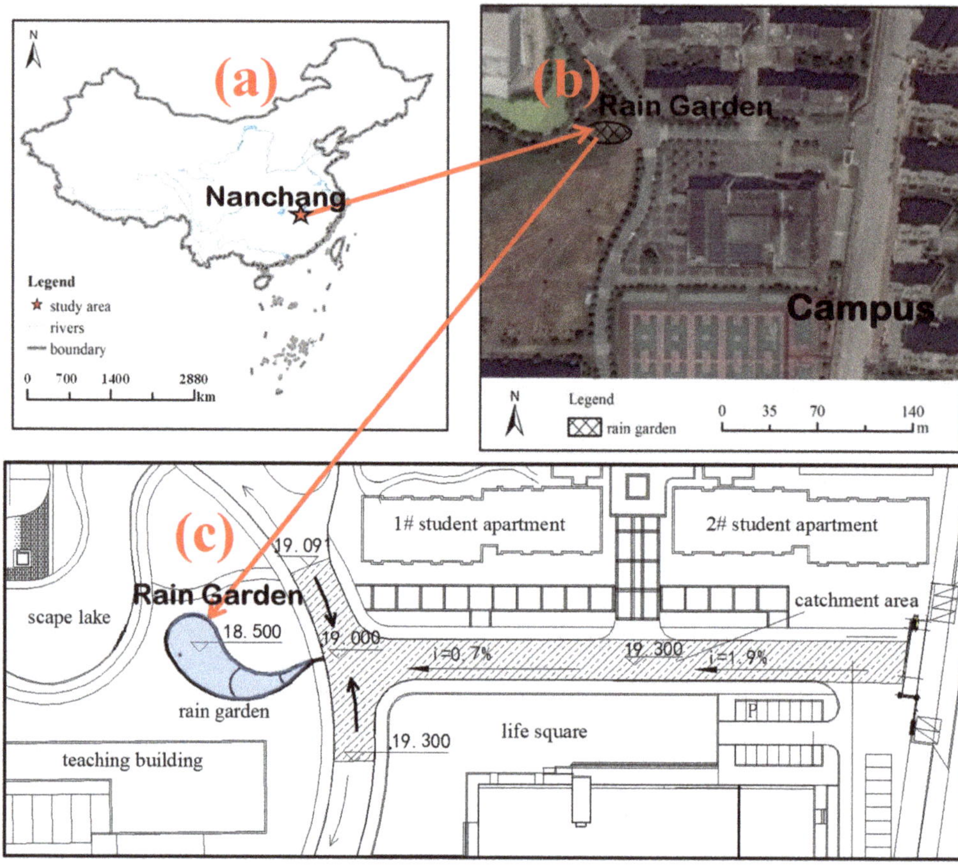

Figure 1. Location map of the rain garden (**a**) The location of Nanchang in China; (**b**) The location of the rain garden in Nanchang; (**c**) The location and surrounding schematic of the rain garden.

2.2. Design Parameters for the Rain Garden

A full-scale rain garden (Figure 1b) was constructed at the Nanchang University campus in July 2016 for the collection and treatment of road runoff. The catchment area of the test rain garden (Figure 1c) was 1533.24 m^2, obtained by measuring the size of the surrounding pavement draining into the rain garden. According to the Design Specifications of China's Outdoor Drainage Design Code [28], the surface type of the catchment area was mainly hard concrete pavement, and the runoff coefficient was set to be 0.9. The volume capture ratio of annual rainfall in the Nanchang area was 60–85% by the Sponge City Construction Technology Guide of Nanchang City. Due to the frequent occurrence of the rainy season, the control target was set at 85%, and the design rainfall depth was 38.9 mm, with the average recurrence interval (ARI) of 5 years. The total design runoff volume (V) was calculated by the volumetric method Equation (1):

$$V = 0.001FH\psi, \quad (1)$$

where F is the runoff catchment area (m^2); ψ is the runoff coefficient; and H is the design rainfall depth (mm).

The water balance method was used for the surface area of the test rain garden [27]. Firstly, it assumed that runoff from the catchment area would entirely flow into the rain garden. When the amount of runoff exceeded the capacity of storage and infiltration, the total runoff balance of the rain garden was calculated, as shown by Equation (2). Secondly, the method ignored evaporation from the rain garden during the calculation time period. In addition, in the design of the rain garden, the effluent could be assumed to be zero. Finally, the area of the rain garden could be calculated by Equation (6), which was derived from Equations (2)–(5). Parameters set in the equations were shown in Table 1.

(a) Design storage capacity:

$$V = G + V_w + W_s, \tag{2}$$

where G is the medium void storage (m^3); V_w is the aquifer storage (m^3) and W_s is the permeation during rainfall (m^3).

(b) Mediumvoid storage:

$$G = n \cdot A_f \cdot d_f, \tag{3}$$

where n is the average porosity of the filter media layer; d_f is the filter media layer thickness (m) and A_f is the rain garden area (m^2).

(c) Aquifer storage:

$$V_w = (1 - m)A_f \cdot h_m, \tag{4}$$

where m is the proportion of plants cross-sectional surface area in the surface area of the aquifer and h_m is the maximum water depth of the standing water aquifer (m).

(d) Permeation:

$$W_s = \frac{60K \cdot (d_f + h) \cdot A_f \cdot T}{d_f}, \tag{5}$$

where K is the permeability coefficient of planting soil ($m \cdot s^{-1}$); h is the average water depth of the aquifer (m) and T is the rainfall duration (min).

(e) The rain garden area:

$$A_f = \frac{V \cdot d_f}{n \cdot {d_f}^2 + (1 - m)h_m \cdot d_f + 60K \cdot T \cdot (d_f + h)}, \tag{6}$$

Table 1. Parameters selection and reference of Equations (2)–(5).

Parameters	Value	SU	Reference
n	0.3	-	[29]
d_f	0.25	m	Section 2.6
m	0.2	-	[30]
h_m	0.2	m	Section 2.6
K	1.5×10^{-6}	$m \cdot s^{-1}$	laboratory test
h	0.1	m	half of h_m
T	120	min	[31]

2.3. Water Sample Collection

The rain garden was constructed in August 2016. The inlet and outlet of the rain garden were monitored during the period from September 2016 to January 2018. Water quality and flow sampling points were set up at the inlet and the perforated under-drain pipe of overflow well. Automatic flow-monitoring equipment was used to collect data, which will help determine the detailed hydrological and water quality processes at the rain garden.Data collected included runoff volume and discrete samples for water quality. In a whole rainfall event, according to the duration of rainfall, the sampling intervals were 5, 10,

15 and 20 min until the flow was very small or non-detectable. The water quality samples were tested for total nitrogen (TN), nitrate (NO_3-N), ammonium (NH_3-N), total phosphorus (TP), chemical oxygen demand (COD) and total suspended solids (TSS). Sample testing was undertaken according to the test methods specified in the Standard Methods for the Examination of Water in China [32–36].

2.4. Data Analysis

During each rainfall event, the cumulative mass of pollutants in the inflow and outflow were calculated by taking the integral of the product of concentrations and flow rates, as shown in Equation (7). If the value was positive, it meant that the system retained pollutant mass. If the value was negative, it meant the system exported/leached pollutant mass. To undertake the detailed investigation of the treatment performance, the removal efficiencies of pollutant load and the even mean concentration (EMC) reduction were both calculated, as shown in Equations (8) and (9):

$$\text{Total pollutant mass} = \int_0^t C(t)Q(t)dt, \quad (7)$$

$$\text{Pollutant load removal } \% = \left[1 - \frac{\int_0^t C_{out}(t)Q_{out}(t)dt}{\int_0^t C_{in}(t)Q_{in}(t)dt}\right] \times 100\%, \quad (8)$$

$$\text{Pollutant EMC reduction } \% = \left[1 - \frac{\int_0^t C_{out}(t)Q_{out}(t)dt/V_{out}}{\int_0^t C_{in}(t)Q_{in}(t)dt/V_{in}}\right] \times 100\%, \quad (9)$$

where C_{in}(t) and C_{out}(t) are the influent or effluent concentrations of each pollutant at time t ($mg \cdot L^{-1}$); Q_{in}(t) and Q_{out}(t) are the influent or effluent flow rates at time t ($L \cdot s^{-1}$) and V_{in} and V_{out} are the influent or effluent volume (L). Limits of integration refer to time 0 (runoff initiation) and time t (time at which runoff ceases).

2.5. Storage Capacity and the Rain Garden Area

The rain garden was expected to not only alleviate the local flooding, but also effectively improve the water quality of a nearby landscape lake on campus. Figure 2 shows the photos of the site before, during and after the construction of the rain garden.

Figure 2. Pictures of the rain garden before, during and after construction.

The rain garden surface area required was at least 204.9 m^2 as calculated by Equation (5) when the design storage capacity was 54.78 m^3, calculated by Equation (1). A two-stage front pool was designed between the inlet and the rain garden (Figure 3), taking consideration of calculation results, terrain features of the site and creating some visual effects. The areas of the front pool No.1 and front pool No.2 were 15 m^2 and 64 m^2, respectively. The surfaces of the two-stage front pool were covered with turf, and the interior was filled with red soil only without a gravel drainage layer. The main function of the two-stage front pool was to provide preliminary runoff and erosion control. The filter media layer of the rain garden was filled with combination substrates. The outflow of the rain garden was collected by the PVC perforated pipe at the bottom and eventually flowed into the campus landscape lake.

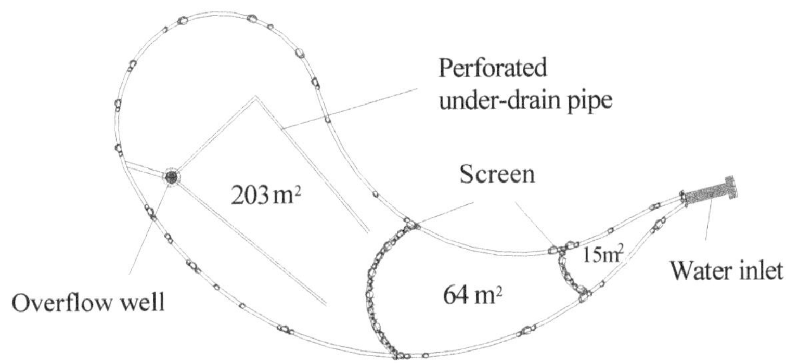

Figure 3. Plane layout of the rain garden.

Based on the design layout of the rain garden, a variety of plant communities were set up in the rain garden and the surroundings. Plants were an important part for the rain garden, which could retain water and certain pollutants. Native plants were the best choice in most cases since they were adapted for local environmental conditions and required less care. Additionally, plants should be able to tolerate periodic inundation. The principles of economic benefit, local conditions and diversity were followed and major plant species were chosen, such as *Canna generalis*, *Lythrumsalicaria*, *Cyperus alternifolius*, *Irispseudacorus* and *Miscanthus sinensiswere* [26].

2.6. Inlet and Cross-Section Design of the Rain Garden

Inlet design is a critical part of a rain garden. When the road elevation is higher than the surface of the rain garden, road runoff would flow into the rain garden and be tested. For the Nanchang site, the minimum elevation of the catchment area was 19.0 m, where catch-basins were available to collect road runoff. After the transformation (Figure 4), when the rainfall was light, road runoff could be completely collected by the rain garden.

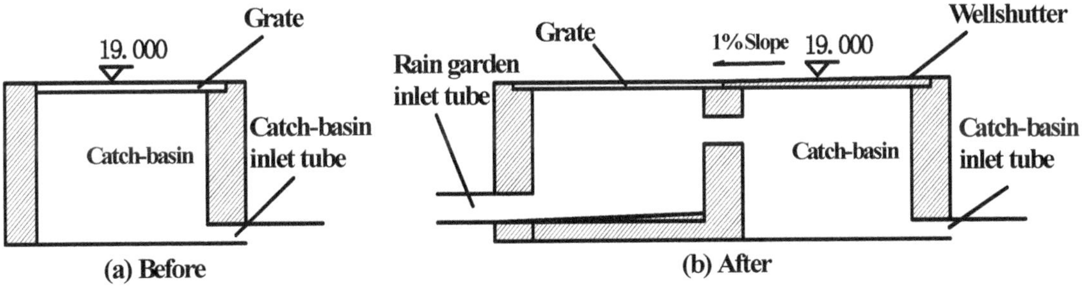

Figure 4. Before and after the inlet reconstruction.

The rain garden had some specific design features that would enhance runoff infiltration and temporary storage in underlying soil layers, which would help reduce both the total runoff volume and its peak flow [37,38]. As the elevation of the landscape lake was 17.65 m, which was 1.35 m lower than the inlet. It was necessary to strictly control the structural thickness of the rain garden. Details of the design features are shown in Figure 5. It should be noted that the design of the rain garden requires attention to the following points.

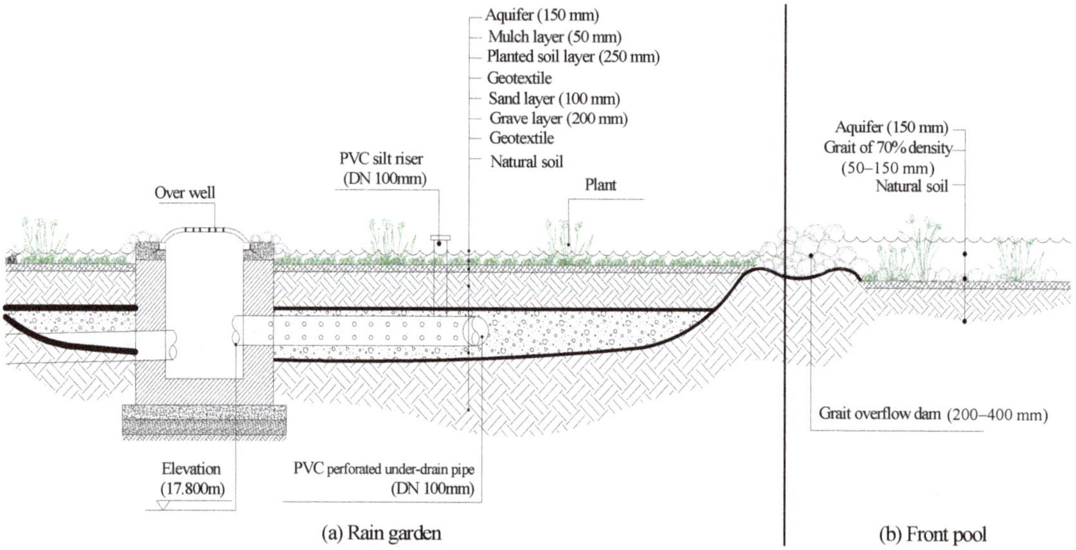

Figure 5. Section structure of the rain garden.

(a) The aquifer was mainly for storage runoff and precipitation of TSS.

(b) The mulch layer was covered with bark of 50 mm deep, which could maintain soil moisture [22,39]. Moreover, a suitable microbial environment was built between the bark and soil layer, which was propitious to the microorganisms on the degradation of organic matter and reduce runoff erosion of the topsoil.

(c) The filter layer required good permeability to provide a suitable growth environment for plants. Its depth depended on the type of soil and plants. When herbs were used, its depth was about 250 mm. As the clay content of red soil was above 40%, its permeability coefficient was only 1.5×10^{-6} m·s^{-1}. Runoff could not infiltrate as soon as possible or might even spillover if red soil was used as the planting soil without being amended. Therefore, the filter media layer was filled with a mix of 30% sand, 10% compost and 60% red soil as combination substrates, which provide better osmotic properties and organic matter. The permeability coefficient of the amended media layer was determined to be 1.48×10^{-5} m·s^{-1}.

(d) The sand filter layer, with a depth of 100 mm, prevented the soil substrate from sinking and blocking the perforated drain.

(e) The gravel drainage layer was 200 mm in depth. There were two perforated under-drain pipes, 150 mm in diameter with a drilling diameter of 15 mm to 20 mm [40,41]. The perforated pipes were used for the timely discharge of the filtered water. The particle size of the gravel was 20–30 mm, which was greater than the perforation aperture. The middle of the perforated under-drain pipe had a 100 mm-diameter silt riser, which was used to regularly remove sediment in the perforated under-drain pipe.

3. Results and Discussion

Impermeable surfaces in urbanized environments accelerate surface water runoff during rainfall events, decrease infiltration [42], reduce aquifer replenishment and degrade the water quality of aquatic ecosystems receiving pollutant-laden rain runoff, thus accelerating the issue of water pollution further [43]. Urban stormwater runoff represents a great challenge to modern water pollution management [44,45].

3.1. Runoff Reduction and Pollutant Removal

Ten rainfall events were monitored to determine the characteristics of the stormwater runoff entering the rain garden facility and evaluate its performance in terms of pollutant removal and volume reduction. Rainfall depth ranged between 6.3 and 30.9 mm with a mean value of 19.2 mm.

In particular, the planting media thickness and soil porosity in the rain garden were significant indicators of overflow in native soils with lower seepage rates [14]. Based on monitoring ten rainfall events, the rain garden had a good retention capacity and there was no overflow during the monitoring period, indicating that the rain garden worked well. The data matrix on runoff control is given in Table 2.

Table 2. Runoff reduction and pollutant removal efficiency data for every rainfall event.

Rainfall Event	Rainfall/mm	Runoff Reduction/%	EMC Reduction/%						Load Removal/%					
			NH_3-N	NO_3-N	TN	TP	COD	TSS	NH_3-N	NO_3-N	TN	TP	COD	TSS
2016.09.11	11.3	78.5	43.8	44.0	35.7	59.8	6.9	35.6	86.8	86.5	85.9	90.8	80.5	86.0
2016.10.22	26.6	71.5	13.1	−22.5	−58.9	−4.1	9.6	73.9	75.2	65.1	54.7	70.3	74.2	92.6
2016.11.23	6.3	85.6	52.6	−93.7	40.9	70.4	−58.5	59.3	93.2	38.1	91.5	95.7	77.2	94.1
2016.12.21	10.5	64.2	−0.6	3.5	−6.8	14.1	46.3	68.7	64.0	65.6	61.8	69.3	80.8	88.8
2017.03.12	22.6	87.3	33.7	−7.4	−13.1	−21.4	55.1	89.5	92.0	83.1	84.6	82.9	92.0	98.5
2017.04.09	23.5	83.6	19.9	14.1	42.0	61.7	14.7	92.6	86.8	85.9	90.5	93.7	86.0	98.8
2017.05.08	22.5	80.5	−7.9	−43.8	−62.4	−4.4	36.4	83.3	78.9	71.9	68.3	79.6	87.6	96.8
2017.06.06	30.9	81.8	47.4	44.0	27.6	27.6	−159.6	87.8	90.4	89.8	86.8	86.8	52.7	92.3
2017.11.17	27.3	87.7	7.5	32.3	−90.5	29.4	−104.4	55.6	94.9	96.3	90.8	96.0	90.5	97.2
2018.12.14	12.2	68.6	68.8	77.7	68.1	−16.6	23.8	36.3	90.2	93.0	90.0	63.4	76.1	80.0
max	30.9	87.7	68.8	77.7	68.1	70.4	55.1	92.6	94.9	96.3	91.5	96.0	92.0	98.8
min	6.3	64.2	−7.9	−93.7	−90.5	−21.4	−159.6	35.6	64.0	38.1	54.7	63.4	52.7	80.0
mean	19.2	78.9	27.8	4.8	−1.7	21.6	−13.0	68.3	85.3	77.5	80.5	82.9	79.8	92.5
SD	8.5	6.6	21.4	37.6	44.6	28.1	56.7	17.2	7.5	13.9	11.3	9.8	7.8	4.6

Overall, the total runoff control rate ranged from 64.20% to 87.70%, and the average runoff control rate was 78.9%, which achieved the Sponge City Construction Standards for Nanchang [31]. Field performance assessment demonstrated that this rain garden effectively cut inflow volumes through the filter media. This has important implications for the management of urban waterways, where increased flows are a key stressor [38].

The rain garden had the best removal efficiency for TSS, followed by NH_3-N, TP and TN. TSS, TP and nitrogen showed different removal characteristics, which could be attributed to different treatment mechanisms [7,46,47]. Rain gardens can remove nutrients and hydrocarbons from stormwater via several mechanisms [7]. Nutrients are removed by several mechanisms: filtration, adsorption, sedimentation, ion exchange, chemical precipitation, biological decomposition and plant uptake [12]. Pollutants such as TSS and TP would be primarily removed by physical processes while nitrogen would be primarily removed by biochemical processes, such as denitrification [15]. TSS was removed via the physical filtration of the particulates and colloids during percolation through the filter media. The rain garden was consistently effective in removing TSS irrespective of the rainfall sizes, runoff volumes and influent loads' amounts and treatments [23]. The rain garden was effective at treating phosphorus regardless of soil type [12].TP removal efficiency was highly dependent on the filter media. The red soil was effective in TP reduction since the content of phosphorus in red soil was relatively low. Moreover, the red soil contained a large amount of Fe_2O_3 (amorphous iron oxide), Al_2O_3 (aluminum oxide) and kaolinite, which were conducive to the adsorption and fixation of TP [48–50]. The red soil had four kinds of parent materials; details of them are shown in Table 3 [51].

Table 3. Parent materials of red soil and its compositions proportion.

Parent Material	Quaternary Red Clay	Granite	Arenite	Pelite
Proportion of red soil/%	4.1	17.1	11.6	13.2
Organic matter/%	0.7	1.4	0.9	1.5
TP/%	0.06	0.09	0.06	0.06
SiO_2/%	73.3	44.6	71.7	73.3
Fe_2O_3/%	5.7	13.7	7.0	6.6
Al_2O_3/%	15.7	37.4	17.4	16.4
Kaolinite/%	38.6	43.7	38.9	32.1

The removal efficiencies of nitrogen and COD in the rain garden fluctuated greatly. The water quality pollutant-load reduction fluctuated, which was consistent with relevant research results in bioretention tanks [52]. Soil media and plants played a vital role in the pollutant removal processes of rain gardens [7]. Plants were significant to treatment after media saturation. The extent of plants that assimilated pollutants was largely dependent on root structure, runoff detention time and the ability of plants to acquire pollutants from the media [5,30,53]. It was noteworthy that Table 2 also showed negative values for pollutant reduction percentages, particularly for EMC reduction inNO_3-N, TN and COD. This explained the occurrence of nutrient leaching which could be attributed to the flushing of runoff retained in the filter media layer from the preceding rainfall event containing elevated pollutants due to the evapotranspiration. Furthermore, nutrients presented in the rain garden could also contribute to pollutant leaching. Various plant-based mechanisms and chemical processes such as adsorption, reduction, sedimentation, cation-exchange capacity, complexation and so forth were involved in the removal of contaminants from stormwater [7].

The removal efficiency of pollutant load for NH_3-N, NO_3-N, TN, TP, COD and TSS increased by 57.4%, 72.7%, 82.2%, 61.2%, 92.7% and 24.2% compared to EMC reduction separately. The removal efficiency of the pollutant load was generally higher than the EMC reduction. This might be due to the fact that runoff volume control was taken into consideration in the calculation of pollutant load removal. In addition, the concentration of pollutants of campus runoff was generally lower than that of urban roads and parking lots, resulting in less obvious EMCs removal efficiency of the rain garden. It could be observed that there were wide differences in the efficiencies of pollutants' removal among different rainfall events due to a number of factors such as plants, rainfall patterns and soil condition [25]. The removal effectiveness had been shown to be reliant upon the rainfall patterns (e.g., length of wet and dry weather) and temperature [54,55].

3.2. Time Variation of Pollutant Concentrations

In order to discuss the migration of runoff pollutants in the rain garden, the variation in concentration of each pollutant with the change in time was analyzed by sampling data from 10 January 2017 (Figure 6). The average residence time between the start of the influent and the appearance of effluent was approximately 100 min, which included the total flow-through time in the two-stage front pool and the infiltration time in the rain garden.

As shown in Figure 6, the concentration of pollutants varies with rainfall time and showed some common characteristics, and each concentration of inflow pollutants decreased with rainfall duration and finally tended to flatten out, which was due to the initial scouring effect. The concentrations of these pollutants were higher before inflow in the early 20 min, and the pollution of rain water was more serious at the initial stage. The pollutant load was always heavy in the initial stage of the runoff [9]. There were significant fluctuations in the concentration of outflow pollutants except for TSS, which was due to different degrees of the initial scouring effect of different pollutants [56]. The fluctuations in the outflow were generally lower than those in the inflow.

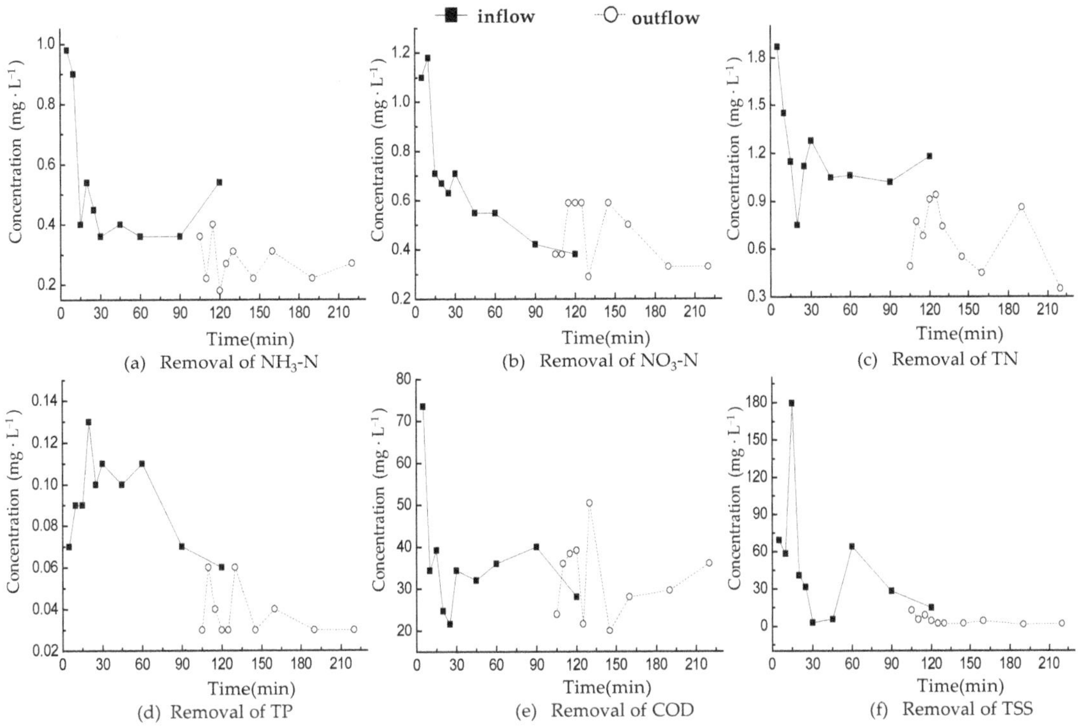

Figure 6. The time-variation of pollutant concentrations in inflow and outflow.

The average concentrations of NH_3-N, TN, TP and TSS in the outflow were relatively good, and the concentration of these pollutants decreased gradually and then tended to be stable. After adsorption and filtration by the rain garden, the concentrations of NH_3-N in the outflow were relatively steady, for the optimal sponge had an excellent treatment effect on NH_3-N in rainwater while ensuring rapid infiltration [57]. The concentrations of NO_3-N and COD in the outflow were unstable, and greater concentrations appeared in the early stage of the outflow. The concentrations of NO_3-N and TN fluctuated, which was related to the fact that the removal of NO_3-N in the rain garden was easily affected by various factors, and nitrogen retention may have occurred there [12]. Because NO_3-N is a part of TN, the fluctuation of the concentrations of NO_3-N will also cause the concentrations of TN fluctuate to some extent.

Among the water quality indicators, the COD concentration fluctuated the most. The concentrations of COD in 30 min decreased gradually before inflow, and the average concentration of COD was 25.30 mg·L^{-1} in the later stage of inflow. Even the COD concentration in the effluent was higher than that in the influent at the initial stage of operation. This was due to the poor stability of the rainwater garden at the initial stage of the operation. The microbial activities and organic secretions released by the plant roots in the rainwater garden system entered the effluent, resulting in a higher COD concentration in the effluent.

Despite variation in inflow concentrations, pollutant concentrations in the effluent were relatively constant, although an initial spike was sometimes observed forNO_3-N, and COD. It could be seen that the range of outflow pollutant concentrations were lower compared to the inflow concentrations, suggesting a level of reliability in treatment [38].

3.3. Limitations or Directions for Further Research

Rain gardens can retard surface runoff, reduce and delay flood peaks effectively and play a major role in rehabilitating the water cycle. The control effects of rain garden on pollutants could be improved by by-passing some initial runoffs.

Although stormwater cannot be treated completely without conventional sewage systems in urban areas, rain gardens can decrease the dependence on these. Stormwater infiltration and redistribution by rain gardens are also potentially significant ecosystem services and impart value to vacant land that presently has little or no value [58].

This is the first study presenting treatment performance results on rain garden in red soil area of Nanchang city at the field-scale. Ultimately, the results of this paper provide key insights into the design and operating conditions of rain garden, especially for the future reliable treatment of stormwater. However, in order to fully validate the rain garden studied, long term operational monitoring needs to be put in place to provide assurance that Sponge City construction planning and design objectives are being continuously met. More data will be obtained with auto-sampling, which is typically necessary for performance monitoring and maintenance. A large set of additional data may be provided for further simulation and model analysis, and finally for the development of a validation framework for stormwater treatment systems.

4. Conclusions

In this study, the special features in the design of a rain garden and the modification of the filter media layer play an important role in the field performance of a rain garden in a red soil region. Rainfall characteristics and catchment partition were important parameters in designing the rain garden. The construction of rain gardens in red soil regions, such as Southern China, should pay close attention to the permeability of the filter media layer and the architectonics of the rain garden.

The average runoff control rate obtained by this study was 78.9%, which achieved the Sponge City Construction Standards for Nanchang. The efficiency of runoff pollutant load removal generally was higher than the EMC reduction rates. The rain garden showed the best removal in TSS, followed by NH_3-N and TP. Under the same average recurrence interval (ARI) the mean load removal of TSS, NH_3-N, TP, TN, COD and NO_3-N were 92.5%, 85.3%, 82.9%, 80.5%, 79.8% and 77.5%, respectively. The red soil was effective in TP reduction. On the other hand, the removal efficiency of NO_3-N, TN and COD were negative at times, showing pollutant leaching.

The results of the study indicated favorable storage/infiltration functions in the field performance of this rain garden, the potential to control more than 70% of storm runoff and its effectiveness at pollutants' load removal. The results of this study could provide a good reference for the construction of rain gardens in a red soil region. Therefore, the application of this rain garden may be recommended in other red soil urban areas.

Author Contributions: Conceptualization, R.Z., C.C. and Z.L.; methodology, W.L. (Wei Liu) and Y.L.; validation, W.L. (Wei Liu), W.L. (Wencai Le) and Y.L.; formal analysis, Y.L., C.Y. and W.L. (Wencai Le); resources, R.Z.; data curation, C.C.; writing—original draft preparation, C.C. and W.L. (Wei Liu); writing—review and editing, C.C. and R.Z.; visualization, C.Y.; supervision, Z.L.; project administration, R.Z.; funding acquisition, R.Z. All authors have read and agreed to the published version of the manuscript.

Funding: This research was funded by The National Key R&D Program of China (No. 2016YFC0401500), the National Science Foundation of China (No. 51169019) and Jiangxi Provincial Department Science and Technology (No. 20171BBG70080).

Data Availability Statement: The data presented in this study are available in the article.

Acknowledgments: All individuals included in this section have consented to the acknowledgement.

Conflicts of Interest: The authors declare no competing financial interest, and the funders had no role in the design of the study; in the collection, analyses, or interpretation of data; in the writing of the manuscript, or in the decision to publish the results. We certify that it is our original work, and we have participated sufficiently in the work to take responsibility for the appropriateness of the experimental design and method, and the analysis and interpretation of the data. All authors have reviewed the final version of the manuscript and approve it for publication. This manuscript has not been published nor is it being considered for publication elsewhere.

References

1. Chahal, M.K.; Shi, Z.; Flury, M. Nutrient leaching and copper speciation in compost-amended bioretention systems. *Sci. Total Environ.* **2016**, *556*, 302–309. [CrossRef] [PubMed]
2. Ding, W.; Wu, J.; Tang, R.; Chen, X.; Xu, Y. A Review of Flood Risk in China during 1950–2019: Urbanization, Socioeconomic Impact Trends and Flood Risk Management. *Water* **2022**, *14*, 3246.
3. Zhang, Y.; Xu, H.; Liu, H.; Zhou, B. The Application of Low Impact Development Facility Chain on Storm Rainfall Control: A Case Study in Shenzhen, China. *Water* **2021**, *13*, 3375. [CrossRef]
4. Bak, J.; Barjenbruch, M. Benefits, Inconveniences, and Facilities of the Application of Rain Gardens in Urban Spaces from the Perspective of Climate Change-A Review. *Water* **2022**, *14*, 1153. [CrossRef]
5. Zhou, P.L.; Han, J.Q.; Zhang, H.X. A Review of Researches on Plant Configuration and Decontamination Efficiency of Rain Gardens in China. *IOP Conf. Ser. Earth Environ. Sci.* **2020**, *510*, 032029. [CrossRef]
6. Takaijudin, H.; Ghani, A.A.; Zakaria, N.A. Challenges and developments of bioretention facilities in treating urban stormwater runoff; A review. *Pollution* **2016**, *2*, 489–508.
7. Sharma, R.; Malaviya, P. Management of stormwater pollution using green infrastructure: The role of rain gardens. *Wires Water* **2021**, *8*, 1507. [CrossRef]
8. Liu, J.; Sample, D.J.; Bell, C.; Guan, Y.T. Review and research needs of bioretention used for the treatment of urban stormwater. *Water* **2014**, *6*, 1069–1099. [CrossRef]
9. Guo, C.; Li, J.K.; Li, H.E.; Zhang, B.; Ma, M.H.; Li, F. Seven-Year Running Effect Evaluation and Fate Analysis of Rain Gardens in Xi'an, Northwest China. *Water* **2018**, *10*, 944. [CrossRef]
10. Laukli, K.; Vinje, H.; Haraldsen, T.K.; Vike, E. Plant selection for roadside rain gardens in cold climates using real-scale studies of thirty-one herbaceous perennials. *Urban For. Urban Green.* **2022**, *78*, 127759. [CrossRef]
11. Hess, A.; Wadzuk, B.; Welker, A. Evapotranspiration in Rain Gardens Using Weighing Lysimeters. *J. Irrig. Drain. Eng.* **2017**, *143*, 04017004. [CrossRef]
12. Wadzuk, B.; DelVecchio, T.; Sample-Lord, K.; Ahmed, M.; Welker, A. Nutrient Removal in Rain Garden Lysimeters with Different Soil Types. *J. Sustain. Water Built Environ.* **2021**, *7*, 04020018. [CrossRef]
13. Tu, M.C.; Traver, R. Performance of a Hydraulically Linked and Physically Decoupled Stormwater Control Measure (SCM) System with Potentially Heterogeneous Native Soil. *Water* **2019**, *11*, 1472. [CrossRef]
14. Bethke, G.M.; William, R.; Stillwell, A.S. Rain Garden Performance as a Function of Native Soil Parameters. *J. Sustain. Water Built Environ.* **2022**, *8*, 04021021. [CrossRef]
15. Mangangka, I.R.; Liu, A.; Egodawatta, P. Performance characterisation of a stormwater treatment bioretention basin. *J. Environ. Manag.* **2015**, *150*, 173–178. [CrossRef]
16. Shafique, M.; Kim, R. Low impact development practices: A review of current research and recommendations for future directions. *Ecol. Chem. Eng.* **2015**, *22*, 543–563. [CrossRef]
17. Abduljaleel, Y.; Demissie, Y. Identifying Cost-Effective Low-Impact Development (LID) under Climate Change: A Multi-Objective Optimization Approach. *Water* **2022**, *14*, 3017. [CrossRef]
18. Zhang, Y.; Zhao, W.; Chen, X.; Jun, C.; Hao, J.; Tang, X.; Zhai, J. Assessment on the Effectiveness of Urban Stormwater Management. *Water* **2021**, *13*, 4. [CrossRef]
19. Trowsdale, S.A.; Simcock, R. Urban stormwater treatment using bioretention. *J. Hydrol.* **2011**, *397*, 167–174. [CrossRef]
20. Wang, J.j.; Li, T. Discussion on design essentials of rain gardens and its application in Shanghai. *Environ. Sci. Tech.* **2013**, *7*, 164–167.
21. Shao, Z.Y.; Li, S.; Lv, B.; Chai, H.X.; Ao, L.G.; Zhang, X.Y.; Li, W.Q.; He, Q. Analysis of the sediment remobilization phenomenon in a rain garden using CSTR theory. *J. Water Clim. Chang.* **2018**, *9*, 356–366. [CrossRef]
22. Guo, J.C.Y.; Luu, T.M. Operation of Cap Orifice in a Rain Garden. *J. Hydrol. Eng.* **2015**, *20*, 1061. [CrossRef]
23. Shrestha, P. Effects of different soil media, vegetation, and hydrologic treatments on nutrient and sediment removal in roadside bioretention systems. *Ecol. Eng.* **2018**, *112*, 116–131. [CrossRef]
24. Eckart, K.; Mcphee, Z.; Bolisetti, T. Performance and implementation of low impact development—A review. *Sci. Total Environ.* **2017**, *607–608*, 413–432. [CrossRef]
25. Paule-Mercado, M.A.; Lee, B.Y.; Memon, S.A. Influence of land development on stormwater runoff from a mixed land use and land cover catchment. *Sci. Total Environ.* **2017**, *599–600*, 2142–2155. [CrossRef]
26. Zhou, C.Y.; Huang, W.; Qiu, S.Y.; Liu, Z. A quantitative study on the amount of water-retaining agent based on adhesive-modified red bed weathered soil. *B Eng. Geol. Environ.* **2021**, *80*, 3139–3150. [CrossRef]

27. Zheng, F.W.; Rao, W.B.; Chu, X.D.; Bai, H.; Jiang, S.Y. Chemical and sulfur isotopic characteristics of precipitation in a representative urban site, South China: Implication for anthropogenic influences. *Air Qual. Atmos. Heal.* **2020**, *13*, 349–359. [CrossRef]
28. *GB50014-2021*; National Standards of the People's Republic of China, Standard for Design of Outdoor Wastewater Engineering. China State Bureau of Standards: Beijing, China, 2021.
29. Li, J.K.; Li, F.; Li, H.E.; Guo, C.; Dong, W. Analysis of rainfall infiltration and its influence on groundwater in rain gardens. *Environ. Sci. Pollut. Res.* **2019**, *26*, 22641–22655. [CrossRef]
30. Li, J.Q.; Mao, L.L.; Mao, K.; Li, B.H.; Li, H.Y.; Che, W. Case study on rain garden storage in filtration system for disposal of roof runoff. *China Water Wastewater.* **2010**, *26*, 129–133. (In Chinese)
31. *Nanchang Sponge City Construction Planning and Design Guidelines*; Nanchang Housing and Urban-Rural Development Bureau: Nanchang, China, 2017; (2022 revised).
32. *GB11893-1989*; National Standards of the People's Republic of China, Water Quality-Determination of Total Phosphorus-Ammonium Mlybdate Spectrophotometric Method. China State Bureau of Standards: Beijing, China, 1989.
33. *GB7479-87*; National Standards of the People's Republic of China, Water Quality-Determination of Ammonium-Nessler's Reagent Colorimetric Method. China State Bureau of Standards: Beijing, China, 1987.
34. *GB11894-89*; National Standards of the People's Republic of China, Water Quality-Determinatlon of Total Nitrogen-Alkaline Potassium Persulfate Digestion-UVSpectrophotometric Method. China State Bureau of Standards: Beijing, China, 1989.
35. *GB11901-89*; National Standards of the People's Republic of China, Water Quality-Determination of Suspended Substance-Gravimetric Method. China State Bureau of Standards: Beijing, China, 1989.
36. *HJ/T346-2007*; The Environmental Protection Industry Standards of the People's Republic of China, Water Quality-Determination of Nitrate-Nitrogen-Ultraviolet Spectrophotometry. China State Bureau of Standards: Beijing, China, 2007.
37. Zhang, L.Y.; Oyake, Y.; Morimoto, Y.; Niwa, H.; Shibata, S. Flood mitigation function of rain gardens for management of urban storm runoff in Japan. *Landsc. Ecol. Eng.* **2020**, *16*, 223–232. [CrossRef]
38. Hatt, B.E.; Fletcher, T.D.; Deletic, A. Hydrologic and pollutant removal performance of stormwater biofiltration systems at the field scale. *J.Hydrol.* **2009**, *365*, 310–321. [CrossRef]
39. Chen, Y.; Liu, M.Y.; Duan, L.H. Analysis on design essentials of rain gardens and its application. *J. Environ. Eng.* **2017**, *12*, 6–10. (In Chinese)
40. Meng, Y.Y.; Yin, R.X.; Zhang, S.H. Study on design of drainage system with bioretention facilities. *China Water Wastewater* **2015**, *9*, 135–138. (In Chinese)
41. Li, J.K.; Liu, F.; Li, Y.J. Simulation and design optimization of rain gardens via DRAINMOD and response surface methodology. *J. Hydrol.* **2020**, *585*, 124788. [CrossRef]
42. Katsifarakis, K.; Vafeiadis, M.; Theodossiou, N. Sustainable drainage and urban landscape upgrading using rain gardens. Site selection in Thessaloniki, Greece. *Agric. Agric. Sci. Proc.* **2015**, *4*, 338–347. [CrossRef]
43. Malaviya, P.; Singh, A. Constructed wetlands for management of urban stormwater runoff. Crit. *Rev. Environ. Sci. Technol.* **2012**, *42*, 2153–2214. [CrossRef]
44. Hu, M.; Shealy, T. Overcoming status quo bias for resilient stormwater infrastructure: Empirical evidence in neurocognition and decision-making. *J. Manag. Eng.* **2020**, *36*, 04020017. [CrossRef]
45. Vo, P.T.; Ngo, H.H.; Guo, W.; Zhou, J.L.; Listowski, A.; Du, B.; Bui, X.T. Stormwater quality management in rail transportation—Past, present and future. *Sci. Total Environ.* **2015**, *512–513*, 353–363. [CrossRef] [PubMed]
46. Jeon, M.; Guerra, H.B.; Choi, H.; Kwon, D.; Kim, H.; Kim, L.H. Stormwater Runoff Treatment Using Rain Garden: Performance Monitoring and Development of Deep Learning-Based Water Quality Prediction Models. *Water* **2021**, *13*, 3488. [CrossRef]
47. Zhang, R.; Zhou, W.; Field, R.; Tafuri, A.; Yu, S.L.; Jin, K. Field test of best management practice pollutant removal efficiencies in Shenzhen. China. *Front. Environ. Sci. Eng. China* **2009**, *3*, 354–363. [CrossRef]
48. Wu, Y.C.; Zou, Z.W.; Huang, C.X.; Jin, J. Effect of Biochar Addition on Phosphorus Adsorption Characteristics of Red Soil. *Front. Environ. Sci.* **2022**, 10. [CrossRef]
49. Gou, X.M.; Cai, Y.; Wang, C.Q.; Li, B.; Zhang, Y.; Tang, X.Y.; Shen, J.; Cai, Z.H. Effects of different long-term cropping systems on phosphorus adsorption and desorption characteristics in red soils. *J. Soil Sediments* **2020**, *20*, 1371–1382. [CrossRef]
50. Mai, Y.; Huang, G. Hydrology and rainfall runoff pollutant removal performance of biochar-amended bioretention facilities based on field-scale experiments in lateritic red soil regions. *Sci. Total. Environ.* **2020**, *761*, 143252. [CrossRef]
51. Zhao, Q.G.; Xie, W.M.; He, X.Y. *Red Soil of Jiangxi*; Jiangxi Science & Technology Press: Nanchang, China, 1988.
52. Li, Y.J.; Fu, H.; Zhang, J.Y.; Zhang, Z.X.; Li, J.K. Study of pollutant accumulation characteristics and microbial community impact at three bioretention facilities. *Environ. Sci. Pollut. Res.* **2021**, *28*, 44389–44407. [CrossRef]
53. Bonnie, J.G.; Tim, D.F.; Belinda, E.H. Interactions between design, plant growth and the treatment performance of storwater biofilters. *Ecol. Eng.* **2017**, *105*, 21–31.
54. Fowdar, H.; Payne, E.; Schang, C.; Zhang, K.F.; Deletic, A.; McCarthy, D. How well do stormwater green infrastructure respond to changing climatic conditions? *J. Hydrol.* **2021**, *603*, 126887. [CrossRef]
55. Li, P.; Liu, J.; Fu, R.; Liu, X.; Zhou, Y.Y.; Luan, M. The performance of LID (low impact development) practices at different locations with an urban drainage system: A case study of Longyan, China. *Water Pract. Technol.* **2015**, *10*, 739–746. [CrossRef]

56. Alyaseri, I.; Zhou, J.P.; Morgan, S.M.; Bartlett, A. Initial impacts of rain gardens' application on water quality and quantity in combined sewer: Field-scale experiment. *Front. Environ. Sci. Eng.* **2017**, *11*, 19. [CrossRef]
57. Jing, Y.P.; Li, J.; Mei, Y.M.; Liu, X.; Yu, X.L.; Hu, X.Z.; Song, F.F.; Lu, M.M. Design and performance of urban sponges in red soil: Improvement of physical and chemical properties. *J. Water Clim. Chang.* **2021**, *12*, 371–383. [CrossRef]
58. Shuster, W.D.; Dadio, S.; Drohan, P.; Losco, R.; Shaffer, J. Residential demolition and its impact on vacant lot hydrology: Implication for the management of stormwater and sewer system overflows. *Landsc. Urban Plan.* **2014**, *125*, 48–56. [CrossRef]

water

MDPI

Article

Study on Multi-Objective Optimization of Sponge Facilities Combination at Urban Block Level: A Residential Complex Case Study in Nanjing, China

Mingkun Xie [1,2], Yuning Cheng [3,*] and Zengchuan Dong [2,*]

1 College of Civil and Transportation Engineering, Hohai University, Nanjing 210024, China
2 College of Hydrology and Water Resources, Hohai University, Nanjing 210024, China
3 Department of Landscape Architecture, School of Architecture, Southeast University, Nanjing 210096, China
* Correspondence: 101004222@seu.edu.cn (Y.C.); zcdong@hhu.edu.cn (Z.D.)

Abstract: Urban block-scale sponge system design needs address how to specify the optimal approach to combine the number of areas and types of sponge facilities for diverse land conditions and sponge system design objectives, while ensuring sponge performance and economic efficiency. With the gradual application of multi-objective optimization algorithms in the design of sponge cities, multi-objective combinatorial problem solving for sponge facilities based on optimization algorithms is more accurate and efficient than traditional design methods based on the designer's experience. This study utilizes a residential complex in Nanjing as a practical example, selects six types of typical sponge facilities to construct a multi-objective optimization combination model for sponge facilities, and employs the SPEA-2 algorithm to determine the optimal combination of sponge facility types and quantities. Finally, 186,754 combinations of sponge facilities were calculated. For the three sponge objectives of optimal performance and economy for stormwater infiltration and storage, optimal performance and economy for runoff pollution control, and optimal average overall performance for stormwater infiltration, runoff pollution control, and economy, a number of combinations of sponge types and numbers were obtained.

Keywords: multiobjective optimization; sponge city planning and design; urban block; sponge facility combination

Citation: Xie, M.; Cheng, Y.; Dong, Z. Study on Multi-Objective Optimization of Sponge Facilities Combination at Urban Block Level: A Residential Complex Case Study in Nanjing, China. *Water* **2022**, *14*, 3292. https://doi.org/10.3390/w14203292

Academic Editors: Haifeng Jia, Dafang Fu, Wei-Shan Chen and Jiangyong Hu

Received: 3 September 2022
Accepted: 10 October 2022
Published: 18 October 2022

1. Introduction

Under natural conditions, the underlying surface has good permeability and precipitation can infiltrate directly into the soil and participate in the hydrological cycle [1]. However, the increase in impermeable underlying surface in cities has gradually led to an increase in the amount of runoff volumes and a decrease in the amount of naturally infiltrated stormwater, causing many urban water environment problems [2]. Examples include urban flooding, lack of water in urban green spaces, etc. [3,4]. Sponge cities aim to increase the resilience of cities to rainfall by optimising the urban underlying surface and thereby increasing the natural infiltration and storage capacity of rainwater [5]. Now, China's sponge cities have moved from the pilot exploration stage to the systematic demonstration stage [6].

Given the scarcity of land resources in built-up urban areas, how to optimize the combination of sponge facilities to maximize ecological and economic benefits has become an important issue for sponge city planners. The selection and combination of sponge facility types is an important step in the planning and design process of urban sponge systems [7,8]. The performance and cost of sponge facilities vary, how to reasonably select the type and number of sponge technology facilities based on site conditions and sponge city policy requirements requires consideration of multiple design objectives such as runoff control, pollution control and stormwater resource utilization [9].

The problem of optimising the combination of sponge or eco-stormwater management facilities has been extensively researched by many scholars in similar fields. Advanced experiences can provide us with a reference. These concepts mainly include the Low Impact Development (LID), Green Infrastructure (GI) and Best Management Practices (BMPs) etc. [10]. Different combinations of facilities will create different stormwater management effects. For example, in terms of stormwater runoff control [11], stormwater pollution control [12] and in terms of economic performance [13]. Therefore, depending on the objectives of the project, different types of LID-BMP-GI facilities are combined in different practical projects. The combination of facility types chosen for the different scales of the project also varies considerably [14–16]. In recent years, multi-objective optimisation models have become more widely used in the sponge city [17,18].

A multi-objective optimization model-based approach provides methods and approaches to address the multi-objective combination of sponge facilities [15,19]. The design variables, objective functions, constraints, software tools and solution algorithms are all significantly different in the construction of multi-objective optimisation models due to the different purposes and scenarios for which different combinations of stormwater management facilities are studied [15,20–23].

The aims of this study are: (1) to construct a multi-objective optimisation model for a typical combination of sponge facility type and scale at the urban plot scale. (2) To explore the application and solution of parametric design software (e.g., grasshopper), which is more commonly used by urban planners, to the multi-objective optimisation model. (3) Apply the constructed multi-objective optimisation model for sponge facility combinations to a case study in Nanjing, China, and apply it empirically to solve for sponge facility types and scales under different optimal combination scenarios.

2. Methodology

2.1. The Multi-Objective Optimization Model

In real-world cases, people are often faced with decision problems that consist of multiple interacting and conflicting goals that need to be weighed to obtain the best solution to the problem. A multi-objective optimization problem is when we are faced with a decision problem with more than one objective [24].

In general, multi-objective optimization problems have conflicting objectives, and it is very unlikely that multiple objectives will be optimal at the same time. Thus, the solution to a multi-objective optimization problem is not unique, but rather there is a set of many Pareto optimal solutions. The set of Pareto optimal solutions constitutes the Pareto front [25]. In Bi-objective optimization problems, the Pareto front is usually a two-dimensional curve, while in multi-objective optimization problems, the Pareto front is a hypersurface [26]. The set of Pareto optimal solutions is a set of possibilities, but in practice it is necessary for the decision maker to select one or more of the set of Pareto optimal solutions as the optimal solution to the multi-objective optimization problem. The decision maker can select one or several optimal solutions from among the many Pareto fronts based on the visualisation. If more than one optimal solution is selected, the values of the objective function corresponding to the multiple optimal solutions are compared and analysed in relation to the application scenario in which the problem needs to be solved. The mathematical expression for the multi-objective optimization problem is as Equation (1) and the components of a multi-objective optimization model as Table 1.

$$\begin{cases} opt\ f(x) = [f_1(x), f_2(x), \ldots, f_i(x)] \\ \quad s.t.\ x \in X \\ \quad X \subseteq R^n \end{cases} \tag{1}$$

where $opt\ f(x)$ is the objective function that maximises or minimises the objective function; $f_1(x), f_2(x), \ldots, f_i(x)$ is the i single objective function; R^n is the set of objective function constraints.

Table 1. Components of a multi-objective optimization model.

Elements of the Model	Description
Decision variables	Constructing a choice of options for describing the characteristics of a system (process) in a mathematical model, where each different set of values taken for the design variables corresponds to a solution value for the problem.
Constants	Known constants to be considered in model construction.
Objectives	A function constructed according to the objective problem to be solved, usually requiring a maximum or minimum value.
Constraints	Constraints that need to be met to establish decision variables.

2.2. Model Framework

The multi-objective combination optimization model for sponge facilities is mainly used to solve the problem of selecting and scaling combinations of different sponge facilities. This means that different types and numbers of sponges are required to achieve both optimal stormwater management performance and economic cost objectives per unit area of urban land. A logical framework for a multi-objective optimization model of sponge facilies combination is shown in Figure 1.

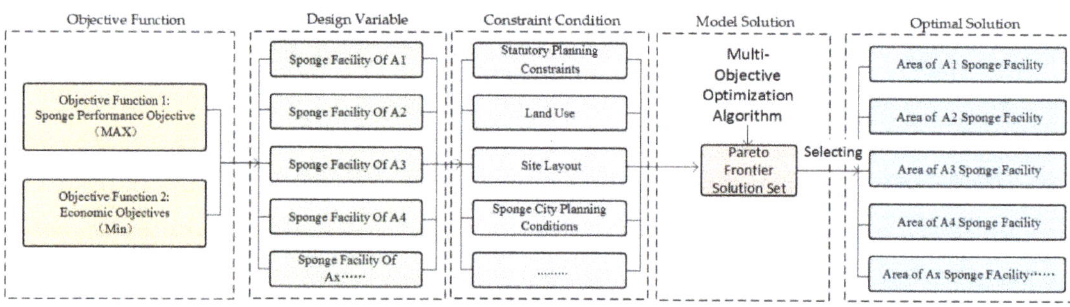

Figure 1. A logical framework for a multi-objective optimization model of sponge facilities combination.

The modelling steps are as follows:

- First, determine the sponge city design objective functions under different site conditions.
- Second, determine the number of design variables and constraints, while obtaining model constants such as sponge capacity attributes and economic cost per unit sponge facility.
- Again, establish a list of multi-objective optimization model for sponge facilities combination.
- Finally, the optimization algorithm is used to solve the model list to obtain the Pareto solution set, and the optimal solution for the combination of sponge facilities is selected according to the project situation.

2.3. Model Components

2.3.1. The Objective Functions

For different site conditions, the objective function can consist of three objectives: rainwater infiltration and storage, rainwater harvesting and utilisation, and runoff pollution control, with the overall objective of optimal sponge performance at the lowest economic cost. Its mathematical formula is expressed as:

$$\mathrm{F} = opt\ f(x) = [f_1(x), f_2(x), \ldots, f_i(x)] \quad (2)$$

$$\mathrm{Max}\ f_1(x),\ \text{and}\ \mathrm{Min}\ f_2(x) \quad (3)$$

where F is the overall objective of the optimisation; $f_1(x)$ is the sponge performance objective function of the sponge facility; and $f_2(x)$ is the economic cost objective number.

(1) Sponge performance objective function

Under the control scale requirements for sponge systems, three indicators are used to measure the sponge performance of different types of sponge facilities, namely rainwater infiltration capacity, rainwater resource storage capacity and runoff pollution control capacity, one or more of which can be selected as the overall sponge performance objective function depending on site conditions.

- Rainwater infiltration and storage effect

The rainwater infiltration and storage effect includes the infiltration and storage volume of rainfall on the site by the sponge facilities, with reference to the formula for calculating rainwater infiltration and storage facilities in the the *Technical Guide for Sponge City Construction* [27], The mathematical expression for rainwater infiltration and storage effect is:

$$V_s = \sum_{m=1}^{n} (K_m J A_{im} T_{sm} + A_{sm} h_m n k_m) \tag{4}$$

where, V_s is rainwater infiltration volume (m^3); K is soil infiltration coefficient ($m{\cdot}s^{-1}$); take the value can refer to Table A2; J is the hydraulic gradient, usually taken as J = 1; A_i is the effective infiltration area (m^2); horizontal infiltration surface according to the projected area, vertical infiltration according to the effective water level height 1/2 calculation, oblique infiltration surface according to the effective water level height 1/2. T_s is infiltration time (s), usually take 2 h; A_s is effective storage area (m^2); h is effective rainwater storage depth (m); nK is facility fill porosity, no fill to take 1; n is the number of different sponge facility types.

- Rainwater harvesting and utilization effect

The rainwater harvesting capacity effect of a sponge facility is determined by its unit storable volume and unit area, expressed by the mathematical formula:

$$V_r = \sum_{m=1}^{n} v_m A_m \tag{5}$$

where V_r is the volume of rainwater collected and utilised (m^3); v_m is the volume of water stored per unit area of the sponge facility (m^3/m^2); A_m is the area of the sponge facility (m^2); n is the different sponge facility types.

- Runoff pollution removal effect

The pollution control objectives of sponge cities are mainly reflected in the removal effects of sponge facilities on SS, COD, BOD, TN and TP in urban runoff surface source pollution. In urban runoff SS is significantly correlated with several other water quality indicators, and the *Technical Guidelines for Sponge City Construction* issued by the Ministry of Housing and Urban-Rural Development of China selects SS indicators as runoff pollution control indicators, so this study also adopts SS as a measure of pollution removal capacity of sponge facilities, with the formula expressed as:

$$V_p = \sum_{m=1}^{n} k_{pm} A_m P \tag{6}$$

where V_p is the amount of pollution removed by the sponge facility (t); k_{pm} is the removal rate of SS by different sponge facilities (%); A_m is the area of different sponge facility types (m^2); P is the annual pollution load of site runoff (t/m^2); and m is the different sponge facility types.

(2) Economic objective function

The total cost of the project's sponge facilities is expressed as the sum of the product of the area of the different facility components and the cost per unit area of such facilities, where the cost per unit area of the sponge facilities needs to be determined according to the local market and facility section structure.The mathematical expression of which is:

$$E = \sum_{m=1}^{n} A_m m_m \quad (7)$$

where E is the total construction cost of the proposed sponge facility(yuan); A_m is the area of the type of facility (m^2); m_m is the cost per unit area of the mth type of facility (yuan); and n is the different sponge facility types.

2.3.2. The Decision Variables

The variables in the construction of the multi-objective optimization model for sponge facilities combination are the number of areas for different types of facilities

With reference to several types of sponge facilities that are more frequently used in China's sponge city practice, this study selects six types of typical sponge facilities for application, namely horizontal green space without water storage modules, horizontal green space with water storage modules, water-storing sunken green space, permeable hard surface, designed water body and green roofs. The details are shown in Table 2.

Taking the area A of each type of sponge facility as a decision variable and using Ai to denote the planned design area of different facilities, where i denotes the type of facility (i = 6 for horizontal green space, horizontal green space with water storage module, water-storing sunken green space, permeable hard surface, designed water body, green roof), the variables involved in this study are listed in the Table 3.

2.3.3. The Constraints

In actual sponge city project design practice, constraints are imposed by site planning and land use and other factors on the area that can be laid out for different types of facilities, mainly including the following categories.

(1) Constraints of the total area

This means that the sum of the area of each sponge facility type on the site should be less than or equal to the total area of the planned and designed site.

$$A1 + A2 + A3 + A4 + A5 + A6 \leq A_{TA} \quad (8)$$

where A1–A6 is the sum of the area of each type of sponge facility on the site; A_{TA} is the total area of the planned and designed site.

(2) Constraints of the site green space ratio

The total area of green space type facilities such as horizontal green space and sunken green space must be less than the total area of green space on the site, in accordance with the green space ratio restrictions in the site plan.

$$A1 + A2 + A3 \leq A_{GS} \quad (9)$$

where A1 is the area of the horizontal green space without water storage modules; A2 is the area of the horizontal green space with water storage; A3 is the area of the sunken green space on the site.

Table 2. Description of the characteristics of 6 types of typical sponge facilities.

S/N	Name	Description	Section Illustration
1	Horizontal Green Space Without Water Storage Modules	Low cost, low infiltration and runoff pollution control, low stormwater storage capacity	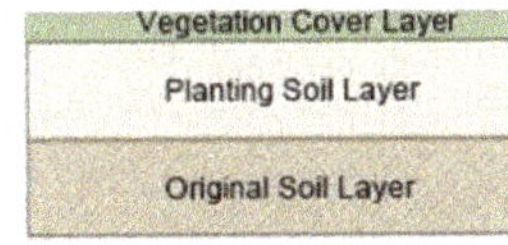
2	Horizontal Green Space With Water Storage Modules	Infiltration and runoff pollution control advantages of horizontal green space, but space saving, high rainwater harvesting efficiency and high cost	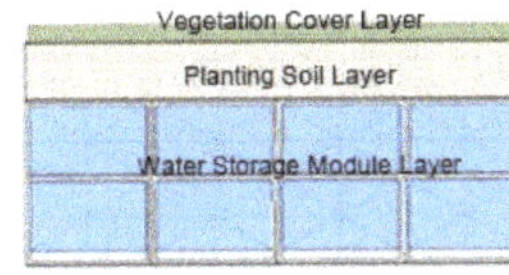
3	Sunken Green Space	Low cost with a certain volume of water storage and pollution control function for rainwater runoff, but the actual storage volume is insufficient	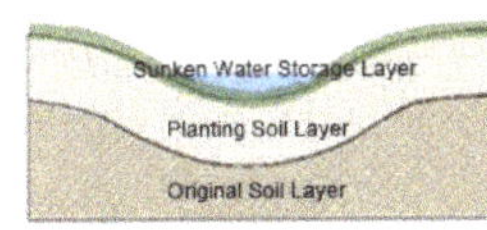
4	Permeable Hard Surface	Effective stormwater infiltration and runoff pollution control, insufficient stormwater storage capacity and high costs.	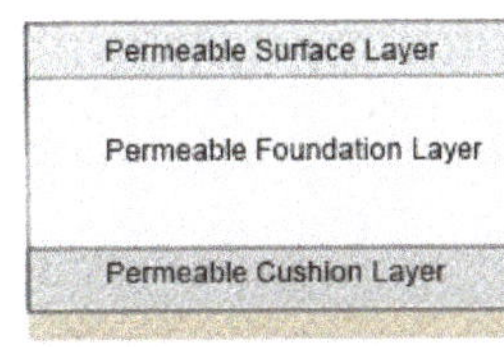
5	Green Roof	Only be used on building roofs, with less scope for application and higher costs.	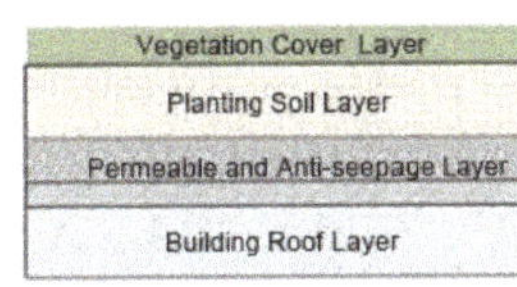
6	Designed Water Body	High rainwater storage capacity, low pollution control and low cost of construction.	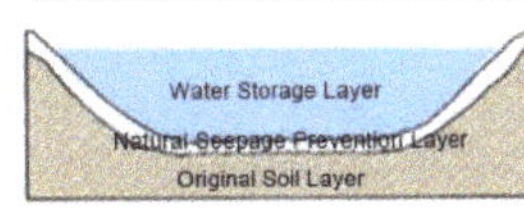

Table 3. Composition of variables for a multi-objective optimization model for sponge facilities combination.

Serial Number	Design Variable Symbol	Description	Unit
1	A1	Area of horizontal green space	m^2
2	A2	Area of horizontal green space with water storage modules	m^2
3	A3	Area of water-storing sunken green space	m^2
4	A4	Area of permeable hard surface	m^2
5	A5	Area of designed water body for water storage	m^2
6	A6	Area of green roof	m^2

(3) Runoff Control Constraints for Sponge City Construction

In order to meet the design storage volume requirements corresponding to volume capture ratio of annual rainfall and runoff of the sponge city, the sum of infiltration volume, rainwater harvesting volume and runoff pollution treatment volume in the function should be greater than or equal to the design storage volume under the design rainfall corresponding to the annual volume capture ratio of the site.

$$V_s + V_r + V_p \geq V_{\mathrm{DSV}} = \sum_{m=1}^{n} (K_m J A_{im} T_{sm} + A_{sm} h_m n k_m + v_m A_m + k_{pm} A_m P) \geq V_{\mathrm{DSV}} \quad (10)$$

where V_s is the infiltration volume; V_r is the rainwater harvesting volume; V_p is the runoff pollution treatment volume on the site. V_{DSV} is the design storage volume.

(4) Constraints of the Site Building Density Rate

According to the plot ratio requirements in the site plan, the area of the green roof must be smaller than the site building footprint.

$$\mathrm{A6} \leq \mathrm{A_B} \quad (11)$$

where A6 is the area of the green roof on the site; $\mathrm{A_B}$ is the planned building footprint.

(5) Constraints of the Hard Surface Area

The permeable hard surface provided on the site shall be less than equal to the sum of the areas of all hard surfaces such as roads and squares on the site.

$$\mathrm{A4} \leq \mathrm{A_{HS}} \quad (12)$$

where A4 is the area of permeable hard surface sponge facilities on the site; $\mathrm{A_{HS}}$ is the total area of hard surface such as roads and squares on the site.

(6) Constraints of the Water Surface Rate

If the site water surface ratio constraint is stipulated in the detailed site plan, consideration needs to be given to the fact that the proportion of the site occupied by designed water body for storage should be less than or equal to the water surface ratio requirement in the site plan.

$$\mathrm{A5} \leq \mathrm{A_{WS}} \quad (13)$$

where A5 is the area of the designed water body in the site and A is the planned site water surface rate footprint.

(7) Non-negative Constraints

Non-negative area for each type of sponge city facility

$$\mathrm{A1, A2, A3, A4, A5, A6} \geq 0 \quad (14)$$

where A1–A6 is the sum of the area of each type of sponge facility on the site.

2.3.4. The Constants

The constants are the basic attribute parameters for the different types of sponge facilities. The values of the basic constants for rainwater infiltration and storage, rainwater harvesting and utilisation, runoff pollution control and unit cost can be found in Tables A1–A5, listed in the Appendix A.

2.4. Model Solution

2.4.1. Algorithms

In recent years, evolutionary algorithms based on simulating the evolutionary process of natural organisms have become an important method for solving multi-objective optimization problems. Compared to traditional mathematical planning methods evolutionary algorithms have the characteristics of being informative, adaptable and scalable. At present, the more commonly used evolutionary algorithms include Genetic Algorithms, Simulated

Annealing Algorithms, The Strength Pareto Evolutionary Algorithm, Ant Colony Optimization Algorithm etc. [28]. The Non-Dominated Sorting Genetic Algorithm (NSGA-II) and The Strength Pareto Evolutionary Algorithm (SPEA-2) are the most widely used and influential multi-objective evolutionary algorithms, and are the reference for performance comparisons of other evolutionary algorithms. Compared to the first generation of traditional genetic algorithms, the advantages of the enhanced evolutionary algorithm include reduced computational complexity of the initial evolutionary algorithm, faster computational speed, improved accuracy of the optimization results and reduced computational effort [29,30]. In this study, the Strength Pareto Evolutionary Algorithm (SPEA-2) was selected as the algorithmic solution tool for the multi-objective optimization sponge facility combination model.

2.4.2. Software Tools

In this study, Rhinoceros and Grasshopper, which are widely used in the field of urban planning and design, are used as the basic software platforms for the calculation of optimization algorithms [31–34]. Grasshopper is a parametric design plug-in for Rhinoceros with visual programming capabilities, but it cannot be run independently of Rhinoceros itself, and can generate results through operator command operations, and can also realise circular iterative operations based on the written algorithms, greatly improving the efficiency of planning and design personnel [35].

This study uses the Octopus multi-objective optimization plug-in. Octopus is a multi-objective optimization plug-in based on the Grasshopper developed by the University of Applied Arts in Vienna, Austria, and Bollinger+Grohmann (Frankfurt, Germany). It provides designers with a quick and easy-to-use multi-objective parametric design plug-in for computational design and is widely used in the field of planning and design [36,37].

3. Case Study

3.1. Overview of the Study Area

3.1.1. Location

The study area is located in the southeast of Nanjing, with a total planned area of approximately 18.13 square kilometres for the Shangfang District urban area. The No. 2 residential complex selected for this study is located on the southern side of Shangfang District, with a total area of 3.77 hectares (Figure 2).

3.1.2. Weather

The study area is located in a subtropical monsoon climate, with an average annual rainfall of 1038.7–1124 mm and an average annual rainfall day of 124.2 days. Most of the annual rainfall is concentrated in June to August, accounting for more than 50% of the annual rainfall, with June to July being the rainy season.

3.1.3. Underlying Surface

The current topography of the study area is relatively flat, with over 80% impervious area and the main land types are impervious building roofs and road squares, along with some infiltrated bare earth wasteland. The current status of the site has a large proportion of highly developed industrial, residential and storage land attributes. The vegetation is poorly protected and the water system in the area is underdeveloped, with most of the original water system being filled in and some areas poorly drained.

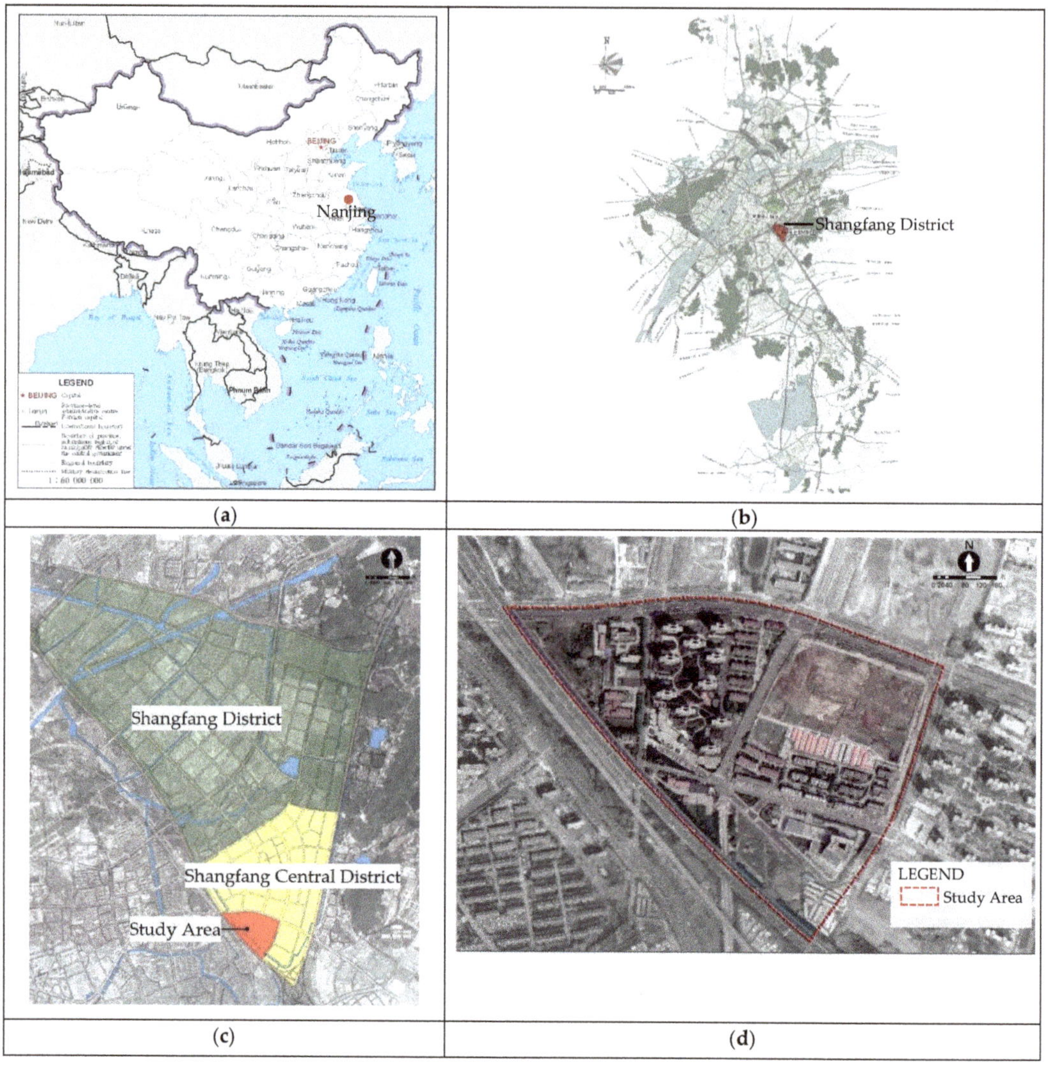

Figure 2. (**a**) Location of Nanjing in China, Source: Ministry of Natural Resources of China; (**b**) Location of Shangfang District in Nanjing; (**c**) Location of the study area in the Shangfang District; (**d**) Satellite image of the study area, Source: Google Maps 2018-10.

3.1.4. Planning and Policy Status

According to the Nanjing Urban Master Planning (2007–2020), the Jiangning District Urban and Rural Master Planning (2010—2030) and the Nanjing Jiangning Shangfang Group Control Detailed Planning, the current land use in the study area has been greatly adjusted, with a large proportion of industrial land replaced by residential and commercial land. The commercial land use is mainly concentrated in the northwest and southeast of the site in two separate plots 02–06 and 02–19, and the green space in the area has been reorganized and a separate green space has been planned in the west, so that the green space rate has been improved to a large extent (Table 4).

Table 4. Statistics of the urban planning underlying surface classification of the study area.

Serial Number	Underlying Surface Classification	Planning Indicators (m^2)
1	Total site area	377,133
2	Area of building area	41,260
3	Area of road and open space	229,343
4	Area of green space	106,530
5	Area of water surface	9600

3.1.5. Sponge City Scale Control Requirements

Based on the requirements for runoff control in this area in the Nanjing Sponge City Construction Pilot City Implementation Planning and the Nanjing Sponge City Special Planning, the total annual runoff control rate for the study area sponge city is 85%, corresponding to a design rainfall of 38.8 mm, We obtained the control index parameters of annual runoff control rate, design rainfall amount, surface source pollution control rate, rainfall field control rate and design of storage volume for the control unit in which the study area is located through the upper planning. The details are shown in Table 5.

Table 5. Study area sponge city design scale and control requirements.

Planning Control Index	Annual Runoff Control Rate (%)	Design Rainfall Amount (mm)	Surface Source Pollution Control Rate (%)	Rainfall Field Control Rate (%)	Design of Storage Volumes (m^3)
Values	79.52	29.7	55	87.3	750–1000

3.2. Optimization Objectives

By analysing the site topography, land use type and runoff characteristics, the main water environment issues faced in the study area as follow:

(1) A large proportion of impervious surface area and a low water surface ratio resulting in a high volume of runoff from the site.

(2) The site is highly developed and surface runoff pollution is more serious.

(3) A high proportion of green space is planned for the site, with high water demand for vegetation and high long-term maintenance and management costs.

By integrating the water environment issues that need to be addressed at the site, rainwater infiltration and collection, runoff pollution control and economic objectives are selected as the site sponge system design optimization objectives, and a multi-objective optimization model for sponge facility.

3.3. Constraint Settings

Based on the control requirements for boundary conditions, green area ratio, building density, water body area and square road area in the detailed control planning of the study area, as well as the scale control requirements in the sponge city planning (Table 5), In conjunction with Section 2.3.3 the constraints, we obtained the relevant constraints for the model.

3.4. Model List

After setting the objective function and constraints, we obtained a list of elements for a multi-objective optimization model for sponge facilities in the study area, with each element specified as follows in Table 6.

Table 6. A list of elements for a multi-objective combined optimization model for sponge facilities in the study area.

Design Variables	A1: Area of horizontal green space			
	A2: Area of horizontal green space with water storage module			
	A3: Area of water-storing sunken green space			
	A4: Area of permeable hard surface			
	A5: Area of designed water body for water storage			
	A6: Area of green roof			
Objective Functions	Overall objective function	Sub-objectives	Formula	Description
	opt{$f1x$, $f2(x)$, $f3(x)$} Max $f1(x), f2(x)$ *and Min* $f3(x)$ (Maximum sponge efficiency and lowest economic cost)	$f1x$ Rainwater infiltration and storage objective	Max $f1(x) = \{\sum_{m=1}^{n}(K_m J A_{im} T_{sm} + A_{sm} h_m n k_m) + v_m A_m\}$ n = 6	Max $f1(x)$ The larger the rainwater infiltration storage capacity the better the objective
		$f2(x)$ Runoff pollution control objective	$f2(x) = \sum_{m=1}^{n} k_{pm} A_m p$	Max $f2(x)$ The greater the runoff pollution removal capacity the better the objective.
		$f3(x)$ Economic objective	$f3(x) = \sum_{m=1}^{n} A_m m_m$	*Min* $f3(x)$ The lower the economic cost, the better the objective.
Constraints	Constraints of the total area		A1 + A2 + A3 + A4 + A5 + A6 $\leq$ 377,000	
	Constraints of the site green space ratio		A1 + A2 + A3 $\leq$ 106,530	
	Runoff Control Constraints for Sponge City Construction		1.7248A1 + 2.764A2 + 2.39A3 + 1.89A4 + 0.34A5 + 1.344A6 $\geq$ 750	
	Constraints of the hard surface area		A4 $\leq$ 229,343	
	Constraints of the Site Building Density Rate		A6 $\leq$ 41,260	
	Constraints of the Water Surface Rate		A5 $\leq$ 9600	
Optimization Algorithms	The Strength Pareto Evolutionary Algorithm-2 (SPEA-2)			

4. Results

4.1. Parameter Settings

The six decision variables A1–A6 were connected to the G-side of the Octopus plug-in in the form of a Number slider. The three objective functions of rainwater infiltration, runoff pollution control and economic objectives are entered into the O-side of the Octopus plug-in using the objective function equation in Evaluate. The rainwater infiltration function and pollution control are evaluated for the maximum value and the economic objective for the minimum value. At the same time, constraints are set between the six decision variables Number slider, through the setting of constraints can improve the speed of the model calculation, reduce the number of invalid scenario simulation calculation. The sponge objective function and the pollution control objective function are maximum values, and as the default design of the Octopus plug-in program is to find the minimum value, the sponge objective function and the pollution control objective function formula output data need to be negative (i.e., the output formula is multiplied by −1).

The Octopus parameters were set as follows: Optimization algorithm: SPEA-2, population size: 100, maximum number of iterations: 500, elite rate 0.5, mutation probability 0.1, mutation ratio 0.5, crossover rate 0.8, and Grasshopper cell connection method detailed in Figure 3.

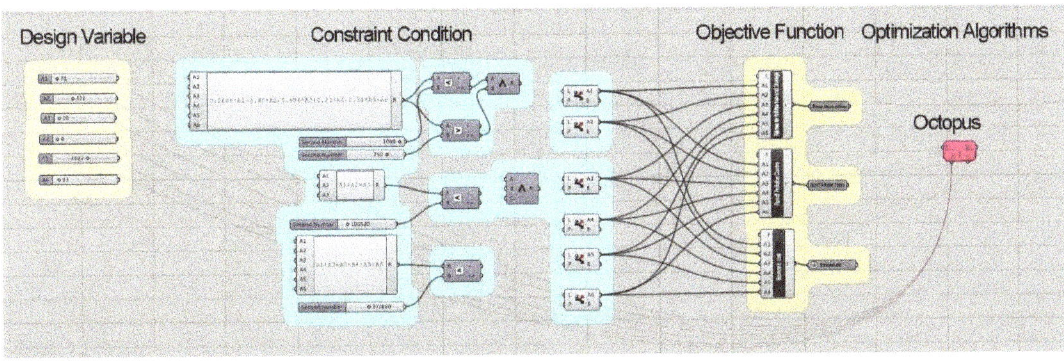

Figure 3. Connection and illustration of the Grasshopper operator for multi-objective combinatorial optimization of sponge facilities.

4.2. Validity Verification

After 500 iterations of the Octopus plug-in, the results of the three objective functions were extracted and analysed, showing that the values of rainwater infiltration and runoff pollution removal capacity gradually increased with the number of iterations and the economic cost values gradually decreased with the number of iterations, indicating that the objective functions were gradually optimised with the iterative operation of the algorithm.

The resultant values gradually stabilise at the 10th iteration of the rainwater infiltration objective function (Figure 4) and at around 370 iterations of the pollution control objective function and they gradually reach a maximum value at the 10th iteration of the rainwater infiltration objective function. In order to verify whether the optimization algorithm converges towards the optimization objective, 500 generations of the evolutionary process of the Pareto optimal solution of concentrated rainwater infiltration storage, pollution control objective function value per generation maximum value, and economic cost objective function minimum value for comparison can be seen from Figures 4–6. In the 400th generation around the three objective functions, basic unity basically remains unchanged and the optimization objective gradually gains better convergence, meeting the optimization requirements.

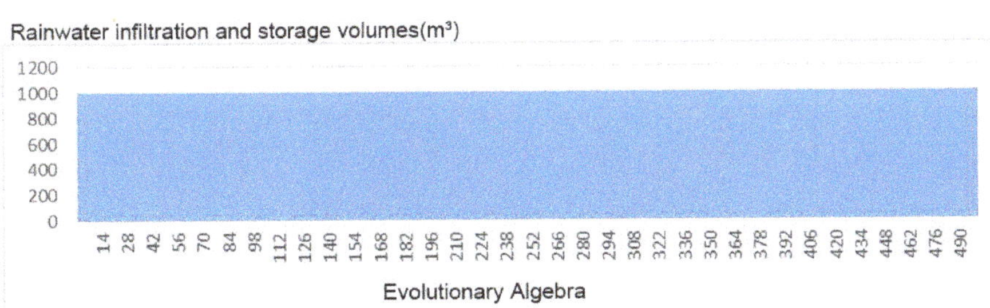

Figure 4. Evolutionary process of maxima in Pareto-optimal solutions for rainwater infiltration and storage objectives.

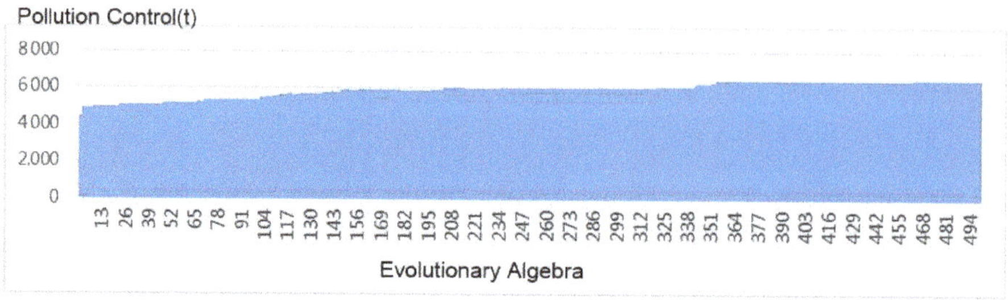

Figure 5. Evolutionary process of maxima in the Pareto-optimal solution of pollution control objectives.

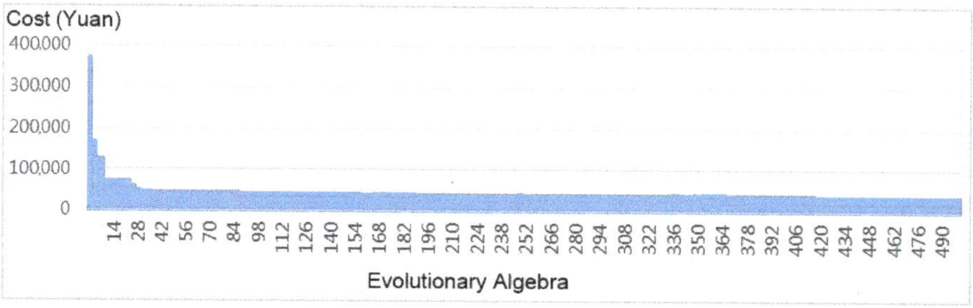

Figure 6. Minimal value evolutionary process in the Pareto optimal solution of economic cost objectives.

4.3. Analysis of the Results

A total of 186,754 optimal solution sets were obtained after 500 generations of the algorithm. 123 sets of optimal solutions obtained in the 500th generation were selected (Figure 7), and this solution space can provide designers with a rich choice of solutions. Based on the design objectives of this case, the optimal solution for rainwater infiltration and storage, the optimal solution for pollution control, the optimal solution for economic efficiency and the comprehensive average optimal solution are compared and analysed (Figure 8).

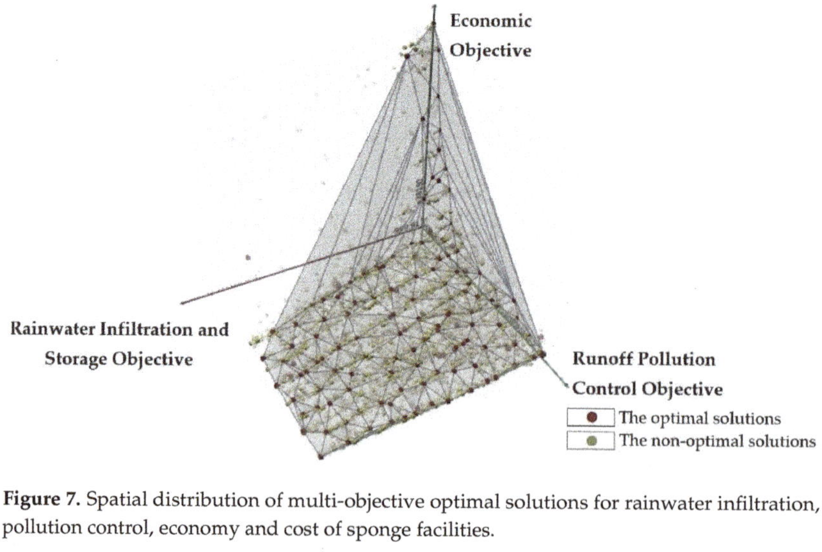

Figure 7. Spatial distribution of multi-objective optimal solutions for rainwater infiltration, runoff pollution control, economy and cost of sponge facilities.

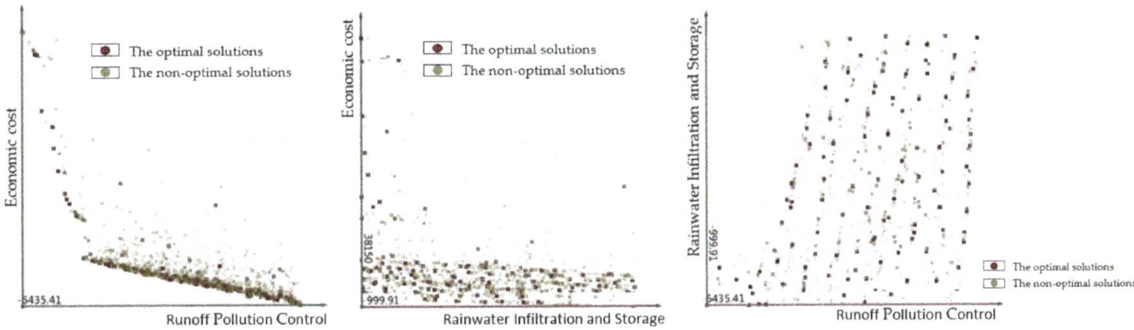

(a) (b) (c)

Figure 8. (**a**) Pareto front distribution of runoff pollution control (x-axis) and economic cost (y-axis); (**b**) Pareto front distribution of rainwater infiltration and storage (x-axis) and economic cost (y-axis); (**c**) Pareto front distribution of runoff pollution control (x-axis) and rainwater infiltration and storage (y-axis).

4.3.1. Optimal Solution for Rainwater Infiltration, Storage and Economic Objectives

In the spatial distribution of the Pareto optimal solution set, the green X-axis represents the infiltration capacity of the combination of sponge facilities, and the closer the value is to the origin, the better the infiltration capacity of the combination of facilities. The green Y-axis is the economic cost of the combination of facilities, the closer to the origin the lower the economic cost of the combination of facilities. As shown in Figure 9, the value with the strongest rainwater infiltration capacity and the lowest economic cost is the optimal set of blue circles in the bottom left corner, with a rainwater infiltration capacity of 998 m^3 and an economic cost of 52,440 yuan.

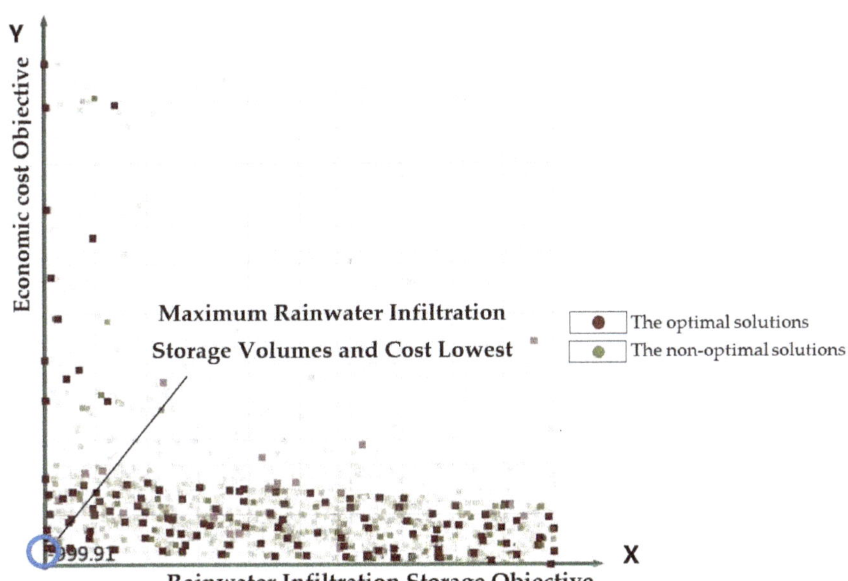

Figure 9. Rainwater infiltration and storage and economic objective are both optimal.

4.3.2. Optimal Solution for Runoff Pollution Control and Economic Objectives

In the spatial distribution of the Pareto optimal solution set, the green *X*-axis represents the pollution control capacity of the combination of facilities, and the *Y*-axis represents the economic cost level, the higher the runoff pollution control capacity of the combination of facilities, the closer the *X*-axis numerical optimal solution is to the position of the origin, as shown in Figure 10, the blue circle in the upper left corner is the optimal target parameter value for runoff pollution control, which corresponds to a runoff pollution treatment capacity of 6345 t. The value of the model variable when the runoff pollution control objective is optimal as shown is Table 7.

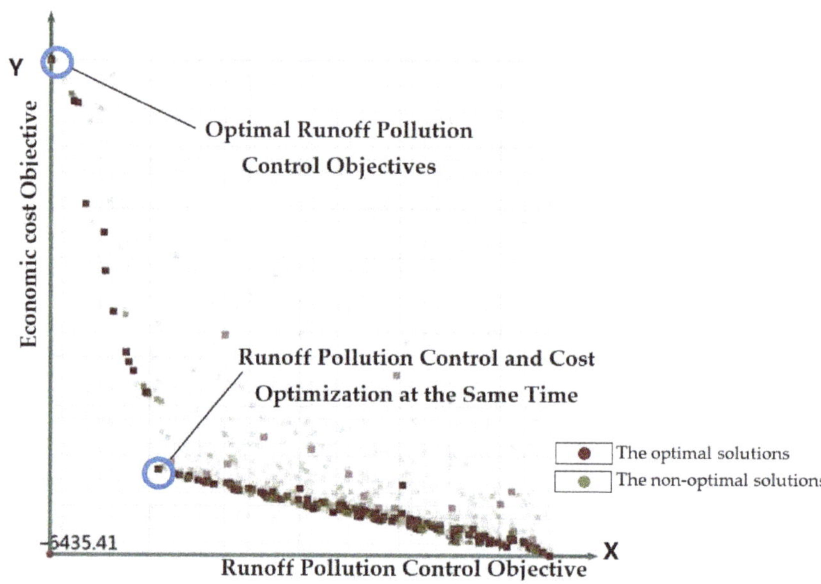

Figure 10. Optimal parameters for runoff pollution control.

Table 7. The value of the model variable when the runoff pollution control objective is optimal.

Optimization Objective	Objective Values			Area of Sponge Facilities (m^2)					
	Rainwater Infiltration and Storage Volumes (m^3)	Runoff Pollution Control (t)	Economic Costs (yuan)	A1	A2	A3	A4	A5	A6
Optimal runoff pollution control (regardless of economic factors)	999	6435	453,730	2173	0	4	1938	2	6
Optimal runoff pollution control and lowest economic cost	995	5258	109,340	3579	0	1	1	0	33

4.3.3. Optimal Solution for All Objectives

As can be seen from Figure 11, the Pareto solution set for each objective should be as close as possible to the origin of the coordinates and located at the outer convex position of the solution set set surface. The selection of the integrated optimal value takes into account the factors of rainwater infiltration and storage, pollution control and cost, and compromises between the optimal values. 4 combinations of sponge facilities are selected from the three-objective Pareto optimal solution set as the optimal solution alternatives

(Table 8). We offer four sets of optimal decision options and the final decision should be made by the designer, project implementer and stakeholders. Of the four solutions, solution 2 can be chosen if the best infiltration, storage and decontamination performance is considered, and solution 4 can be chosen if the lowest economic cost is desired.

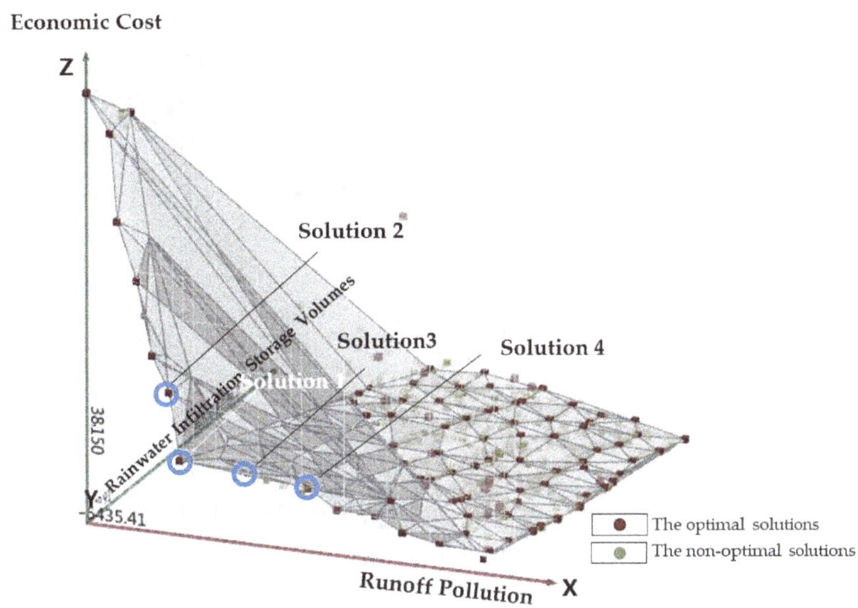

Figure 11. Optimal parameters for all objectives.

Table 8. The value of the model variable when the comprehensive objective is optimal.

S/N	Facility Combination Solution	Area for Various Types of Facilities (m^2)						Rainwater Infiltration and Storage Volumes (t)	Runoff Pollution Control (t)	Economic Costs (yuan)
		A1	A2	A3	A4	A5	A6			
1	Solution 1	3579	0	1	1	9	34	999	5260	110,110
2	Solution 2	3243	1	1	365	11	46	999	5400	174,590
3	Solution 3	2905	0	4	5	2	202	986	4517	98,890
4	Solution 4	2131	0	2	43	10	405	991	3723	93,820

4.3.4. Target Interval Selection and Priority for Specific Facility Types

There are also special cases in the selection process of sponge facilities, where application scenarios may require the selection of a combination of options within a certain target range, for example the selection of a combination of facilities within a certain price range. In addition to this, decision makers often require a single type of facility to be selected as the main facility, for example a demonstration area with a green roof. In this case, the combination solution can be selected from the set of Pareto solutions derived from arithmetic, the optimal solution for all facility combinations can be derived as a data table, a target-variable correspondence can be established, and then the number of facility combinations corresponding to a certain target interval value or a certain type of facility quantity dominated by conditional statements can be selected.

In this case, a total of 61,920 sets of optimal solutions were generated, corresponding to a total of 371,520 facility area parameters. See Table 9 for a schematic representation of facility combinations dominated by a single type of facility.

Table 9. The value of design objective for the maximum area of a particular type.

S/N	Facility Combination Solution	Area for Various Types of Facilities (m^2)						Rainwater Infiltration and Storage Volumes (m^3)	Runoff Pollution Control (t)	Economic Costs (yuan)
		A1	A2	A3	A4	A5	A6			
1	Maximum Area of A1	3632	0	19	2	2	13	999	5345	112,450
2	Maximum Area of A2	209	439	345	719	138	165	999	5400	174,590
3	Maximum Area of A3	1101	9	1330	19	43	18	986	4208	209,770
4	Maximum Area of A4	2131	0	2	43	10	405	991	3723	93,820
5	Maximum Area of A5	809	10	708	100	858	105	999	4946	568,780
6	Maximum Area of A6	3300	0	21	2	0	975	999	6400	450,570

5. Discussion

After calculation, a total of 186,754 combinations of sponge facilities were obtained for the study case. This solution space provides designers with a rich choice of sponge facility combinations, allowing them to flexibly select sponge facility combinations according to the site water environment problems to be solved and the design objectives.

Although good computational results were obtained with the SPEA-2 algorithm via the Octopus plug-in, the accuracy of the data results and the applicability of the software tool are issues that deserve further research and discussion. Therefore, in this study, the NSGA-II algorithm was selected for the comparative validation of the multi-objective combinatorial optimization model for sponge facilities, and the choice of software tools was also discussed.

5.1. Comparison of Algorithms

In the case of consistent objective functions and constraints, the NSGA-II algorithm of MATLAB software was used to solve the case study model, and from the spatial analysis of the final obtained solution set distribution, the rainwater infiltration storage objective frontier solution set range was between 757–1000 m^3, the runoff pollution control objective function frontier solution set range was distributed between 3155–5028 t, and the economic cost The target solution set is distributed between RMB 216,430 and RMB 276,219. Four integrated optimal solutions were selected from the relevant solution sets, and the calculation results of each solution and the corresponding A1-A6 variables are detailed in Table 10.

Table 10. A comprehensive optimal solution based on the NSGA-II algorithm.

S/N	Facility Combination Solution	Area for Various Types of Facilities (m^2)						Rainwater Infiltration and Storage Volumes (m^3)	Runoff Pollution Control (t)	Economic Costs (yuan)
		A1	A2	A3	A4	A5	A6			
1	Solution 1	2197	0	241	1	399	15	858	3678	127,712
2	Solution 2	2261	1	233	754	336	3	999	5011	275,823
3	Solution 3	2259	1	204	237	459	4	917	4083	177,862
4	Solution 4	2203	1	175	22	636	8	907	3590	143,711

The results obtained based on the NSGA-II algorithm are compared with the optimization results of the SPEA-2 algorithm. In terms of the number of optimal solution sets, both algorithms obtain a rich set of Pareto front solutions, which can provide a rich choice of sponge facility type combinations for site scale sponge system design decisions; in terms of the optimization objective values obtained, both optimization algorithms obtain optimal solution sets within the constraint range for the rainwater infiltration and storage objective; in terms of the pollution control objective the SPEA-2 algorithm obtained results in the range of 3500–5500 t, while the NSGA-II algorithm obtained target calculated values in the range of 3155–5028 t, from the total obtained optimal target value analysis, the SPEA-2 algorithm obtained target optimal value than NSGA-II to about 10% higher; economic cost objectives SPEA-2 algorithm obtained optimal value solution set distribution in the range of 90,000–360,000 yuan. The Pareto front solution obtained by SPEA-2 is also superior to the NSGA-II algorithm. Due to the stochastic nature of the genetic algorithm, the results obtained by the author for the NSGA-II algorithm after several operations are still relatively close to the final values of the study, and the results are basically representative of the algorithm's ability to solve the multi-objective combination model for sponge facilities.

By comparing the two algorithms, the objective optimal values obtained by the two algorithms are generally close to each other, and the distribution characteristics of the number of different types of sponge facilities corresponding to the same objective values are also more consistent, and both algorithms can meet the needs of sponge city design well. However, the SPEA-2 algorithm achieves better results than the NSGA-II algorithm for the multi-objective combination of sponge facilities.

5.2. Selecting Software Tools

The study used two optimization algorithms with different software tools, Grasshopper and MATLAB with different characteristics.

(1) Octopus is a multi-objective optimization plug-in based on the Grasshopper parametric design platform, and its interface is more user-friendly and easier for urban planners to get used to. Whereas mathematical software like MATLAB mostly requires the user to have programming skills and the learning time cost will be higher.

(2) In terms of data visualization, Octopus provides a more convenient way to visualize the process and result data, while MATLAB-based optimization algorithms can often only obtain images and data of the final optimal solution, which is more cumbersome than grasshopper for recording and visualizing historical data of the evolutionary process.

5.3. Comparison with Related Research

We compared this study with related studies on the application of multi-objective optimisation models in sponge cities in recent years, and the differences between this research work and other studies in terms of research objectives, scale, methodology and tools can be clearly seen. Meanwhile, we illustrate the highlights of relevant research. Please see Table 11 for more details. The relevant study also provides a good reference for our future research works.

Table 11. Comparison of this study with related research.

References	Objectives	Scale	Methodology	Tools	Highlights
Te Xu, Haifeng Jia et al. (2017) [15]	LID-BMPs planning, LID-BMP chain layout optimization	Block-scale, site-scale	Multi-objective optimization	SWMM-based methodology, NSGA-II algorithm	Coupling MOEA to SWMM and LID-BMP chain layout, optimization was combined with block-scale scenario analysis
Kun Zhang, Ting Fong May Chui (2018) [9]	selected, designed, and allocated for LID-BMP-GI	From site to catchment scale	Strategic planning cycle	Spatial allocation optimization tools (SAOTs)	Spatial allocation of LID-BMP-GI practices is illustrated. Strategic planning cycle
Yang Yu, Yongchao Zhou et al. (2022) [22]	LID spatial allocation optimization	Neighborhood scale	Integrated hydrological computing enginean with optimization algorithm	SWMM &MATLAB, PICEA-g algorithm	LID spatial allocation optimization couples SWMM & MATLAB, PICEA-g algorithm
Joong Gwang Lee, Ariamalar Selvakumar et al. (2012) [14]	SUSTAIN-based approach to optimising applications in BMPs	Watershed-scale	Optimization module	SUSTAIN, NSGA-II algorithm	Details of the SUSTAIN model
Zijing Liu, Haifeng Jia et al. (2022) [38]	Decision-making framework for GI layout	City scale	An adaptive GI layout decisionmaking System	Arcgis	Considering Site Suitability and Weighted Multi-Function Effectiveness:

Table 11. *Cont.*

References	Objectives	Scale	Methodology	Tools	Highlights
Jingwei Hou, Moyan Zhu et al. (2020) [20]	Optimal spatial priority scheme of urban LID-BMPs	City scale	Multi-objective model	ArcMap, ADEA (Adaptive differential evolution algorithm)	Includes different investment periods
Zijing Liu, Changqing Xu et al. (2022) [21]	Multiobjective optimization of green-grey coupled infrastructures	Block-scale	Multiobjective evaluation framework, Intelligent optimization algorithm	SWMM NSGA-II algorithm	Integrating socioecological indexes, Grey-green infrastructure coupling
Our works	Sponge facilities combination	Block-scale	Multi-objective optimization	Octopus, Grasshopper, SPEA-2	Six typical sponge facilities, Application of Grasshopper with SPEA-2 algorithm

Meanwhile, this study also has some limitations, for example: (1) In the selection of design variables for the model, we have selected only six typical types of sponge facilities, other than that other types of sponge facilities are not involved. (2) We did not consider additional social influences, such as landscape aesthetics and social behavioural preferences, in our objective setting. (3) The analysis of application scenarios for selecting the best solution and optimal solution for the model can be further extended.

6. Conclusions

This study focuses on the combination of typical sponge facility areas and types at the urban block scale to achieve optimal stormwater management performance and economic benefits for different site conditions and sponge system objectives, provided that the sponge system scale is determined. Six typical sponge facilities were selected for multi-objective optimization modelling and applied in practice with a case study located in Nanjing, China. By comparing the SPEA-2 and NSGA-II algorithms, both algorithms can meet the computational requirements of sponge city planning and design, but from the experimental results, it is observed that the SPEA-2 algorithm is superior for the application of multi-objective optimisation combination models for sponge facilities.The mathematical models, software tools and empirical cases involved in this study can provide references for sponge city practice and research.

In further research, more types of sponge facilities can be selected as design variables, and a corresponding multi-objective optimisation model database for sponge facilities can be established to further expand its scope of application. In terms of model application scenarios, two types of application scenarios can be categorised, one for completed urban areas and the other for new urban areas in the future; at the same time, the multi-objective optimisation algorithm can be coupled with the urban hydrological model to achieve more complex and refined optimisation results.

Author Contributions: Conceptualization, M.X. and Y.C.; methodology, M.X.; software, M.X.; validation, M.X.; formal analysis, M.X.; investigation, M.X.; resources, Y.C.; data curation, Y.C.; writing—original draft preparation, M.X.; writing—review and editing, M.X.; visualization, M.X.; supervision, Y.C. and Z.D.; project administration, Y.C.; funding acquisition, Y.C. All authors have read and agreed to the published version of the manuscript.

Funding: This research was funded by National Natural Science Foundation of China, grant number 51838003 and the Fundamental Research Funds for the Central Universities, grant number B220201015.

Data Availability Statement: Not applicable.

Conflicts of Interest: The authors declare no conflict of interest.

Appendix A

Table A1. Constants implications of multi-objective optimal combination models for sponge facilities.

Scheme	Constants	Description of Constants	Unit
1	s	Infiltration storage capacity per unit area of facility	m^3
3	r	Water collected per unit area of facility	m^3
4	p	Runoff pollution removal per unit area of facility	t
5	e	Facility cost per unit area of facility	Yuan/m^2

Table A2. Reference infiltration coefficients for different soil types [39].

Materials	Grain Size (mm)	Weight (%)	Permeability Coefficient K (m/s)
Clay	-	-	$<5.7 \times 10^{-8}$
Silty clay	-	-	$5.7 \times 10^{-8} \sim 1.16 \times 10^{-6}$
Powdered soil	-	-	$1.16 \times 10^{-6} \sim 5.79 \times 10^{-6}$
Silt	>0.075	>50	$5.79 \times 10^{-6} \sim 1.16 \times 10^{-5}$
Fine sandy clay	>0.075	>85	$1.16 \times 10^{-5} \sim 5.79 \times 10^{-5}$
Medium sand	>0.25	>50	$5.79 \times 10^{-5} \sim 2.31 \times 10^{-4}$
Homogenised medium sand	-	-	$4.05 \times 10^{-4} \sim 5.79 \times 10^{-4}$
Coarse sand	>0.50	>50	$2.31 \times 10^{-4} \sim 5.79 \times 10^{-4}$
Round gravel	>2.00	>50	$5.79 \times 10^{-4} \sim 1.16 \times 10^{-3}$
Pebbles	>20.0	>50	$1.16 \times 10^{-3} \sim 5.79 \times 10^{-3}$
Slightly fractured rock	-	-	$2.31 \times 10^{-4} \sim 6.94 \times 10^{-4}$
Rocks with many fissures	-	-	$>6.94 \times 10^{-4}$

Table A3. Porosity reference for different soil types.

Type of Soil	K (mm/h)	ψ (mm)	Φ (Fractions)	FC (Fractions)	WP (Fractions)
Sandy Soil	120.4	4.9022	0.437	0.062	0.024
Loamy Sandy Soil	29.972	6.096	0.437	0.105	0.047
Sandy Loamy Soil	10.922	10.9982	0.453	0.190	0.085
Loamy Soil	3.302	8.89	0.463	0.232	0.116
Silty Loamy Soil	6.604	16.9926	0.501	0.284	0.135
Sandy Clay Loam Soil	1.524	21.9964	0.398	0.244	0.136
Clay Loamy Soil	1.016	21.0058	0.464	0.310	0.187
Chalky Clay Loam Soil	1.016	27.0002	0.471	0.342	0.210
Sandy Clay Soil	0.508	24.003	0.430	0.321	0.221
Chalky Clay Soil	0.508	29.0068	0.479	0.371	0.251
Clay Soil	0.254	32.004	0.475	0.378	0.265

* K is saturated hydraulic conductivity, mm/h, ψ is flat suction head, mm, φ is porosity, fraction, FC is water yield capacity, fraction, WP is withering point, fraction [40].

Table A4. Reference for runoff pollution removal capacity of different sponge facilities.

Name of coefficient	Horizontal green space	Horizontal Green Space with Water Storage Modules	Water-storing Sunken Green Space	Permeable Hard Surface	Designed Water Body for Water Storage	Green Roof
K	65	85	60	85	0	75
P	Annual average SS of stormwater runoff from urban areas					

* K is the removal rate of SS from runoff by different sponge facilities (%), P is annual average SS of stormwater runoff from urban areas [41].

Table A5. Reference unit cost for different types of sponge facilities.

Serial Number	Types of Sponge Facilities	Rource of Values	Unit Area Cost (yuan/m^2)	Notes
1	Horizontal green space	*Technical Guide for Sponge City Construction*	30–50	
2	Horizontal green space with water storage modules	Prices in Nanjing	800–1200	Depth of 0.8–1 m
3	Water-storing sunken green space	*Technical Guide for Sponge City Construction*	40–50	Average depth 100–200 mm
4	Permeable hard surface	*Technical Guide for Sponge City Construction*	60–200	
5	Designed water body for water storage	Prices in Nanjing	80	Average depth 1 m
6	Green roof	*Technical Guide for Sponge City Construction*	100–300	

References

1. Yang, D.; Yang, Y.; Xia, J. Hydrological cycle and water resources in a changing world: A review. *Geogr. Sustain.* **2021**, *2*, 115–122. [CrossRef]
2. Zhang, J.; Song, X.; Wang, G.; He, R.; Wang, X. Development and challenges of urban hydrology in a changing environment: I: Hydrological response to urbanization. *Adv. Water Sci.* **2014**, *25*, 594–605.
3. Liu, J.; Shao, W.; Xiang, C.; Mei, C.; Li, Z. Uncertainties of urban flood modeling: Influence of parameters for different underlying surfaces. *Environ. Res.* **2019**, *182*, 108929. [CrossRef] [PubMed]
4. Cheng, Y.N.; Xie, M.K. Being both Opposite and Complementary: Urban Road Sponge System Practice Based on Digital Technology—Taking Nanjing Tianbao Street Ecological Road as the Example. *Chin. Landscape Archit.* **2017**, *33*, 5–13.
5. Hamidi, A.; Ramavandi, B.; Sorial, G.A. Sponge City—An emerging concept in sustainable water resource management: A scientometric analysis. *Resour. Environ. Sustain.* **2021**, *5*, 100028. [CrossRef]
6. Yin, D.; Xu, C.; Jia, H.; Yang, Y.; Sun, C.; Wang, Q.; Liu, S. Sponge City Practices in China: From Pilot Exploration to Systemic Demonstration. *Water* **2022**, *14*, 1531. [CrossRef]
7. Gao, J.; Li, J.; Li, Y.; Xia, J.; Lv, P. A Distribution Optimization Method of Typical LID Facilities for Sponge City Construction. *Ecohydrol. Hydrobiol.* **2020**, *21*, 13–22. [CrossRef]
8. Li, Q.; Wang, F.; Yu, Y.; Huang, Z.; Li, M.; Guan, Y. Comprehensive performance evaluation of LID practices for the sponge city construction: A case study in Guangxi, China. *J. Environ. Manag.* **2019**, *231*, 10–20. [CrossRef]
9. Zhang, K.; Chui, T.F.M. A comprehensive review of spatial allocation of LID-BMP-GI practices: Strategies and optimization tools. *Sci. Total Environ.* **2017**, *621*, 915–929. [CrossRef]
10. George's County, Maryland Department of Environmental Resources Programs and Planning Division. *Low Impact Development Design Strategies: An Integrated Design Approach*; U.S. Environmental Protection Agency: Washington, DC, USA, 1999.
11. Brown, R.A.; Line, D.E.; Hunt, W.F. LID Treatment Train: Pervious Concrete with Subsurface Storage in Series with Bioretention and Care with Seasonal High Water Tables. *J. Environ. Eng.* **2012**, *138*, 689–697. [CrossRef]
12. Jia, H.; Wang, X.; Ti, C.; Zhai, Y.; Field, R.; Tafuri, A.N.; Cai, H.; Yu, S.L. Field monitoring of a LID-BMP treatment train system in China. *Environ. Monit. Assess.* **2015**, *187*, 373. [CrossRef] [PubMed]
13. Xu, C.; Hong, J.; Jia, H.; Liang, S.; Xu, T. Life cycle environmental and economic assessment of a LID-BMP treatment train system: A case study in China. *J. Clean. Prod.* **2017**, *149*, 227–237. [CrossRef]
14. Lee, J.G.; Selvakumar, A.; Alvi, K.; Riverson, J.; Zhen, J.X.; Shoemaker, L.; Lai, F.-H. A watershed-scale design optimization model for stormwater best management practices. *Environ. Modell. Software* **2012**, *37*, 6–18. [CrossRef]
15. Xu, T.; Jia, H.; Wang, Z.; Mao, X.; Xu, C. SWMM-based methodology for block-scale LID-BMPs planning based on site-scale multi-objective optimization: A case study in Tianjin. *Front. Environ. Sci. Eng.* **2017**, *11*, 48–59. [CrossRef]
16. Zhang, Z.; Gu, J.; Zhang, G.; Ma, W.; Zhao, L.; Ning, P.; Shen, J. Design of urban runoff pollution control based on the Sponge City concept in a large-scale high-plateau mountainous watershed: A case study in Yunnan, China. *J. Water Clim. Chang.* **2021**, *12*, 201–222. [CrossRef]
17. Jiang, C.; Li, J.; Gao, J.; Lv, P.; Yao, Y.; Li, H. Advances in research of optimal rainwater infrastructure configuration in sponge city construction. *J. Hydroelectr. Eng.* **2021**, *40*, 19–29.
18. Sun, H.; Li, L.; Tian, Y.; Zhang, T.; Zuo, W.; Cai, G.; Zhang, F. Sponge city planning and design based on multi-objective optimization and comprehensive evaluation. *Acta Sci. Circumstantiae.* **2020**, *40*, 3605–3614.
19. She, L.; Wei, M.; You, X.-Y. Multi-objective Layout Optimization for Sponge City by Annealing Algorithm and Its Environmental Benefits Analysis. *Sustain. Cities Soc.* **2021**, *66*, 102706. [CrossRef]
20. Hou, J.; Zhu, M.; Wang, Y.; Sun, S. Optimal spatial priority scheme of urban LID-BMPs under different investment periods. *Landsc. Urban Plan.* **2020**, *202*, 103858. [CrossRef]

21. Liu, Z.; Xu, C.; Xu, T.; Jia, H.; Zhang, X.; Chen, Z.; Yin, D. Integrating socioecological indexes in multiobjective intelligent optimization of green-grey coupled infrastructures. *Resour. Conserv. Recycl.* **2021**, *174*, 105801. [CrossRef]
22. Yu, Y.; Zhou, Y.; Guo, Z.; van Duin, B.; Zhang, W. A new LID spatial allocation optimization system at neighborhood scale: Integrated SWMM with PICEA-g using MATLAB as the platform. *Sci. Total Environ.* **2022**, *831*, 154843. [CrossRef] [PubMed]
23. Zhang, G.; Hamlett, J.M.; Reed, P.; Yong, T. Multi-Objective Optimization of Low Impact Development Designs in an Urbanizing Watershed. *Open J. Optim.* **2013**, *2*, 95–108. [CrossRef]
24. Mytilinou, V.; Kolios, A.J. A multi-objective optimization approach applied to offshore wind farm location selection. *J. Ocean Eng. Mar. Energy* **2017**, *3*, 265–284. [CrossRef]
25. Branke, J.; Deb, K.; Miettinen, K.; Sowiński, R. *Multiobjective Optimization: Interactive and Evolutionary Approaches*; Branke, J., Ed.; Springer: Berlin, Germany, 2008. [CrossRef]
26. Deb, K. *Multi-Objective Optimization Using Evolutionary Algorithms*; Ross, S., Weber, R., Eds.; John Wiley & Sons: New York, NY, USA, 2001; p. 15.
27. Beijing University of Civil Engineering and Architecture. *Technical Guide for Sponge City Construction*; The Ministry of Housing and Urban-Rural Development; China Architecture & Building Press: Beijing, China, 2015; pp. 49–53.
28. Eiben, A.E.; Smith, J.E. Multiobjective Evolutionary Algorithms. In *Introduction to Evolutionary Computing*; Natural Computing Series; Springer: Berlin, Germany, 2015. [CrossRef]
29. Deb, K.; Pratap, A.; Agarwal, S.; Meyarivan, T.A.M.T. A fast and elitist multiobjective genetic algorithm: NSGA-II. *IEEE Trans. Evol. Comput.* **2002**, *6*, 182–197. [CrossRef]
30. Zitzler, E.; Laumanns, M.; Thiele, L. SPEA2: Improving the strength pareto evolutionary algorithm. *Tech. Rep. Gloriastrasse.* **2001**, *103*, 95–100.
31. Roudsari, M.S.; Pak, M. Ladybug: A parametric environmental plugin for grasshopper to help designers create an environmentally-conscious design. In Proceedings of the BS 2013: 13th Conference of the International Building Performance Simulation Association, Chambéry, France, 26–28 August 2013.
32. Shan, R. In Integrating Genetic Algorithm with Rhinoceros and Grasshopper in Whole Building Energy Simulation. In Proceedings of the Grand Renewable Energy 2014, Tokyo, Japan, 27 July–1 August 2014.
33. Shi, X.; Yang, W. Performance-driven architectural design and optimization technique from a perspective of architects. *Autom. Constr.* **2013**, *32*, 125–135. [CrossRef]
34. Cubukcuoglu, C.; Nourian, P.; Sariyildiz, S.; Tasgetiren, M.F. A discrete event simulation procedure for validating programs of requirements: The case of hospital space planning. *SoftwareX* **2020**, *12*, 100539. [CrossRef]
35. Bahdad, A.A.S.; Fadzil, S.F.S.; Onubi, H.O.; BenLasod, S.A. Sensitivity analysis linked to multi-objective optimization for adjustments of light-shelves design parameters in response to visual comfort and thermal energy performance. *J. Build. Eng.* **2021**, *44*, 102996. [CrossRef]
36. Lakhdari, K.; Sriti, L.; Painter, B. Parametric optimization of daylight, thermal and energy performance of middle school classrooms, case of hot and dry regions. *Build. Environ.* **2021**, *204*, 108173. [CrossRef]
37. Toutou, A.; Fikry, M.; Mohamed, W. The parametric based optimization framework daylighting and energy performance in residential buildings in hot arid zone. *Alexandria Eng. J.* **2018**, *57*, 3595–3608. [CrossRef]
38. Liu, Z.; Yang, Y.; Hou, J.; Jia, H. Decision-Making Framework for GI Layout Considering Site Suitability and Weighted Multi-Function Effectiveness: A Case Study in Beijing Sub-Center. *Water* **2022**, *14*, 1765. [CrossRef]
39. Ministry of Housing and Urban-Rural Construction of the People's Republic of China; General Administration of Quality Supervision, Inspection and Quarantine of the People's Republic of China. *Technical Code for Rainwater Management and Utilization of Building and Sub-District*, 1st ed.; China Architecture & Building Press: Beijing, China, 2017; pp. 8–9.
40. Rawls, W.J.; Brakensiek, D.L.; Miller, N. Green-ampt Infiltration Parameters from Soils Data. *J. Hyd. Engr.* **1983**, *109*, 62–70. [CrossRef]
41. Ministry of Housing and Urban-Rural Construction of the People's Republic of China. *Technical Guide for Sponge City Construction: Technical Guide for Sponge City Construction*, 1st ed.; China Architecture & Building Press: Beijing, China, 2014; p. 46.

water

MDPI

Article

Comparison of the Transition to More Sustainable Stormwater Management in China and the USA

Yitong Zhao *, Mackay Price and Sam Trowsdale

Department of Environmental Science, University of Auckland, Auckland 1010, New Zealand; wpri344@aucklanduni.ac.nz (M.P.); s.trowsdale@auckland.ac.nz (S.T.)
* Correspondence: yzhb454@aucklanduni.ac.nz; Tel.: +64-273-094-241

Abstract: This paper presents a comparative cross-nation study of the transition to more sustainable stormwater management (SSWM) in the United States and China. Multi-level perspective and multiphase models are used to examine the transition dynamics and reflect on how transition theory explains the change within federal and socialist context. Instead of simply differentiating the two countries' transition patterns by using terms such as bottom-up or top-down, we consider the importance of changes at all three levels of the system. The main difference between the transition process in the United States and China is the extent to which niche level innovations are developed, especially in the type of actors and activities investigated. The analysis suggests that the Chinese transition is less radical, while the U.S. pathway exhibits signs of reconfiguration, dealignment and realignment. Developing learning networks across sectors and actors to spread knowledge and experience appears to be the next major challenge for the Chinese Sponge City initiative. Despite the feasibility of transition theory for transition comparison, the author suggests its usage with caution and critical reflection to avoid the risk of embedding the mindset of 'catch-up' and convergence.

Keywords: stormwater management; transition; multi-level perspective

Citation: Zhao, Y.; Price, M.; Trowsdale, S. Comparison of the Transition to More Sustainable Stormwater Management in China and the USA. *Water* **2022**, *14*, 1960. https://doi.org/10.3390/w14121960

Academic Editor: Sajjad Ahmad

Received: 26 May 2022
Accepted: 17 June 2022
Published: 19 June 2022

1. Introduction

Urban stormwater management is traditionally based on the principles of a command-and-control technocracy whereby water is transported away from a city as quickly as possible using built infrastructure to avoid flooding. Building on the sustainability debate that came to the fore in the 1990's, there was a widespread aspiration to shift to more sustainable stormwater solutions. These took off at the turn of the 21st Century with strategies such as Water Sensitive Urban Design (WSUD), Sustainable Urban Drainage Systems (SuDs), and Low Impact Development (LID) and gained global traction as the world urbanized and looked to do so in an environmentally friendly way [1]. In a similar way, the Sponge City Initiative (SCI) was proposed by the Chinese government in 2012, envisioning cities with the capacity to infiltrate, drain and filter stormwater freely, improving resilience to floods, droughts, and contamination, hooking into some of the major problems in the contemporary environmental crisis discourse [2].

The SCI is a national strategy to achieve the goals of new-type urbanization (xinxingchengzhenhua) and the harmonious development of human and nature [3]. SCI calls for more integrated urban water management rooted in the physical and hydrosocial water cycle [4,5] and a planning and design strategy for sustainable urban development that promotes rainwater systems integrated with strategic ecosystem conservation and restoration or remediation [6]. Its implementation requires a comprehensive integration of multiple aspects from policies, designs, to social communication and other subsystems [7] and in many ways contrasts to the techno-centric six-word principles (of the sponge city) that have been frequently adopted by the government, academy and industry (infiltration, detention, storage, purification, usage, and drainage).

Perhaps the most familiar change to stormwater management in the SCI is the encouragement of LID as an alternative approach (albeit largely technical) to hard-engineering stormwater systems, primarily for flood mitigation and pollution control. Borrowing the terminology from the U.S., LID refers to on-site natural and engineered infiltration and storage techniques such as pervious paving, rain-gardens, and greenroofs. However, widespread implementation of new approaches is not easy because established systems tend to privilege previous technologies and practice, known as path dependency.

Addressing such persistent societal problems (often environmental) is a central topic of socio-technical transition research. From a socio-technical perspective, sustainability transitions cannot be achieved through 'green' technology alone but requires a broader systemic societal change, whereas technical evolution can be used as an entrance point for studying societal changes [8,9]. Sustainable stormwater management (SSWM) transition, in-line with other subjects, refers to a long-term, purposeful, multi-dimensional, and fundamental transformative process that involves technological, social, institutional, and economic changes in the stormwater management system [10,11].

The multi-level perspective (MLP) serves as the core analytical framework for studying socio-technical transition dynamics. It views transitions as a non-linear process that results from the interplay of developments at three functional levels: landscape, regime, and niche [12]. Landscape usually refers to various external trends (for example, demography, macroeconomy, political culture, societal concerns) and shocks (for example, wars, crisis, accidents) that affect transition. Niches are protective spaces where innovative activity can take place. The regime corresponds to the incumbent system, encompassing the dominant rules and practices that guide activities in particular directions. Based upon MLP, different transition pathways can be outlined, for example according to the timing and the nature of the three levels' interaction [13]. Historical transition pathways that trace the emergence of sustainable regimes could inspire ideas of transition management [14], meaning that sustainability transitions could be facilitated by purposive intervention (governance).

Reflecting its origins within European-capitalist social contexts, socio-technical transition theories have been applied to examine numerous western developed cases from energy, agro-food, transportation, and the water sector. Several researchers have used a socio-technical approach to provide insights on 'what characterizes and shapes the SSWM transition'. Some focus on the competition processes between technology choices. For example, four barriers that affect the adoption of alternative techniques in French stormwater management (difficulty in behavior adaptation, unclear advantages over traditional sewerage method, legal constraints, and the reluctant manufacturers) were examined using transition theory [15]. Some highlight the role of actors, for example, the importance of actor-networks launched by a small group of actors across various sectors and the networked bridging organizations throughout the transition of Melbourne stormwater quality management [16]. Others have explored the timing and types of actions. Ref. [17] focused on the link between technical attributes, governance and transition stage within a Brussels case study, concluding that soft actions such as manuals and legislation prevailed in the early stage of transition, while decentralized processes and collaboration between formal and informal networks were important for the diffusion of later actions. The development of transition pathways allows comparison between transitions with different contexts. For example, the comparison of WSUD development of Netherlands and Australia concluded that both countries have similar transformation pathways followed by dealignment and realignment [18].

In China, the application of socio-technical theories is mainly limited to the energy and transport fields. A handful of exceptions in the water management sector exist: Ref. [19] used the MLP to understand the dynamics and forces that can induce a leapfrogging development of wastewater treatment; the endogenous (regional) and exogenous (international) innovation processes in early transition to on-site water recycling in Chinese cities [20,21] focus on the water regime transition of the social-ecological system at the county level, whereas the transition process is deemed to be driven by landscape pressure and internal instability within the regime but ignoring the effect of niche innovations. To the best of our knowledge, there is no such published literature that specifically focuses on the SSWM transition process although numerous studies have acknowledged that the Chinese urban water and stormwater management is experiencing transition through the implementation of SCI [22–27]. Most of these studies focused on recent policy changes and practices or identified opportunities and constraints of taking up the new initiative in a general approach.

It is still unclear to what extent western-centric transition theories describe the transition being experienced in China and, if they do, what does it tell us about the stormwater transition known as the SCI? There is a need to assess the cross-cultural robustness of transition theories as transitions differ by nature, scope, and driving forces [28]. Moreover, how is the case in China different from that of other countries, and importantly, does this lead to unique challenges?

This article applies a socio-technical analytical approach to historically reconstruct stormwater management trajectories in the United States and China. The U.S. is selected because it is well-studied in terms of SSWM and is the origin of LID. It is also arguably a major international source of experience for the SCI. These two countries have similar scales of economies, vast territories, and various climates. They also have distinct political, legal, organizational, and cultural characteristics and history. Therefore, the comparison of their stormwater transition process in this study aims to:

1. identify the current stage of the two countries' SSWM transition;
2. identify unique opportunities and challenges for the Chinese SCI;
3. reflect on the feasibility of applying socio-technical transition theory to compare transitions in federal and socialist societies.

2. Rational and Approach

This study begins with an overview of the SSWM transition stories of the two countries. A descriptive, country-based approach is adopted. The historical analysis is performed at the country-scale to describe the nationwide transition under different social systems. Drawing on various literature data sources, including policies, regulations, guidance documents, history articles, media reports, and peer-reviewed journal articles, the transition stories are told from the rise of social awareness around the need to modify modern reticulated systems to address growing environmental and social issues. For the U.S., post-World War II is considered the first key node of the transition, when rapid urban expansion and booming industrial sectors greatly increased domestic and industrial effluents, intensified drainage pressure, and resulted in serious water pollution. For China, the timeline was drawn after the founding of the People's Republic of China (1949) when the recovery of urban drainage systems began, alongside a renewed focus on cities and industry. Since it is unrealistic to involve every issue that is relevant to water and environment, only those considered important for stormwater transition are discussed.

A socio-technical transition study using MLP usually needs to assign different elements to the three levels (landscape, regime, and niche). In this study, the regime refers to the stormwater management paradigm at the national level. Examples of elements may include dominant stormwater practices, relevant policy, legislation (rules), guidelines, institutional arrangements, etc. The niche level mainly includes local practice (for example, local innovative policy, local legislation, social activities), and technological innovations. The landscape factors can be represented by slow-changing trends such as urbanization, population growth and mobility, climate change, macroeconomics, macropolitics, social values, culture patterns, and disruptive shocks such as environmental crisis, war, and disasters. For ease of interpretation, the timelines of major changes at the three levels (not fixed) are illustrated in Figures 1 and 2. While it should be noted that these elements are not bound to their level but may evolve to other levels, for example, the local social activities may scale up to large-scale social movements which could be regarded as a landscape pressure to the regime. The boundary between may not be very distinct, for example, the policy and market can sometimes be regarded as part of the regime and other times as landscape factors. In a word, the assignment of elements is context specific. Therefore, in the text we are not going to separately describe the development at each level (in the way that some transition studies have) but we will examine them in the specific context.

	Major landscape change	**Tangible regime response**	**Key niche activities**
Predevelopment	Post war (1945–) prosperity (industry boom, suburbanization, population growth)	The Federal Water Pollution Control Act 1948; Large–scale hydro–engineering construction (since 1930s)	Discontent among public towards environmental degradation; states' actions on quality protection
	Environmental movement enlarged from niches (since 1950s)	The Water Quality Act 1965, setup of the Federal Water Pollution Control Administration (1965–1970)	Influential publications (e.g., *Silent Spring*), setup of environmental organizations, public protests
	Fire disaster on the Cuyahoga River (1969)	Setup of EPA (1970); The Clean Water Act (1972)	
Take–off	UN Conference on the Human Environment(1972)	The Endangered Species Act 1973	Growing public concern on stormwater ecological impact
		National Urban Runoff Program (1979–1983); Amendments to the CWA (1987)	First usage of the term 'LID' in land use planning (1977); first state rule requiring treatment of stormwater, Florida (1982); statewide infiltration standards, Maryland (1984)
	Earth Summit 1992 (for climate change and sustainable development	Phase I stormwater regulations (1990); Regulations of TMDL	Dam removal movement gradually began (since 1990s)
		Phase II stormwater regulations (1999); release of LID manual for national use (1999); LID literature review by EPA (2000)	First LID manual, Maryland (1997)
Acceleration	Failure of the structural flood defence, New Orleans (2009)	Memo: Use of GI in NPDES Permits and Enforcement (2007)	Green Alley Programs, Chicago (2005); calls from scientists and practitioners for consistent, national rule (2009)
		EPA released questionnaires on stormwater practices (since 2010s); revision of MS4 rules (2014); setup of National Municipal Stormwater Alliance (2016)	

Figure 1. The transition trajectory of the U.S. stormwater management.

	Major landscape change	**Tangible regime response**	**Key niche innovations**
	Planned economy, *danwei* system	Urban drainage system recovery (since 1950s, suspended from 1958 and restart in 1980s); Large-scale hydro-engineering	
	Guanting reservoir pollution (1972); UN Conference on the Human Environment (1972) Reform and opening up (since 1978)	The Water Pollution Prevention and Control Law 1984; The Environmental Protection Law 1989	
Predevelopment	Earth Summit 1992 (for climate change and sustainable development discourse); Pollution incident of Huai River (1994)	The Ninth Five-Year Plan for water pollution prevention and control (since 1996)	Government-civil environment activities (e.g., China Environmental Protection Century trip) Sponge city concept for flood control was first mentioned (2003); local institutional innovation of River Chief
	Drinking water crisis in Southern China (2007)	Regional and Watershed Restriction Approvals (2007); Three Red Lines (2011)	Stormwater research burgeoned; First LID demonstration site, Shenzhen (planning since 2011); sponge-like city was proposed on the 2011 Two Sessions
	Beijing 712 storm event (2012); the strategic decision of ecological civilization construction (2012)	The strategic decision on the construction of ecological civilization (2012); Sponge city construction was requested by President Xi (2013); Outline of the Integrated Plan for Urban Drainage (Rainfall) and Flood control (2013)	
Take-off	Conferred legislative power on cities (2015); special provision on public participation in environmental protection (2015)	National technical guidance for sponge city construction (2014); selection of sponge city pilots (2015, 2016); The Strategy of Yangtze River Protection (2016); Assessment Standard for sponge city construction effect (2018)	China Economic Weekly report on urban flood issues of sponge city pilots (2016); public doubts over SCI implementation; Shutdown of dams and hydropower stations, Zhangjiajie (since 2019)

Figure 2. The transition trajectory of the Chinese stormwater management.

Based upon the historical timeline, the transition phases and pathways of the two countries are analyzed. The multi-phase model is often used to identify different transition stages. Four phases were identified: predevelopment, where the regime is relatively stable while some niches may emerge under landscape pressure; take-off phase, where change starts to build up and the regime begins to shift; acceleration phase, in which structural changes occur in a visible way; and stabilization, when the new system reaches equilibrium again [29]. These stages have general descriptions in some of literature [13,29–31], making comparisons requires indicators developed specifically for SSWM transition to make the evaluation criteria consistent (Table 1). The indicators for entering stabilization are somewhat open to debate since the implications for SSWM or sustainability are subjective and dynamic, while such uncertainty does not affect the comparison in this study as transitions in both countries are considered have a long way to reach stabilization (as discussed below).

Table 1. Indicators of transition phases.

Transition Phase	Indicators in This Study
Predevelopment	• The stormwater management regime is centered on drainage issues. End-of-pipe and hard engineering logic is dominant. • Niches may emerge by the introduction of innovative technologies, local practices and activities.
Take-off	• The regime incorporates some of the innovative niches to reorient its stormwater practices. • Broader objectives of stormwater management come into sight.
Acceleration	• Varied actors and stakeholders are involved and a learning and supporting network is shaping between them. • Innovative knowledge and experience is well spread. • The new stormwater regime starts to provide comprehensive benefits across socio-cultural, economic, and ecological sectors beyond stormwater quantity and quality issues.
Stabilization	Open to debate

The typology given by [13] is a common reference tool to identify transition pathways: transformation, reconfiguration, technological substitution, and de/realignment. Under a transformation pathway, underdeveloped niche innovations fail to break through to wider levels whereby actors gradually modify the direction of development trajectories and innovation activities. In reconfiguration, niche innovations are well developed and are being incorporated to trigger subsequent adjustments and change to the regime's basic architecture. During technological substitution, niche innovations have developed to replace the regime; and under de/realignment, landscape pressure creates space within the regime whereby niche innovations co-exist and compete for extended periods until one of them replaces the regime.

3. Stormwater Management Trajectories

3.1. United States

3.1.1. Predevelopment: Suburbanization and Raising Discontent of Water Pollution

Before WWII, industrialization and urbanization had progressed in the U.S. for more than a century (if regarding the textile industry pioneered in 1790s as the beginning). The new round of urban expansion set off by the post-war prosperity expressed itself differently: revitalized factories relocated to rural areas to reduce costs. Nuclear families (a family group consisting of parents and their children), promoted as the ideal family structure to stimulate the economy, were encouraged to move to suburbs as the proliferation of automobiles enabled further travel distant. As a result, farmland, forest, and undeveloped green space were converted to transportation infrastructure and estates, which contributed to increased surface runoff and pollution.

These land use changes were reflected in the deterioration of water bodies, strengthened the landscape pressure while on the other hand provided unique niches for raising public environmental awareness. When people moved to the suburbs, interests in outdoor recreational activities that closely related to nature grew as one of the romantic fantasies of nuclear family life. Ironically, environmental degradation quickly dispelled such fantasies. Suburban dwellers, mostly upper-middle class individuals, later became the major force of the national environmental organizations and actively participated the massive environmental movement from 1950s to 1970s [32]. The 1950s and 1960s saw a series of influential publications (for example, Silent Spring), the setup of nonprofit organizations (for example, Nature Conservancy), and public protests (for example, against construction of the Echo Park Dam) on environmental protection. Some local officials and communities which faced with significant pollution hazards began to take active measures such as establishing local water boards and financing clean-up programs. It should be noted, however, that this was built on the earlier efforts of conservation movement (1890–1920), the prevalence of preservationism and wilderness since the middle of the 19th century. In other words, it took a rather long process for public perceptions towards their relationship with nature and the environment to realize such transition.

The stormwater regime at the time still focused on drainage and sanitation issues and retained a strong end-of-pipe, hard engineering logic. There was a dominant and pervading belief in science and technology, which made engineers the core urban solver [33]. The construction of large-scale hydro-engineering projects, especially dams since the Great Depression (1930s), went further; these were driven by the demand of economic stimulation and extended political influence [34]. The Federal Water Pollution Control Act 1948 (FWPCA) firstly required a cooperation between federal and state entities to address declining water quality. However, the Act only 'encouraged' pollution control without federal supervision [35]. Federal efforts mainly limited in assisting the construction of large-scale grey infrastructure such as sewers and treatment plants [36]. States retained primary power and responsibility of water pollution management. By 1966, all states had passed some type of water pollution legislation, but enforcement varied greatly [37]. For many state governments, the priority of economic development was still unassailable. Consequently, the federal encouragement and local clean-up efforts could do little to regulate rivers and streams especially those issued with industrial permits under state legislation.

3.1.2. Take-Off: Ecology Concern and Stormwater Quality Legislation

The 1969 fire disaster on the Cuyahoga River, Cleveland, which was caused by accumulated industrial pollution from storm water overflows and raw sewage discharges, was a notable shock to the regime, triggering widespread distrust of the ability of state and local governments to effectively manage water quality [38]. It is generally believed that this incident directly led to the establishment of the Environmental Protection Agency (EPA) in 1970 and the revision of FWPCA in 1972 (the Clean Water Act, CWA) [39]. While the 1969 Cuyahoga event was not the first and even not the most serious river-fire disaster, the significant regime change was considered more deeply promoted by the relatively fully developed niche activities and the already destabilized regime.

The 1972 FWPCA explicitly separated point and non-point source pollution, the former became subject to the National Pollutant Discharge Elimination System (NPDES) permit, however, non-point source pollution was still not specifically addressed. Pushed by the prevalent environmental activism that aimed to protect the nation's eco-heritage from extinction, the Endangered Species Act of 1973 was passed. The Act offered a new window for bringing stormwater quality into the regime as bio-indices for quality evaluation were introduced, whereby a higher standard of water quality was required for habitat reservation [40]. Under this Act, a growing number of biologists and environmental groups expressed their concern over the impact of stormwater on ecosystem health. In 1979, EPA launched the National Urban Runoff Program (NURP) to investigate the impact of stormwater on species' habitats, whereby its final report formally determined stormwater as a pollution source. The report highlighted the effectiveness of detention basins, retention ponds, and wetlands on capturing contaminants from stormwater runoff. These findings in turn spurred increasing interests on nature-based solutions to stormwater management. During the same period, several states and municipalities started formulating stormwater ordinances demanding on-site storage and detention (for example, Florida, Pennsylvania). Maryland took the lead by creating a statewide infiltration program during the mid 1980s, which became the origin of LID.

Building upon the results of NURP, the next decade saw significant changes in the U.S. stormwater regime. The amendments of CWA in 1987, section 402(p) in particular, brought stormwater under the NPDES permit (National Research Council 2009). States were required to develop and implement nonpoint pollution management programs. Combined sewer overflow (CSO) was identified as a point source, whereby its control was guided under the National Combined Sewer Overflow Control Strategy in 1989 and its revised version in 1994. In response to the revised CWA, the Municipal Separate Storm Sewer System (MS4) program was launched in two phases (Phase I, 1990-; Phase II, 1999-) to require a NPDES permit for stormwater discharge and submission of a stormwater management or pollution prevention plan.

Spaces for niches to grow were also intentionally created through these regime changes. In 1992, the Total Maximum Daily Load (TMDL) limit describing the maximum pollutant that a water body can receive was incorporated into CWA. Most states were responsible for developing TMDLs and submitting them to EPA for approval. For water that did not meet the minimum criteria, cleanup plans that were essentially watershed-based were required. This promoted the establishment of collaborative watershed groups (for example, partners of Chesapeake Bay program). Moreover, public participation was mandated in specific provisions: the Phase II rule defined six elements as minimum control measures that contributed to successful stormwater programs: public education and outreach, public involvement/participation, illicit discharge detection and elimination, construction site runoff control, post construction runoff control, and pollution prevention/good housekeeping [41].

3.1.3. Nowadays: Exploration of Multiple Benefits from Stormwater Management

While the MS4 programs did not strictly require the implementation of specific runoff control measures, it did create momentum for LID adoption. In order to help local governments and developers comply with stormwater legislation and regulations, the EPA along with organizations such as the Natural Resources Defense Council and the LID center made a concerted effort to spread the stormwater experience of Maryland. The diffusion of LID was accelerated via the release of manuals that guided its national application [42], case reports containing evidence of its cost-effectiveness [43,44], the establishment of websites that assembled policies and training resources of sustainable stormwater issues (for example, http://water.epa.gov (accessed on 12 June 2020)), and databases that offered precedent design and performance of stormwater devices (for example, the Best Management Practices database). LID for flood mitigation was also emphasized, especially after the flood defense failure in New Orleans in 2005, which again highlighted the deficiency of hard engineering flood control approaches and the necessity for transition [45]. The dismantling of dams since the end of 1990s also signaled the erosion of the past hard-engineering regime.

Green infrastructure (GI), a concept that originated from landscape planning, shares similar logic with LID in mimicking natural hydrological processes. The popular Green Alley programs in several U.S. cities (for example, Chicago and Los Angeles), although initially stormwater-focused, were acclaimed to provide ecological and cultural services [46]. A memo released by EPA and the Energy Independence and Security Act of 2007 endorsed GI as a wet weather infrastructure solution [47]. Federal agencies were required to reduce stormwater runoff via LID and GI from federal projects [48].

Despite LID having been increasingly used in the U.S. since the 1990s and generally positive feedback has been received from previous practice, LID adoption and implementation at the national level is not as expected [48–51]. The enforcement and outcomes of stormwater programs are also highly variable [52]. Multiple reasons have been identified such as technical infeasibility, lack of financial support, uncertainty of LID cost-effectiveness, challenges in operation and maintenance, conflict of stormwater regulations and other local ordinances, etc. While some of these challenges might be common to all countries, the large degree of discretion left for states to have their own governing system (metric, review processes, institutions setup, etc.) and legal codes is the most distinct characteristic of the

U.S. system which may be beneficial for local-based decision-making but complicates policy transfer between cities and states, making both tracking compliance to federal regulations difficult. Although there have been strong calls for consistent nation-wide rules [52] and commitments by EPA to enact them, in 2014, the EPA deferred on such actions, insisting on leveraging existing requirements and playing a supporting role in the implementation of local stormwater programs.

The regime actors decided to move to a more cooperative manner instead of a more top-down approach. Since 2010, the EPA started to distribute thousands of questionnaires to MS4s, NPDES permitting authorities, and developers, to collect specific information about their stormwater practices, oversight, and finance issues to inform how could EPA improve its stormwater program [53]. The Nonpoint Source Outreach Toolbox has been continuously updated in recent decades in order to assist state and local agencies and other organizations with educating the public on nonpoint source pollution or stormwater runoff [54]. In 2016, the National Municipal Stormwater Alliance was established to represent MS4 permittees at the national level, providing a unified voice when working with the EPA, states, regional municipalities and other stormwater organizations. In the same year, the EPA issued a revision of the MS4 rules, which resulted in more flexibility for permitting authorities to issue and administer small MS4 permits. A draft guide, toolkit, and technical assistance was released to promote comprehensive, community-wide planning approaches to manage stormwater [55].

3.2. *China*

3.2.1. New China: Restoration of Drainage Infrastructure, Rising of Technocrats, and Restricted Exposure to Nature

Stormwater drainage has a long history in China. The well-known Fushou drainage system built in the Northern Song dynasty (960–1127) includes both combined drainage and storage infrastructures, with one of the sections still functioning today [56]. People in the agricultural society viewed domestic sewage as precious fertilizer which usually would not be discharged into stormwater sewers [57]. However, due to a century of war and drastic social changes since 1840, the Chinese urban economy and infrastructure were badly damaged. Therefore, one of the important tasks of urban planning after the founding of the People's Republic of China was the basic sanitation remediation of major cities through sewer construction and retrofitting, as well as filling up significantly polluted river channels for sanitation and urban development [58]. However, this endeavor was soon halted by the Great Leap period (1958–1960) and the Great Cultural Revolution (1966–1976). Following this period saw the emergence of technocrats as social order returned and the revolutionary veterans stepped down from decision-making positions. Drainage recovery work rapidly resumed under the supervision of these technocrats. Given the lack of funding and mature technology in China at the time, most of the drainage systems were built as combined sewers. The design standards were far below that of the U.S. For example, the recurrence interval of most drainage systems in Chinese cities were usually in a range from 0.5 to 5 years (5 to 10 years in the U.S.).

As the economy steadily grew, China's population rose rapidly from 0.54 billion in 1949 to 0.83 billion in 1970. Following the similar route, increased domestic and industrial effluents led to the degradation of water quality in China's aquatic environments. Meanwhile, technocrats kept enlarging their influence due to the continued to prioritize industrialization and economic development of macro policy. Similar to the U.S. hydro-engineering construction boom since the 1930s, multiple plans for large dams and hydropower stations were implemented for flood control, cheaper energy, and economic stimulation. Waterways were modified or filled for further development. This development paradigm was underpinned by the once-popular political slogan 'rendingshengtian', which means 'man can conquer nature'. The 'antagonistic' mindset together with additional impacts of dam construction (for example, population migration) were pushing people away from water and environment both physically and cognitively.

While compared with the U.S. case which finally developed strong enough bottom force to affect the regime, there was little sign of influential niche activities in China at the time. The highly centralized planned economy and the dominant urban society organization approach, as well as the *danwei* (work unit) system that attach people's residence to their workplace are considered as the two major landscape factors for this lack. The Five-year Plan, developed to set national development goals under the planned economy, was over-reliant on pure administrative means, which greatly inhibited niche-level activities. Its profound influence on subsequent environmental management remained for a long time even after it changed to a socialist market economy in the 1980s. The *danwei* is typically enclosed by walls, which implies security and has a unique social identity for Chinese people [59]. Such values contrast with western desire for outdoors life (recall the nuclear family fantasies). Most social and recreational activities took place in the *danwei*, which restricted the exposure of city dwellers to nature and therefore minimized public concerns towards degraded environmental water quality.

3.2.2. Evolving Water Quality Management: Tightening Regulation and Point Pollution Source Focused

A major perceived shock that opened a window for the water quality regime was the 1972 pollution event of the Guanting Reservoir, Beijing, which killed numerous fish and threatened the health of locals [60]. This event coincided with the first instance of Chinese participation in the 1972 United Nations Conference on the Human Environment, which was the first world conference to make the environment a major issue. In response to this pollution event, a series of local measures were immediately adopted to deal with the reservoir issue, whereas little had been addressed for pollution in other areas due to the near absence of legal and institutional systems of environment protection.

Although the global environmental movement was not as influential in China as it was in the western world, certain worldviews were still assimilated by top leaders, which brought new opportunities for change. The first act on water pollution, the Water Pollution Prevention and Control Law was passed in 1984. The Ninth Five-Year Plan initiated in 1996 aimed to improve the water quality of major rivers and lakes by 2000. The 1990s saw some government-led environmental campaigns (for example, the China Environmental Protection Century demonstration) and the formation of some early environmental groups.

Entering into the 21st century, urbanization in China has already developed to an astonishing scale. The urban population has tripled in the past three decades, driven by factors such as migration policies, rural-urban disparity, and land development for urban use [61]. On the one hand, the centralized managed water system in cities allows people to have access to stable water supply and sanitation. While on the other, such rapid growth has been accompanied by the substantial pollution risk of drinking water resources. The serious eutrophication of major freshwater resources in Southern China, especially the event known as the Wuxi water crisis, served as another catalyst that urged more stringent instruments to be adopted.

In 2007, the Regional and Watershed Restriction Approvals was implemented to limit the number of emerging polluting enterprises from the central government level [62]. The Three Red Lines set for water consumption, water use efficiency, and water environment restoration in 2011 is regarded as the most rigorous regulatory regime for water resources management. Some local institutional innovations became the regime. The River Chief System, first experimented in Changxing county, Zhejiang Province in 2003, was incorporated into law in 2017 to strengthen the responsibility of individual government officials on river health management. These instruments now play a key role in preventing local governments from sacrificing environmental interests, while also consolidating the reliance of top-down administration. However, stormwater was still marginalized at this stage and seldom involved in these point source-focused actions.

3.2.3. Emerging Niches and the Proposal of Sponge City Initiative

As the water quality regime was experiencing incremental transformation, research interest in stormwater sector gradually started to accumulate, which was, to a great extent, attributed to the global flow of relevant knowledge. In 2003, Yu and Li [63], a group of researchers in landscape design who were under the great influence of western landscape perspective proposed the idea of the sponge city and drew an analogy between wetlands and sponges, emphasizing its function in absorbing excess water. They held the view that current planning and development ignored the important role of nature in regulating river flow and flood. Some demonstration sites designed by them were heralded as the predecessor of sponge city experiments (for example, the Qunli wetland park in Harbin). Scientists from Beijing University of Civil Engineering and Architecture led by Che and Li specifically focused on LID utilization for stormwater management [64–66]. The number of studies on runoff quality also grew during this period [67–69].

The Chinese market once offered niches when a number of development projects labeled as 'eco-friendly', 'green', and 'low carbon' cropped up in some cities from 2000s, whereas stormwater retention and water recycling were promoted as one of the 'selling points'. Local governments were supportive because it seemed to be an appealing way to improve city image and attract foreign investment on environmental and ecological grounds. However, without sufficient scientific proof and institutional support, these projects evolved into tools for financial gain rather than environmental protection, with some of these projects being abandoned mid-development (for example, Dongtan Eco-city, Shanghai) [70,71]. A few recognized successful cases do exist, usually developed later around the 2010s in the form of urban planning, such as the early planning of Guangming District in Shenzhen which first introduced the LID concept.

Compared with more invisible stormwater quality, the general public in China is no doubt more familiar with urban flooding issues. Floods in cities across the country had become commonplace. According to People's Daily, 62% of 351 cities in China experienced waterlogging (neilao) from 2008 to 2010 [72]. People started to express their distress over waterlogging through a popular sarcastic joke: 'watch the sea in the city' (chengshikanhai). Under the heat of the complaining, in the 2011 Two Sessions, one of the National People Congress deputies proposed to build 'sponge-like' cities using greenbelts and sloped pavements as a solution to urban flooding. The next year, the Beijing storm disaster on 12 July resulted in at least 79 deaths and about 11.64 billion *yuan* economic loss (approx $1.86 billion USD). These significant casualties and economic losses generated considerable public discontentment towards conventional flood management and lead to growing appeals from scholars to implement stormwater management approaches to improve city resilience to climatic variability. At the 2013 Central Working Conference of Urbanization, President Xi formally put forward the request for nationwide sponge city construction. Although the Beijing storm disaster on July 12th was clearly an important SSWM transition node, it is important to appreciate the other contextual factors which formed the transition, such as the experience from previous niche activities, profession's efforts, changes in public perception, etc.

Urgent demand for flood control is clearly a dominant driver, however, the motivation from ecological sustainability promotion cannot be ignored. Notably, at the end of the same year of the Beijing storm shock, the 18th Communist Party of China National Congress made the strategic decision of 'vigorously promoting the construction of ecological civilization'. Sponge city is regarded as the practice of ecological civilization in urban water management. The Strategy of Yangtze River Protection and the wave of ecological protection in the Yellow River basin established by the central government in 2016 and 2019, respectively, were not only about protecting the river itself, but sought to regulate urban development in cities along the two major rivers. Many projects under the two programs were integrated with local sponge city construction. More comprehensive objectives of sponge city have also been developed in plans and guidance. The Assessment Standard for Sponge City Construction Effect released in 2018 involves requirements for runoff reduction, CSO control, ecology, groundwater, urban heat island, etc. [73]. After reviewing a number of local sponge city plans, the authors concluded that except for the requirement of runoff quantity control, the common controlling indices (kongzhixingzhibiao, in some cities, also referred as mandatory indices) are riparian ecosystem restoration, water surface ratio, and nonpoint source pollution control. The utilization rate of stormwater may also be a controlling index in cities suffering from severe water scarcity due to pollution or drought, for example, in Hefei and Guiyang.

3.2.4. Insufficient Follow-Up Knowledge and Experience Diffusion

Immediately after this conference, a national technical guidance for sponge city construction was published to advise the usage of LID infrastructure to deal with stormwater problems [2]. The Chinese-style pilot projects played a unique role in promoting local stormwater experiments at this early stage. In 2015 and 2016, 30 cities were selected in two batches as sponge city pilots by the ministries of State Council, which implied that they could obtain special financial support and attention on this issue from the central government. All local governments are required to complete the drafting of special plans for sponge cities and devise an integrated plan of urban drainage and flood control [74]. The speed of top-down execution was astonishing (at least on paper), whereby at the end of 2017, almost all major cities have incorporated sponge city in their urban planning documents. The central government also sought to strike a new balance of power and responsibility with local governments. The 2015 amendments to the Legislation Law conferred limited legislative power on all cities, which brought more local wisdom and innovations. For example, Hebi (one of the 30 pilot cities) formulated its first local laws and regulation on the Construction of Circular Economy and Ecological City, which includes a specific chapter for sponge city.

Nevertheless, when the China Economic Weekly [75] reported that 19 of the 30 pilots were still experiencing severe urban flood or waterlogging in rainy seasons, including Beijing, Tianjin, Wuhan, etc., there was widespread skepticism towards the effectiveness of the sponge city solution [76], and criticism among researchers on SCI implementation such as lack of a fully understanding of sponge city concept [77], the trend of 'one model fits all' [78]. The Ministry of Housing and Urban-Rural Development (MHURD) responded by saying that 'sponge cities cannot be built overnight and its construction is still accelerating'. Besides, although SCI was proposed with a mixed of policy goals, the dominant expected effect from sponge city programs was significant improvements in flooding prevention. The timing of the regime shift and its close association with the flood shock created a misleading context that sponge city was only designed for flooding prevention. As a result, public perceptions towards sponge cities can be easily swayed by unsatisfactory outcomes when it rains heavily.

Yet despite these more incidental factors, the institutional arrangements that left little space for public engagement and the lack of cooperative and learning networks might be the root causes of these despairing public perceptions. Although recently the amendment of Environmental Protection Law regulates special provision on public participation in environmental protection [79], which normalized procedures such as public hearings and opinion collection in decision-making, has opened a window for public engagement. So far, citizens are more likely to be passive recipients that are only informed after decisions have been made [80]. In addition, government-led policy propaganda was almost the only source for people to understand the initiative. The failure of central government to diffuse comprehensive knowledge on the implication of sponge cities compounded negative perceptions especially because the public is rarely aware of the benefits of sponge cities in other sectors. SSWM transition can be stymied because these misunderstandings could result in unwillingness to be involved.

3.2.5. Still Firm Incumbent Hard Engineering Approach

LID was not the only component of sponge city projects. In many cities, the construction of large rainwater storage tanks, pumping stations, traditional drainage systems, and deep tunnels projects are listed in local sponge city plans and received substantial investment (for example, Shanghai, Guangzhou, and Wuhan) [77]. They served as a remedy for low-standard drainage systems constructed prior to the sponge city. LID or GI infrastructure is encouraged but not required or regulated. When flood control is emphasized as the major stormwater management goal, it may take a long time for LID to become equally competitive as traditional technologies, unless other innovations or disruptive changes occur. Although recent years also saw the shut-down of some dams and hydropower stations, exemplified by the case in Zhangjiajie, a national park in Hunan province to protect giant salamander habitat, hard engineering approaches to flood management were still held in strong faith at the meso level.

4. Discussion

4.1. Theory Validation

Stormwater management transition in China and the U.S. is in accord with the general multi-level dynamic. At the landscape level, the two countries share similar trends in urban expansion and development, economic development, population growth, climate change, and supranational activities over environmental protection and sustainable development. Flooding and water pollution events play an important role in raising awareness of the necessity of stormwater management transition, as it is usually an underlying social problem that people would not notice until the system fails. These contribute to the destabilization of the incumbent regime and provide momentum for niche activities to grow. Landscape changes do not always lead to positive transitions. For example, compared to the post-war relocation and family structure change in the U.S., the highly centralized production and activity space after liberation in China during the 1950s and 1960s inhibited niche opportunities, especially for the general public. From this perspective, we assert that landscape is a crucial functional level especially in SSWM transition research because its influence ultimately unfolds in the reshaping process of people's perception of urban space and their relationship with the water environment, according to which the transition process can be either hindered or proceeded. This could be the meeting ground between socio-ecological and socio-technical research considering that the former focuses more on the society-nature interaction [81].

In both countries, local experiments on LID or sponge city preceded national actions, although their effect and impact on the regime vary. In the U.S., the driving force from the bottom was prominent due to multiple actors and various niche activities involved in the process, especially social groups and the general public. They served as a strong catalyst of internal niche innovations and then expanded their voice to the macro level, urging the government to make the change. The clear bottom-up characteristics are similar to many western countries that experienced environmental movements from the 1970s. Meanwhile, the regime actions also led to changes on the ground, such as the NURP investigation conducted by EPA that inspired localities stormwater impacts on ecology system and research interests over LID, the MS4 programs that drove innovative management strategies and technologies, the multiple organizations that brought different groups into the conversation and so on.

In the Chinese case, governments and techno-professions were dominant niche actors, whereas the public had little awareness and access to stormwater issues. The incentives of early stormwater experiments were more of an economic gain for developers and local governments compared with improving environmental outcomes. The need for transition was not completely recognized by the regime and neither was the value of local innovations. The opening up of the opportunity window of transition [82], which depends on when the institutional context permits niche innovations to be introduced, was largely delayed until the Beijing flood disaster in 2012. Consequently, the niche development was relatively underdeveloped when the SCI started to be promoted in China compared with LID implementation in the U.S. Lack of additional pressure from grassroots movements made both niche and regime changes less radical. However, these changes were still important in reshaping the institutional context and providing valuable theoretical and practical experience for the proposal of SCI.

Transition is a process of the long-term modulation between the three levels. The significance of the activities at the macro and the micro level should be paid with special attention, rather than just focusing on the resultant regime changes. It may be easy to describe the U.S. and Chinese transition as 'bottom-up', where micro niches cluster and influence the incumbent regime, and 'top-down', where a large and rapid change in landscape leads to regime change. However, these terms should be used with caution because they only describe the surface trigger mechanisms and may in fact hide the idea that the transition is a long-term process with influence from all three levels. There is, of course, a possibility that the transition is the result of being squeezed between top-down and bottom-up influences, which is most likely the case in both countries.

4.2. Transition Pathway

Before the enactment of stormwater quality legislation in the U.S. and the proposal of SCI in China, the water quality regime had experienced gradual, cumulative adjustments from loose regulations to strict effluent discharge restriction, from local affairs to federal or national government supervision, and reorientations of the development path from 'all for economy' to 'ecology matters'. These changes form necessary (but not necessarily sufficient) conditions for transition. Although stormwater management was generally marginalized in urban issues, its concern seems to be only a matter of time as the result of increasing demand for water quality and ecology conditions. However, the length of that time, and the effect of the quality and ecology concern on stormwater transition are different in the two countries.

For the U.S., the provision of ecosystem health is the major driving force of stormwater management transition. The stormwater sector underwent restructuring major legislative changes, followed by the centralization of power in the federal government. The emerging technologies initially adopted to solve locally specific problems were then expanded to a broader geographical range and usage. If the changes in water quality management are considered as moderate pressure to the stormwater regime, then it is fair to say that the transition in the U.S. began with a transformation pathway and followed with observable signs of reconfiguration.

In China, the demand for urban flood control largely drove the proposal of SCI. Admittedly, the Beijing flood disaster is a shock to the regime and the proposal of SCI is a breakthrough of stormwater policies. However, the regime did not experience a shift as radical as that after take-off in the U.S., rather it changed in a more moderate way under the mixture of policy goals and processes which interacted between SCI and other sustainability programs. There are no visible changes to the basic structure of the regime. SCI is still implemented through a state-planning system that is far from complete. Not only does the existing permit system of pollution discharge only target the point source, the urban planning regulations also do not require any order/priority between stormwater discharge and land use [83]. Apart from the duty of municipalities to ensure basic drainage and sanitation conditions (the feature of the incumbent regime), stormwater quality management is essentially non-mandatory.

Differentiated from other technical innovation-based transitions, LID is more of an artificial approach to simulate natural ways of regulating water. There is no doubt that the technical innovations at the niche level are increasingly more developed, at least sufficient to be alternatives for regime actors. However, they are still unable to entirely replace existing drainage systems and may have to co-exist with the previous regime for a long time. The transition itself is therefore not a complete technological substitution in both countries. In addition, the possibility of other particular technologies being fully developed to substitute others should not be excluded.

LID is considered less competitive in China, largely determined by the functional features of this type of technology and the institutional settings that encourage them. While LID is especially useful for reducing pollution from first-flush and mitigating the runoff directly entering into drainage systems, it could be inadequate when dealing with flooding problems in extreme climate events. Compared with the regulatory restrictions on stormwater quality in the U.S. which endow more motivation for LID adoption, the Chinese stormwater management regime is largely centered around quantity. Therefore, traditional stormwater management approaches are still privileged.

4.3. Transition Phases

According to the defined phase border, the predevelopment and take-off phase of the SSWM transition in the U.S. and China can be demarcated around the revision of CWA in 1987 and the sponge city proposal in 2013, respectively. These two actions mark the reorientation of stormwater practices, the beginning of extending stormwater management to more than previous hard-engineering, end-of-pipe, single-purpose drainage. The focus of this section is on whether they have entered into or how close they are to the acceleration phase used in transition theory.

In the U.S., the new regime starts to provide multiple benefits beyond stormwater treatment and there are many indications of a general desire for further changes, either in the promotion of LID or in stormwater governance improvement. Given the experience of local governments, experts, social groups, and the general public in expanding their influence on the other two levels, actors at different levels have relatively strong connections. A support network promoting collective learning is shaping. Under the common effort of the government, research institutions, non-government organizations, industry groups, etc., the adoption of various approaches of knowledge and experience dissemination for LID promotion appears in time, helping smooth the transition from take-off to acceleration. Cross-sector organizations established in recent years bring varied actors together towards the same sustainable goal. The engrained difficulty mainly lies in achieving state/federal balance and policy transfer between different regions to ensure consistent enforcement and outcomes, which remains a core challenge for many environmental governance problems within the federal system.

The Chinese transition history show a relatively limited institutional space for niches to grow. In particular, the general public had less opportunities to participate. A few early social movements to address point source pollution were under government leadership. Although SCI has been successfully proposed in a relatively top-down approach, the negative impacts of lacking multi-stakeholder participation may not become apparent until it hinders the following implementation of the program and the embedding of the emergent stormwater policy. For example, installing LID devices to private properties such as green roofs or rainwater tanks, local facilities maintenance, and moreover, getting financial support from the public could face strong resistance (one suggested funding source is public-private partnership, which proves to be almost failed). There is also a salient vacancy on the follow-up knowledge and experience diffusion after nearly nine-year SCI implementation. Despite several case studies and the release of local guidance, few reviews based on case studies, manuals, or databases that can be influential across varied regions and audience types have been produced. There is a lack of platforms or intermediary organizations that could promote communication, coordination, and cooperation between different institutions, sectors, and regions. This is especially important considering that many cities have little history of relevant knowledge and experience.

This evidence suggests that the SSWM transition in the U.S. is closer to acceleration while that of China is currently at the early phase of take-off. Nevertheless, the Chinese transition history along with its unique socio-institutional system holds its own opportunities. Compared with that in the U.S., the state-planning mechanism may benefit stormwater plans developed with more comprehensive goals and allow the integration of stormwater management and the development of other areas such as economic development. Policy transfer may face fewer obstacles because the basic political, legal, and institutional systems of most regions in China are highly consistent. The Chinese central government has more centralized power and experience in managing areas with geographical and cultural diversity. Local areas have a high compliance with state decisions, whereas the U.S. states may have varied responses to the federal government decisions. However, such an advantage could only be realized with a full recognition of local customs and practical knowledge [84] otherwise there might present the risk of 'one model fits all' and its shortcomings.

5. Conclusions

This study reveals the different features of the transition toward Sustainable Stormwater Management in the U.S. and China. This was conducted in part by applying the multilevel perspective model which showed the requirement for activities at all three levels (landscape, regime, and niche), no matter which social system to which it was applied. The U.S. transition is marked by relatively developed niche activities and strong links between niche and regime and landscape. The civil society serves as a significant actor. In contrast, niches are less developed in China and the governance and politics of the transition do not explicitly incorporate the public as stakeholders. When we considered LID as the focused innovative technology, the U.S. transition pathway shows a more apparent trend of dealignment and realignment. LID is now symbiotic to the regime rather than a substitution in both countries. Based upon the multi-phase model, the U.S. transition is deemed closer to the acceleration phase while the Chinese one is at the early stage of take-off.

The purpose of this study is not to provide a specific answer, but to provide a new way of thinking. The authors believe that the results of the study play a reminder or warning role for stormwater practitioners: the process of seeking the solution to stormwater management problems, such as obvious preference for grey engineering approach, lack of public participation and investment, etc., should not be separated from studying the transition history of the specific social system. Since these problems are formed under the long-standing influence of multi-level elements interaction. Moreover, when brainstorming the more sustainable way, the ideas should not be limited in fields of explicit stormwater engineering area and should attempt to utilize the current landscape and niche characteristics as new niches. Measures that only hit one aspect of one single level are unable to vacillate current regimes.

However, caution and critical reflection are needed when interpreting the comparison since the conclusion that one country is ahead of another may lead to an overly simplistic heuristic that the latecomer needs to facilitate (the same) changes to converge on the earlier adopter. The transition pathway is full of variability, uncertainty, and complexity and is place- and context-based. Essentially, the comparison does not uncover (and in fact, tends to cloak) an important question: is it necessary or possible for a socialist, central-planning society to learn from the route of a capitalist and more decentralized society (and, if so, to what extent and in what way)? It is unclear whether the current transition pathway will lead the U.S. stormwater management toward sustainability. It is also too early to say whether the stormwater quality-led transition with niche-driven radical changes is the best way to approach the environmental crisis in China. From this perspective, applying western-centric transition theory to such a comparison tends to imply that the capitalist transition is the way forward. This may impede countries with a distinct social system (especially eastern, developing, and socialist) from exploring alternative and more environmentally focused development pathways.

Author Contributions: Literature review and historical analysis, Y.Z.; writing—original draft preparation, Y.Z.; writing—review and editing, M.P. and S.T.; supervision, S.T. All authors have read and agreed to the published version of the manuscript.

Funding: This research was funded by China Scholarship Council (201908420296).

Institutional Review Board Statement: Not applicable.

Informed Consent Statement: Not applicable.

Data Availability Statement: Not applicable.

Conflicts of Interest: The authors declare no conflict of interest. The funders had no role in the design of the study; in the collection, analyses, or interpretation of data; in the writing of the manuscript, or in the decision to publish the results.

Abbreviations

SSWM	sustainable stormwater management
LID	Low Impact Development
SCI	Sponge City Initiative
MLP	multi-level perspective
WWII	World War II
FWPCA	Federal Water Pollution Control Act
EPA	Environmental Protection Agency
NPDES	National Pollutant Discharge Elimination System
NURP	National Urban Runoff Program
CSO	Combined sewer overflow
MS4	Municipal Separate Storm Sewer System
TMDL	Total Maximum Daily Load
CWA	Clean Water Act
NRDC	Natural Resources Defense Council
GI	Green infrastructure
MHURD	Ministry of Housing and Urban-Rural Development

References

1. Fletcher, T.D.; Shuster, W.; Hunt, W.F.; Ashley, R.; Butler, D.; Arthur, S.; Trowsdale, S.; Barraud, S.; Semadeni-Davies, A.; Bertrand-Krajewski, J.-L.; et al. SUDS, LID, BMPs, WSUD and more—The evolution and application of terminology surrounding urban drainage. *Urban Water J.* **2014**, *12*, 525–542. [CrossRef]
2. MHURD. Technical Guide for Sponge City Construction. 2014. Available online: http://www.mohurd.gov.cn/wjfb/201411/t20141102_219465.html (accessed on 24 July 2020).
3. State Council. *Guiding Opinions of the General Office of the State Council on Advancing the Construction of Sponge Cities*; State Council: Beijing, China, 2015.
4. Wang, H.; Mei, C.; Liu, J.; Shao, W. A new strategy for integrated urban water management in China: Sponge city. *Sci. China Technol. Sci.* **2018**, *61*, 317–329. [CrossRef]
5. Linton, J.; Budds, J. The hydrosocial cycle: Defining and mobilizing a relational-dialectical approach to water. *Geoforum* **2014**, *57*, 170–180. [CrossRef]
6. Jiang, Y.; Zevenbergen, C.; Ma, Y. Urban pluvial flooding and stormwater management: A contemporary review of China's challenges and "sponge cities" strategy. *Environ. Sci. Policy* **2018**, *80*, 132–143. [CrossRef]
7. Xia, J.; Zhang, Y.; Xiong, L.; He, S.; Wang, L.; Yu, Z. Opportunities and challenges of the Sponge City construction related to urban water issues in China. *Sci. China Earth Sci.* **2017**, *60*, 652–658. [CrossRef]
8. Callon, M. Society in the making: The study of technology as a tool for. In *The Social Construction of Technological Systems: New Directions in the Sociology and History of Technology*; The MIT Press: London, UK, 1987; pp. 83–103.
9. Geels, F.W. Socio-technical transitions to sustainability: A review of criticisms and elaborations of the Multi-Level Perspective. *Curr. Opin. Environ. Sustain.* **2019**, *39*, 187–201. [CrossRef]
10. Markard, J.; Raven, R.; Truffer, B. Sustainability transitions: An emerging field of research and its prospects. *Res. Policy* **2012**, *41*, 955–967. [CrossRef]
11. Geels, F.W. Socio-technical transitions to sustainability. In *Oxford Research Encyclopedia of Environmental Science*; Oxford University Press: Oxford, UK, 2018.
12. Rip, A.; Kemp, R. Technological change. *Hum. Choice Clim. Change* **1998**, *2*, 327–399.
13. Geels, F.W.; Schot, J. Typology of sociotechnical transition pathways. *Res. Policy* **2007**, *36*, 399–417. [CrossRef]
14. Loorbach, D. *Transition Management: New Mode of Governance for Sustainable Development*; International Books: Utrecht, The Netherlands, 2007.
15. Patouillard, C.; Forest, J. The spread of sustainable urban drainage systems for managing urban stormwater: A multi-level perspective analysis. *J. Innov. Econ. Manag.* **2011**, *8*, 141–161. [CrossRef]

16. Brown, R.R.; Farrelly, M.A.; Loorbach, D.A. Actors working the institutions in sustainability transitions: The case of Melbourne's stormwater management. *Glob. Environ. Change* **2013**, *23*, 701–718. [CrossRef]
17. Dobre, C.C.; Vinke-de Kruijf, J.; Moretto, L.; Ranzato, M. Stormwater management in transition: The influence of technical and governance attributes in the case of Brussels, Belgium. *Environ. Sci. Policy* **2018**, *85*, 1–10. [CrossRef]
18. Rijke, J.S.; De Graaf, R.; Van de Ven, F.H.; Brown, R.R.; Biron, D. Comparative case studies towards mainstreaming water sensitive urban design in Australia and the Netherlands. In Proceedings of the 11th International Conference on Urban Drainage (ICUD), Edinburgh, UK, 31 August–5 September 2008.
19. Binz, C.; Truffer, B. Leapfrogging in Infrastructure: Identifying transition trajectories towards decentralized urban water management systems in China. In Proceedings of the 2009 DRUID Conference, Copenhagen, Denmark, 17–19 June 2009.
20. Binz, C.; Truffer, B. Anchoring global networks in urban niches: How on-site water recycling emerged in three Chinese cities. In *Urban Sustainability Transitions*; Routledge: London, UK, 2017; pp. 23–36.
21. Xu, G.; Xu, X.; Tang, W.; Liu, W.; Shi, J.; Liu, M.; Wang, K. Fighting against water crisis in China—A glimpse of water regime shift at county level. *Environ. Sci. Policy* **2016**, *61*, 33–41. [CrossRef]
22. Zevenbergen, C.; Fu, D.; Pathirana, A. Transitioning to Sponge Cities: Challenges and Opportunities to Address Urban Water Problems in China. *Water* **2018**, *10*, 1230. [CrossRef]
23. Zhou, M.; Köster, S.; Che, W.; Wang, X. Cross-boundary Evolution of Urban Planning and Urban Drainage Towards the Water Sensitive "Sponge City". In *Urban Water Management for Future Cities*; Springer: Cham, Switzerland, 2019; pp. 303–329.
24. Chan, F.K.S.; Chen, W.Y.; Gu, X.; Peng, Y.; Sang, Y. Transformation towards resilient sponge cities in China. *Nat. Rev. Earth Environ.* **2021**, *3*, 99–101. [CrossRef]
25. Wang, S.; Palazzo, E. Sponge City and social equity: Impact assessment of urban stormwater management in Baicheng City, China. *Urban Climate* **2021**, *37*, 100829. [CrossRef]
26. Yao, Z.; Bell, S. Tacit knowledge in water management: A case study of Sponge City. *UCL Open Environ.* **2021**. [CrossRef]
27. Zhai, J.; Ren, J.; Xi, M.; Tang, X.; Zhang, Y. Multiscale watershed landscape infrastructure: Integrated system design for sponge city development. *Urban For. Urban Green.* **2021**, *60*, 127060. [CrossRef]
28. Loorbach, D.; Rotmans, J. Managing transitions for sustainable development. In *Understanding Industrial Transformation*; Springer: Cham, Switzerland, 2006; pp. 187–206.
29. Rotmans, J.; Kemp, R.; Van Asselt, M. More evolution than revolution: Transition management in public policy. *Foresight* **2001**, *3*, 15–31. [CrossRef]
30. Kemp, R.; Schot, J.; Hoogma, R. Regime shifts to sustainability through processes of niche formation: The approach of strategic niche management. *Technol. Anal. Strateg. Manag.* **1998**, *10*, 175–198. [CrossRef]
31. Van der Brugge, R.; Rotmans, J. Towards transition management of European water resources. *Water Resour. Manag.* **2007**, *21*, 249–267. [CrossRef]
32. Buttel, F.H. New directions in environmental sociology. *Annu. Rev. Sociol.* **1987**, *13*, 465–488. [CrossRef]
33. Schultz, S.K.; McShane, C. To engineer the metropolis: Sewers, sanitation, and city planning in late-nineteenth-century America. *J. Am. Hist.* **1978**, *65*, 389–411. [CrossRef]
34. Skillen, J.R. Concrete Revolution: Large Dams, Cold War Geopolitics, and the US Bureau of Reclamation. In *The AAG Review of Books*; Oxford University Press: Oxford, UK, 2016; Volume 6, pp. 41–49.
35. Barry, F.J. Evolution of the Enforcement Provisions of the Federal Water Pollution Control Act: A Study of the Difficulty in Developing Effective Legislation. *Mich. L. Rev.* **1969**, *68*, 1103. [CrossRef]
36. Digest of Federal Resource Laws of Interest to the U.S. Fish and Wildlife Service. Available online: https://www.fws.gov/laws (accessed on 14 September 2021).
37. Hines, N.W. Controlling industrial water pollution: Color the problem green. *BC Indus. Com. L. Rev.* **1967**, *9*, 553.
38. Adler, J.H. Fables of the Cuyahoga: Reconstructing a history of environmental protection. *Fordham Envtl. LJ* **2002**, *14*, 89. [CrossRef]
39. Keiser, D.A.; Shapiro, J.S. Consequences of the Clean Water Act and the demand for water quality. *Q. J. Econ.* **2019**, *134*, 349–396. [CrossRef]
40. Coggins, G.C. Conserving Wildlife Resources: An Overview of the Endangered Species Act of 1973. *NDL Rev.* **1974**, *51*, 315.
41. Stormwater Phase II Final Rule: An Overview. 2000. Available online: https://www.epa.gov/sites/default/files/2015-11/documents/fact1-0.pdf (accessed on 7 September 2021).
42. Coffman, L.S.; Goo, R.; Frederick, R. Low-impact development: An innovative alternative approach to stormwater management. In Proceedings of the 29th Annual Water Resources Planning and Management Conference, Tempe, AZ, USA, 6–9 June 1999; pp. 1–10.
43. EPA, U.; Center, L. *Low Impact Development(LID): A Literature Review*; U.S. Environmental Protection Agency: Washington, DC, USA, 2000.
44. Weitman, D.; Weinberg, A.; Goo, R. Reducing stormwater costs through LID strategies and practices. In *Low Impact Development for Urban Ecosystem and Habitat Protection*; Curran: Singapore, 2009; pp. 1–10.
45. Crowe, L.; Rotherham, I.D. Biodiversity and Ecosystem Services in Relation to the Management of Storm Water and the Mitigation of Floods. In *Urban Stormwater and Flood Management*; Springer: Cham, Switzerland, 2019; pp. 159–186.

46. Newell, J.P.; Seymour, M.; Yee, T.; Renteria, J.; Longcore, T.; Wolch, J.R.; Shishkovsky, A. Green Alley Programs: Planning for a sustainable urban infrastructure? *Cities* **2013**, *31*, 144–155. [CrossRef]
47. EPA. Use of Green Infrastructure in NPDES Permits and Enforcement. 2007. Available online: https://www.epa.gov/sites/production/files/2015-10/documents/gi_memo_enforce_0.pdf (accessed on 18 April 2021).
48. McPhillips, L.E.; Matsler, A.M. Temporal evolution of green stormwater infrastructure strategies in three US cities. *Front. Built Environ.* **2018**, *4*, 26. [CrossRef]
49. GAO. *Water Quality: Better Data and Evaluation of Urban Runoff Programs Needed to Assess Effectiveness*; General Accounting Office (GAO): Washington, DC, USA, 2001.
50. Tian, S. Managing Stormwater Runoff with Green Infrastructure: Exploring Practical Strategies to Overcome Barriers in Citywide Implementation. Master's Thesis, University of Nebraska, Lincoln, NE, USA, August 2011.
51. Dhakal, K.P.; Chevalier, L.R. Managing urban stormwater for urban sustainability: Barriers and policy solutions for green infrastructure application. *J. Environ. Manag.* **2017**, *203*, 171–181. [CrossRef] [PubMed]
52. National Research Council. *Urban Stormwater Management in the United States*; National Academies Press: Washington, DC, USA, 2009.
53. EPA. Stormwater Information Collection Request (ICR) Fact Sheet September 2010. Available online: https://www.epa.gov/sites/default/files/2015-11/documents/icr_factsheet_questionnaire.pdf (accessed on 1 May 2021).
54. EPA. NPS Outreach Toolbox. 2017. Available online: https://cfpub.epa.gov/npstbx/index.html (accessed on 12 March 2022).
55. EPA. Community Solutions for Voluntary Long-Term Stormwater Planning. 2016. Available online: https://www.epa.gov/npdes/stormwater-planning (accessed on 10 August 2021).
56. Wu, C.; Qiao, M.; Wang, S. Enlightenment from ancient Chinese urban and rural stormwater management practices. *Water Sci. Technol.* **2013**, *67*, 1474–1480. [CrossRef] [PubMed]
57. Cun, C.; Zhang, W.; Che, W.; Sun, H. Review of urban drainage and stormwater management in ancient China. *Landsc. Urban Plan.* **2019**, *190*, 103600. [CrossRef]
58. Shen, Q.; Peng, S.; Ci, H. Evolution and Future Prospect of Urban Ecological Planning in Modern China. *Urban Plan. Int.* **2019**, *34*, 37–48. [CrossRef]
59. Bjorklund, E. The Danwei: Socio-spatial characteristics of work units in China's urban society. *Econ. Geogr.* **1986**, *62*, 19–29. [CrossRef]
60. Li, Y. *Historical Investigation of the Pollution Incident of Guanting Reservoir in 1970s*; Party School of the Central Committee of CPC: Beijing, China, 2015.
61. Yeh, A.G.; Xu, J.; Liu, K. *China's Post-Reform Urbanization: Retrospect, Policies and Trends*; IIED: London, UK, 2011; Volume 5.
62. Sun, T. *Political and Social Changes and Ecological Environment Evolution in Modern China*; Intellectual Property Publishing House: Beijing, China, 2018.
63. Yu, K.; Li, D. *Way to Urban Landscape-Talk to Mayors*; China Building Industry Press: Beijing, China, 2003.
64. Luo, H.-M.; Che, W.; Li, J.-Q.; Wang, H.-L.; Meng, G.-H.; He, J.-P. Application of rainwater garden to storm and flood control and utilization. *China Water Wastewater* **2008**, *24*, 48.
65. Wang, J.; Che, W.; Yi, H. Low impact development for urban stormwater and flood control and utilization. *China Water Wastewater* **2009**, *14*, 6–9.
66. Zhang, W.; Che, W.; Wang, J.; Wang, S. Management of urban stormwater runoff by green Infrastructures. *China Water Wastewater* **2011**, *27*, 22–27.
67. Li, L.-Q.; Yin, C.-Q.; He, Q.-C.; Kong, L.-L. First flush of storm runoff pollution from an urban catchment in China. *J. Environ. Sci.* **2007**, *19*, 295–299. [CrossRef]
68. Ballo, S.; Liu, M.; Hou, L.; Chang, J. Pollutants in stormwater runoff in Shanghai (China): Implications for management of urban runoff pollution. *Prog. Nat. Sci.* **2009**, *19*, 873–880. [CrossRef]
69. Qin, H.-P.; Khu, S.-T.; Yu, X.-Y. Spatial variations of storm runoff pollution and their correlation with land-use in a rapidly urbanizing catchment in China. *Sci. Total Environ.* **2010**, *408*, 4613–4623. [CrossRef] [PubMed]
70. Hodson, M.; Marvin, S. *After Sustainable Cities?* Routledge: London, UK, 2014.
71. Den Hartog, H. Rural to Urban Transitions at Shanghai's Fringes. *Int. Rev. Spat. Plan. Sustain. Dev.* **2017**, *5*, 54–72. [CrossRef]
72. Wang, W.; Chen, R.; Liu, Y.; Wei, W. From 2008 to 2010, 62% of 351 Cities in China Experienced Waterlogging. *People's Daily*. 24 July 2012. Available online: http://politics.people.com.cn/n/2012/0724/c1001-18580975.html (accessed on 17 December 2020).
73. *GB/T 51345-2018*; Assessment Standards for Sponge City Construction. MHURD: Beijing, China, 2018. Available online: http://www.mohurd.gov.cn/wjfb/201904/t20190409_240118.html (accessed on 3 May 2021).
74. MHURD. *Interim Provisions on Compilation of Special Planning for Sponge City*; MHURD: Beijing, China, 2016.
75. Wang, H. 19 Cities of 30 Pilots Have Experienced Waterlogging This Year. *China Economic Weekly*. 5 September 2016. Available online: http://www.ceweekly.cn/2016/0905/163283.shtml (accessed on 4 May 2021).
76. Sang, Y.-F.; Yang, M. Urban waterlogs control in China: More effective strategies and actions are needed. *Nat. Hazards* **2017**, *85*, 1291–1294. [CrossRef]
77. Qiao, X.-J.; Liao, K.-H.; Randrup, T.B. Sustainable stormwater management: A qualitative case study of the Sponge Cities initiative in China. *Sustain. Cities Soc.* **2020**, *53*, 101963. [CrossRef]

78. Li, H.; Ding, L.; Ren, M.; Li, C.; Wang, H. Sponge City Construction in China: A Survey of the Challenges and Opportunities. *Water* **2017**, *9*, 594. [CrossRef]
79. Froissart, C. From outsiders to insiders: The rise of China ENGOs as new experts in the law-making process and the building of a technocratic representation. *J. Chin. Gov.* **2019**, *4*, 207–232. [CrossRef]
80. Meng, H.-P.; Niu, Z.-R. Present Situation, Problem and Countermeasures of Validity of the Public Participation in China's Environmental Impact Assessment. *J. Yanan Univ. Nat. Sci. Ed.* **2011**, *30*, 78–81.
81. Pant, L.P.; Adhikari, B.; Bhattarai, K.K. Adaptive transition for transformations to sustainability in developing countries. *Curr. Opin. Environ. Sustain.* **2015**, *14*, 206–212. [CrossRef]
82. Olsson, P.; Gunderson, L.H.; Carpenter, S.R.; Ryan, P.; Lebel, L.; Folke, C.; Holling, C.S. Shooting the rapids: Navigating transitions to adaptive governance of social-ecological systems. *Ecol. Soc.* **2006**, *11*, 18. [CrossRef]
83. Wang, T. *Urban Rainwater Management: Theory, Methods and Measures*; Huazhong University of Science and Technology Press: Wuhan, China, 2017.
84. Scott, J.C. Seeing like a state. In *Seeing Like a State*; Yale University Press: London, UK, 2008.

water

MDPI

Review

Application of Analytical Probabilistic Models in Urban Runoff Control Systems' Planning and Design: A Review

Ali Aldrees [1,*] and Salisu Dan'azumi [1,2,*]

1 Department of Civil Engineering, College of Engineering at Al-kharj, Prince Sattam bin Abdulaziz University, Al-Kharj 16278, Saudi Arabia
2 Department of Civil Engineering, Bayero University Kano, Kano 700241, Nigeria
* Correspondence: a.aldrees@psau.edu.sa (A.A.); sdanazumi.civ@buk.edu.ng (S.D.); Tel.: +966-593364065 (A.A.); +234-8038547730 (S.D.)

Abstract: Urban stormwater is known to cause a myriad of problems, ranging from flooding to water quality degradations. This paper provides an extensive review of analytical probabilistic model (APMs) used in the design of urban runoff control systems. APMs are closed-form mathematical expressions representing a long-term system's output performance derived from the probability distribution of the system's input variables. Once derived, the APMs are easy to handle, allow for sensitive analysis, and can be co-opted into optimization frameworks. The implementation of APM in the planning and design of runoff control systems will not only help address the runoff quantity and quality problems of urban stormwater, but will also go a long way in optimizing the benefits derived from the systems. This paper reviews studies that document the negative impacts of urbanization on runoff quantity and quality, and the best management practices (BMPs) used to mitigate the impacts. Three design methodologies used in urban stormwater control systems were reviewed. A detailed review of research on the development and use of APMs in urban stormwater management in various runoff control systems is presented, and recommendations are proffered.

Keywords: best management practices; low-impact development; water-sensitive urban design; blue-green infrastructure; sponge cities

Citation: Aldrees, A.; Dan'azumi, S. Application of Analytical Probabilistic Models in Urban Runoff Control Systems' Planning and Design: A Review. *Water* **2023**, *15*, 1640. https://doi.org/10.3390/w15091640

Academic Editors: Dafang Fu, Haifeng Jia, Jiangyong Hu and Wei-Shan Chen

Received: 20 March 2023
Revised: 15 April 2023
Accepted: 19 April 2023
Published: 22 April 2023

1. Introduction

Urbanization causes a disruption of the natural water cycle. The clearing of land surfaces reduces evapo-transpiration processes that intercept and return rainfall to the atmosphere. Grading the land involves filling depression storage and the removal of topsoil while subsoil is compacted. The construction of impervious surfaces such as roads, roofs and paved walkways reduces infiltration and increases surface runoff. Rainfall that used to infiltrate the ground now runs off the surface at an increased rate, depending on the level of changes made to the land surface. These changes cause an increase in the peak runoff and total volume of runoff. Conversely, the time of concentration of the catchment is decreased, which causes flows across the land surfaces to occur at faster rates. This effect is further aggravated by artificial drainage systems that are designed to convey runoff to rivers as quickly as possible. With the development of impervious surfaces, infiltration into the soil is reduced, thus reducing the quantity of water available to recharge aquifers and feed-in the base-flow during dry weather periods [1–5].

In addition to increasing the runoff quantity, urbanization also affects the runoff quality negatively, by increasing the concentration of pollutants carried by stormwater [6]. As runoff runs over roads, parking lots, rooftops of urbanized areas, it picks up a variety of pollutants and transports them to downstream water bodies. The receiving water body is affected by the cumulative impact of urban activities from the entire watershed, and the resultant changes in stormwater quantity and quality are felt in the downstream waters [7,8].

Several attempts have been made by researchers to quantify the effects of urbanization on stormwater runoff entering receiving water bodies. Todeschini (2016) [9] studied the modifications of stormwater runoff and water quality caused by increased imperviousness in the Bivio Vela industrial catchment in northern Italy. Runoff flows generated from 51 rainfall events of smaller and higher intensities were monitored. The results showed that the conversion of 33% of pervious to impervious surfaces had resulted in a great increase in peak flows, the volume of flows, the number and duration of combined sewer overflows and the pollutants mass discharge. Likewise, Schütte and Schulze (2017) [10] studied the effects of land-use changes due to the proposed urbanization of two sub-catchments of uMngeni catchment (South Africa) on hydrological flows. The study used ACRU software to model the current (2017) and future flows that may arise as a result of the conversion of agricultural lands to impervious surfaces. The results show that increases in impervious surfaces would result in a significant increase in stormflows due to a change in rainfall–runoff conversion caused by the reduced evapotranspiration. Urbanization within a watershed has a number of negative impacts on downstream waters. These impacts include: changes to stream flow and stream geomorphology, degradation of water quality and impact on aquatic habitat [1,4].

Stormwater best management practices (BMPs) are techniques, measures, or structural controls that are used in a given set of conditions to manage the quantity and/or improve the quality of stormwater runoff in the most cost-effective manner [11]. They are designed facilities or modified natural environments that help control the quantity as well as improve the quality of urban stormwater. Urban runoff control systems can be classified in to two BMPs: (1) Methods that are used to reduce the volume of stormwater runoff that will otherwise flow into the receiving water bodies. These methods allow the infiltration of the stormwater into the ground, thereby aiding in groundwater recharge. (2) Methods that remove pollutants from the stormwater through filtration, sedimentation, absorption, biological uptake, etc. [12,13]. However, most stormwater BMPs serve both purposes. The most commonly employed stormwater BMPs include various types of stormwater ponds, filtration practices, vegetated channel practices, wetlands, pervious pavements and rainwater tanks [9]. Green roofs iare also among the most commonly used stormwater BMPs [14–16].

Due to their importance, sustainable stormwater management concepts have been given different names all over the world. Qi et al. (2021) [17] presented a comprehensive list of them, as synonymously adopted all over the world. The list ranges from BMPs to low-impact development, integrated urban drainage systems, sustainable urban drainage systems, stormwater control measures, water-sensitive urban design, resilience cities and sponge cities. Nature-based solutions, green infrastructure, or blue–green infrastructure are other terminologies also used to refer to methods used in mitigating the impact of flood risk related to urban stormwater [18–20]. In its quest to restore its cities following the negative consequences of stormwater runoff due to urbanization, China developed the sponge cities plan in 2013, aimed at promoting source control. The concept uses natural methods to retain rainwater, thereby recharging groundwater, reducing flooding and water pollution problems, and gradually restoring the natural hydrology of the cities. The sponge cities pilot scheme started in 30 cities, and following the successes recorded, the concept is now being adopted at the national level [21,22].

This review paper compiles research on analytical probabilistic models' (APMs') applications to urban stormwater management over the last 35 years, when the models began to debut. A search of the literature was carried out in the SCOPUS and Google Scholar databases using different combinations of the terms: "analytical probabilistic models" AND "stormwater management" OR other BMPs such as "detention ponds, bioretention cells, green roofs, pervious pavements, rain garden, etc." A total of 183 entries were returned by Google Scholar, while SCOPUS returned 45 entries in the first search. The search was repeated and different entries were returned. The entries were filtered and a total of 126 published articles found to be relevant were reviewed. This attempt to compile and review studies on the use of APMs in urban stormwater management is, to the best knowledge of the authors, the first of its kind in this area of research. Figure 1 shows the chronological order of the articles.

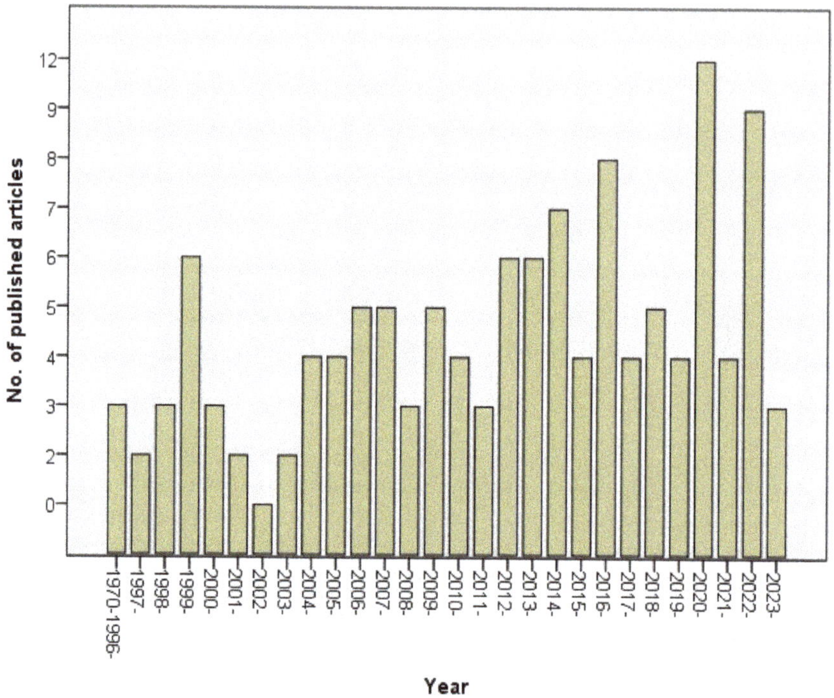

Figure 1. Number of published articles on APMs of stormwater management.

2. Design of Urban Runoff Control Systems

Runoff control systems can be designed based on three approaches: a design storm, continuous simulation and analytical probabilistic models.

The design storm approach uses the statistical analysis of rainfall to establish the IDF curves of an area. A design storm consists of rainfalls of various durations, developed to serve as the input that the runoff control system may experience during its life time [5]. The peak runoff, generated from the design storm, is routed through the runoff control facility in order to estimate the facility's capacity. Many authorities specify design storms in their stormwater regulations. For instance, commonly used regulations specify that post-development peak runoff be less than pre-development peaks for storms having recurrence intervals of 2, 5, and 10 years, with an emergency overflow spillway of 100-year recurrence interval capacity [23]. The duration of the design storm is chosen to be the same as the critical duration or the time of concentration of the catchment. The design storm used for the control system is selected based on risk factors, such as the risk of overflow to downstream conveyance structures and land-use. The assumption being used in the development of the design storm concept is that the recurrence interval of the resulting runoff is the same as the recurrence interval of the rainfall producing the event [24]. This assumption ignores the effect of storm separation time on runoff generation and the temporal variations in storm hyetograph pattern, which are known to affect the behavior of runoff. Due to this, the design storm approach suffers from severe criticisms voiced by many researchers, particularly in the design of stormwater storage systems [24]. Detailed shortcomings of the design approach can be found in [25].

Instead of analyzing the rainfall history to extract concise information, continuous simulation involves the conceptual representation of the catchment and a meteorological input over a longer period. In this case, the effects of storm separation time and the temporal variation of the storm event are captured. The long-term rainfall data are input directly into the continuous simulation software [5]. The result of continuous simulation is the response of the catchment to the rainfall input. However, despite the numerous advantages of the continuous simulation approach, its major problem is its computational burden, requiring a large number of simulations to evaluate system configuration [26]. The models are also extensive, requiring a large computer memory and time for the system analysis [25]. Many types of computer software have been developed for the continuous simulation of a catchment's response to rainfall input, but SWMM is by far the most widely used software. Todeschini et al. (2012) [27] used SWMM to investigate the effectiveness of design configurations and operating conditions of stormwater detention tanks in combination with flow regulators upstream. The results suggest that the optimum performances were obtained by regulating flows into the detention tanks to values between 0.5 to 1 liters per hectare, and tank volumes of 35 to 50 m^3 per hectare of impervious area, for catchments ranging from 4.8 to 48 hectors. These values gave the maximum annual pollutant mass entrapped by the tank and minimum volume of stormwater sent to downstream treatment plants. Mobley and Culver (2014) [28] used the SWMM model of a 12 ha residential site, near Fort Collins, Colorado, to modify the design of a detention pond such that the ecological flows downstream of the catchment were preserved, while maintaining regulatory flow requirements. Pereira et al. (2019) [29] used PC-SWMM to simulate surface runoff, and predict the impacts of urbanization and the use of detention ponds in an urban sub-catchment in Brasilia. Thirteen different configurations of ponds were tested and the best configuration of ponds was observed to be located near the outlet of the catchment. This configuration was found to reduce flood peak discharge by 10 to 30%, and reduce nutrient load by 40 to 60% and COD by 46%. Continuous simulation, though data-intensive and time-consuming, is most commonly used to design or modify the design of existing runoff control systems.

3. Analytical Probabilistic Models (APMs)

APMs are closed-form analytical expressions of a system's output performance derived from the probability distribution of the system's input variables. The approach was initially proposed by Benjamin and Connel (1970) [30]. The approach was first applied to water resources by Eagleson (1972) [31] in water resources engineering, and later extended to the area of stormwater management by many researchers [25,32]. In APMs of urban storm water management, continuous rainfall data are divided into individual events using a pre-defined minimum storm separation time, and the APM parameters are developed using the rainfall statistics [32–34]. The input variable is rainfall characteristics (i.e., rainfall depth, duration, inter-event time), and the output is the catchment's response to the rainfall input (i.e., runoff event volume, peak discharge, etc.). Using derived probability theory, the probability distribution of output variables can be determined analytically if the probability distributions of the input variables are known and the functional relationship between them is either monotonically increasing or decreasing [35].

The application of APMs in urban stormwater management involves the following stages, as described in [24–26,36–46]:

i. Selection of a case study of urban catchment and determination of its physiographic information (catchment area, slope, proportion of imperviousness, drainage length, depression storage, runoff coefficient);
ii. Discretization of the long-term continuous rainfall data into individual events using the minimum inter-event time definition;
iii. Use of probability distributions, such as Exponential, Weibull, etc., to fit rainfall depth, duration and intervention time;
iv. Development of APM parameters for the rainfall station;

v. Development of APMs for stormwater characteristics such as runoff volume and peak runoff captured by the facility, total runoff, pollutants captured and treated by the facility, volume of spilled runoff, etc.

The APMs, once derived, were validated with other known approaches such as design storm and continuous simulation, and the results were generally found to be in good agreement [5,24,26,36–45]. Guo (2001) [5] assessed the suitability of using APMs in the design of urban flood control detention ponds, alongside other approaches, namely, design storm and continuous simulation, using a hypothetical catchment in Chicago. The results show that all the three approaches generated similar results for the prediction of peak flow of various durations from small urban catchments. Quader and Guo (2006) [46] studied the discrepancy in peak flood estimations between design storm and the APM approach. The effects of sub-catchment aggregation and time of concentration, as represented in the two approaches, were examined. A case study catchment of Cataraqui North in Ontario (Canada) was selected, and MIDUSS software was used for the design storm modeling. The results were found to be in good agreement, with only 25% discrepancy between the two peaks generated using the two approaches, which were attributed to sub-catchment aggregation and the choice of time of concentration.

Exponential PDF is the most widely used distribution to model rainfall characteristics, particularly in Canada and the USA, where the APMs of stormwater management were developed. Hassini and Guo (2016) [47] used long-term rainfall data from seven rain-gauge stations in northern USA to test the validity of using one-parameter exponential distribution in modeling rainfall characteristics (depth, duration and inter-event time). Poisson's and Chi-square tests were used. It was found that exponential distribution fits rainfall characteristics well, and was therefore recommended for APM. The exponential PDF is used in more than 80% of the research papers reviewed herein. However, other distributions were tested to determine their fit to rainfall characteristics in other regions of the world. In this regard, Bacchi et al. (2008) [33] compared the use of one-parameter exponential distribution and two-parameter Weibull distribution to model rainfall characteristics for three stations in Italy. The results indicate that the Weibull distribution fits the Italian climate better. The distribution was combined with rainfall–runoff transformation to derive the PDF for the runoff volume and overflow volume of a storage facility, from which the design failure probability can be obtained. Balistrocchi et al. (2009) [48] also used Weibull distribution to model rainfall characteristics for Milano rainfall stations in Italy, while Generalized Pareto Type 3 distribution was used to model rain storm depth with long durations in Toronto, Canada [49]. Pareto and Gamma-2 PDF were found to work well for rainfall depth and duration in Spain [50]. Weibull distribution was also found to fit rainfall characteristics in Poland [51]. Log-normal distribution was found to fit rainfall characteristics for some stations in Korea [52], while three-parameter exponential distribution was used to model rainfall characteristics in Busan (Korea) as an improvement to single-parameter exponential distribution [53].

In rainfall–runoff process modelling from urban catchments, Guo and Adams (1998a) [36] used the exponential distribution to model the frequency distribution of rainfall depth, duration, and inter-event time, from which a closed-form expression for the average annual runoff event volume and the runoff event volume return period was derived. A hypothetical catchment with a different runoff coefficient and various soils was used to test the model against similar results obtained from a numerical simulation model (SWMM). A close agreement between the analytical model and the simulation model was obtained. Similarly, a close agreement was obtained between the results of runoff event volumes and average annual runoff volume with a specified recurrence interval. Guo and Adams (1998b) [37] also used the expression for runoff event volume and its duration together, with the catchment's average time of concentration, to derive an expression of peak discharge rate, using the assumption of a triangular inflow hydrograph and exponential distribution for rainfall depth. A closed-form analytical expression for the exceedance probability of peak discharge per rainfall event and its return period was determined. The results from the analytical model compared favorably with those obtained using a continuous simulation model, SWMM.

Guo et al. (2012) [54] further improved the APM by incorporating both Hortonian and saturated overland flow mechanisms into the model to cater for the increasing use of low-impact development (LID) techniques in urban watersheds, whose infiltration is always below the natural infiltration. The PDFs of runoff event volume under the two scenarios of infiltration excess and saturation excess runoff were derived separately, and combined to obtain APM expressions for the expected value of the runoff event and its recurrence interval, as well as the average annual runoff volume. The results were compared with continuous simulation from HEC-HMS and there was very good agreement. Hassini and Guo (2017) [55] derived APM expressions for the exceedance probability of peak discharge in a small urban catchment considering triangular and trapezoidal hydrographs. Rainfall data from Sherburne, Minnesota in the USA were used and a hypothetical catchment with different times of concentration, imperviousness and soil types was assumed. Design storm (using HEC-HMS model) was used to predict the peak discharges at different return periods. The peak discharges generated from the developed APM were found to be comparable with those of the design storm. Hassini and Guo (2020) [56] further added the effect of saturation excess and infiltration excess runoff to their previous work [24] to develop APMs for runoff event volume and exceedance probability of peak discharge in a small urban catchment. Rainfall data from seven stations in the USA were used and a test catchment in Hamilton (Canada) was used. The results of the APMs were found to be comparable to those of design storm, with a percent difference ranging from 0.1% to 18%. Hassini and Guo (2021, 2022) [57,58] recently developed a new and more accurate APM that can be used for the design of runoff control systems. An APM for the frequency distribution of runoff event volume was developed considering infiltration and saturation excess runoff generation. The new model can effectively estimate runoff volume with different recurrence intervals, and was found to be very sensitive to changes in soil saturation.

APMs have the ability to model flood routing. Guo and Zhuge (2008) [59] developed analytical probabilistic expressions of flood routing to determine the probability distribution of peak outflow from a channel reach with and without a detention pond in between. The outflow hydrographs were obtained, and the results of the analytical models were compared with those of a single-event design storm using stormwater modeling software—the MIDUSS and SWMM surface runoff routing algorithms. It was shown that the analytical models compared well with the design storm. However, the use of different surface routing models gave variations in the results of about 20%. Guo et al. (2009) [60] developed closed-form analytical probabilistic channel routing equations for determining the flood frequency distribution downstream of a catchment, given the catchments' characteristics and APM parameters derived from rainfall data. The model was verified by comparing its results with those of HEC-HMS continuous simulation using 25 reaches and rainfall data from Halifax, Canada. The results of the flood peak attenuations were presented as a function of storage-delay time and return period, which can be used for watershed and stormwater management purposes. Guo and Markus (2011) [61] enhanced the versatility of APMs applied to small watersheds by incorporating SCS-CN for the determination of rainfall excess from the catchments, and Clark's unit hydrograph for runoff routing. The results of the APM were compared with those of design storm using the HEC-HMS model. Twelve watersheds were used in Chicago under urbanizing conditions, and the results show that the analytical model can be used in stormwater management.

Flood peak estimation is another area wherein APMs were also found to be useful. Guo and Dai (2009) [62] expanded the ability of the analytical model to cater for a larger catchment and the master planning of a drainage system. A probabilistic rainfall areal reduction method was used. Both the APM and design storm approaches were used to obtain rainfall frequencies and flood peaks. A case study catchment of the Ganaraska river watershed, Canada was simulated using the OTTHYMO model, and rainfall data from Toronto Pearson International Airport station were used. The results show the capability of APMs in accurately representing the effects of rainfall characteristics across different geographical regions, and their effects on flood frequency.

In the case of modeling the pollutants build-up and wash-off from urban catchments, Behera et al. (2006) [45] assumed that: (1) rainfall duration, inter-event time, pollutant build-up and wash-off follow an exponential distribution; (2) the wash-off load is uniform over the entire catchment and depends on the runoff volume generated, to derive analytical expressions for the PDF of wash-off load, expected value of pollutant event wash-off load, annual average wash-off load and the long-term average pollutant event mean concentration (EMC). The analytical models were calibrated and verified against values observed in a test catchment, and a good agreement was obtained.

The APMs can be used to screen runoff control alternatives in order to determine additional data requirements. Similarly, the APMs can be used in sensitivity analysis to determine the most important parameters, which makes long-term meteorological computation simpler and more economic, supports decision-making and eases stormwater system design [25]. The APMs are computationally efficient and can be easily implemented in a spreadsheet or computer program, as compared to design storm or continuous simulation [54]. Therefore, analytical models can be used as an alternative to time-consuming continuous simulation.

One of the greatest advantages of APMs over design storm and continuous simulation is the option of co-opting them into an optimization framework. The optimization of runoff control systems can be classified based on the objective function (i.e., runoff quantity, quality and/or cost), uncertainty (deterministic or stochastic), and control approach (static or dynamic) [63]. Genetic algorithm, particle swarm optimization, ant colony optimization, artificial bee colony, simulated annealing, harmony search and cuckoo search are some of the optimization techniques that can be applied to optimize flood control systems [17].

4. APM Application to Urban Runoff Control Systems

A schematic diagram showing various stormwater BMPs is shown in Figure 2. APMs have been applied to a wide variety of stormwater BMPs. The application of APMs to these systems in discussed under this section.

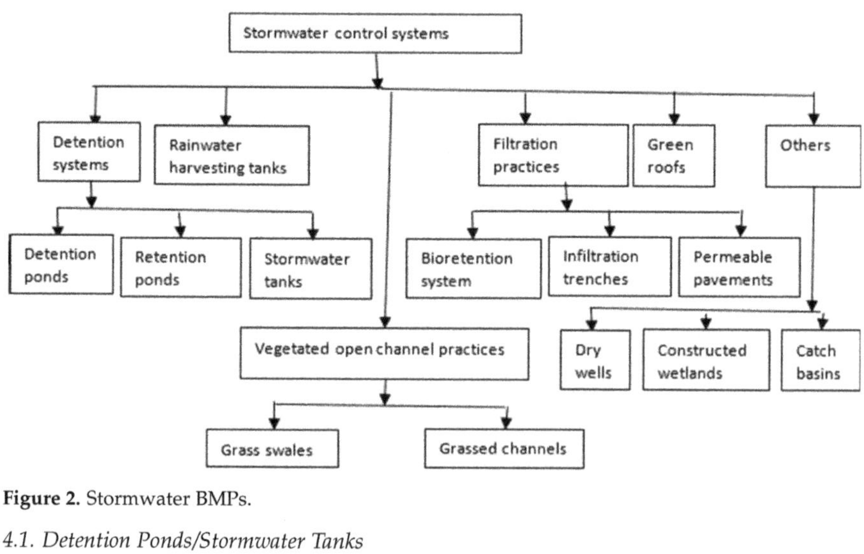

Figure 2. Stormwater BMPs.

4.1. *Detention Ponds/Stormwater Tanks*

Detention ponds involve the temporary storage of runoff in ponds, basins or even underground containers, and are meant to control the quantity as well as quality of urban runoff downstream of a catchment [64,65]. The purpose of stormwater detention is to reduce the flood damages caused by increased runoff due to imperviousness by limiting post-development peak discharges to be less than or equal to pre-development runoff [66], or to a rate based on other criteria specified by the stormwater authorities in charge [67]. Furthermore, stormwater detention improves the quality of stormwater runoff in addition to reducing the peak discharge [68]. The residence time resulting from stormwater detention allows for the suspended particulate matter and adsorbed contaminants to settle [69,70]. As a BMP, detention ponds can help limit the pollutants loaded into receiving water bodies.

Many researchers have dedicated much attention to the application of APMs in detention basins. Papa et al. (1997) [71] derived APM expressions for the pollution control performance of detention ponds for different combinations of active to permanent pool volumes. The results of the study have been compared to those simulated using SWMM software. It was found that the degree of suspended solid removal in both cases was comparable, with a difference of only 5 to 10% in extended dry ponds and 10 to 30% in wet ponds. Guo and Adams (1999a) [42] derived analytical expressions for the probability distribution of peak outflow rate from flood control detention ponds. The derived analytical expressions were used to determine the storage–discharge relationship required to achieve the specified level of flood control at the facility. Using the runoff volume and peak outflow rate presented in [36,37], the runoff rate exceedance probability per rainfall event was derived based on different combinations of storage and outflow. Comparisons were made between the results obtained from the analytical probabilistic model and similar results obtained from SWMM software, and the results were found to be in good agreement. Guo and Adams (1999b) [43] also used the expressions previously developed in [36,37] to derive APM expressions for the long-term performance of a stormwater quality control pond. The expression of flow capture efficiency was derived from the total spill volume, while the volume-weighted average detention time of the basin was derived by taking into account the variable inflow and outflow rates and the inter-runoff event time. The APM expressions describing the detention time and the statistical solution of flow capture efficiency were compared with similar values obtained from SWMM, and the results were found to be in close agreement, thus confirming the validity of the assumptions made in deriving the models.

Li and Adams (2000) [44] used an analytical probabilistic approach to derive runoff quantity and quality control performances for urban runoff control systems. Rainfall was first transformed to runoff, and the runoff transformed to overflow using the derived analytical expressions. The runoff volume was also transformed to runoff pollution mass load using the EMC concept, and was later transformed to total pollution mass discharge load. The APM expressions for fraction of runoff overflow and total pollution mass discharge load were used to derive closed-form APM expressions for the long-term runoff control and long-term pollution control performances of the stormwater storage and treatment systems. Comparisons between the runoff control performances predicted with the analytical model (coded in computer programs called SUDS and EXSUDS) and those obtained using a continuous simulation model STORM were conducted, and the results were in reasonably good agreement.

Analytical expressions for runoff control performances using different forms of rainfall–runoff transformations were developed [24,26,38–41]. Chen and Adams (2005a) [38] modified the rainfall–runoff transformation to consider infiltration rather than a common runoff coefficient, and developed closed-form analytical expressions for runoff control performances, including exceedance probability of a spill volume, expected value of a spill volume, average annual volume and number of spills, and runoff capture efficiency. The performance of the modified analytical model developed was tested against values obtained from continuous simulation using SWMM and the analytical models developed earlier by [25] for rainfall–runoff transformation (called ASTORM models), and good similarities between the three results were obtained. Chen and Adams (2005b) [39] also used the extended version of rainfall–runoff transformation, which divides the catchment into pervious and impervious areas with different depression storages and runoff coefficients, to develop APM expressions for the average annual number and volume of spills and the runoff control efficiencies. The results of the extended analytical model were compared with those from ASTORM and SWMM, and the results were in good agreement, with the extended model outperforming the ASTORM rainfall–runoff conversion model.

Chen and Adams (2006a) [40] used two types of rainfall–runoff transformations, ASTORM and the extended ASTORM, to derive analytical expressions for stormwater quality control based on build-up and wash-off functions. The appropriate models for pollutant build-up and wash-off (designated as Type 1 and Type 2) were chosen, and were combined to formulate the pollutant load model. Finally, the system quality control measures were derived, which are closed-form expressions that can be used to evaluate the long-term system behavior. Comparisons were made between the quality control models developed with observed values, and the values predicted using SWMM gave good estimates of system performance. Chen and Adams (2006b) [41] also used the derived analytical expressions based on three different rainfall–runoff transformations (i.e., ASTORM, Type 1 and Type 2) to derive APM expressions for stormwater quality control measures. In this case, pollutant removal via the extended detention dry pond was assumed to take place primarily through sedimentation, and TSS control was considered as a surrogate measure of other pollutants' removal. Closed-form APM expressions for average annual volume of runoff, average annual number of spills, average annual runoff control and pollutant removal efficiencies were derived. A comparison of the results from Type 1, Type 2 and SWMM was conducted, and the results were in good agreement.

Chen and Adams (2007a) [26] used the ASTORM rainfall–runoff transformation, extended ASTORM and the modified rainfall–runoff transformation to develop analytical expressions for average annual runoff volume from an urban catchment. In the second case, the Horton's infiltration equation was slightly modified, in that the rainfall duration was assumed to be a temporarily averaged constant. Model verification showed that both of the two analytical models compared favorably with results obtained from SWMM. Chen and Adams (2007b) [24] also used rainfall–runoff transformations, and pollutant build-up and wash-off functions, to derive analytical expressions of the cumulative density function (CDF) of pollutants load, as well as the expected value of pollutant EMC and average annual pollutant EMC. In the rainfall–runoff transformation, two types of models were proposed: the lumped parameter rainfall–runoff, and its extended form [39,44]. Two forms of pollutant load model (Type 1 and Type II) were obtained, and the expected pollutants' EMC and average annual pollutants' EMC values were derived. The pollutant load models were compared with observed values, and a good agreement was obtained. However, the Type II load model was found to outperform Type I in the estimation of average annual pollutants' EMC.

Apart from the tremendous contributions made to the development of APM in relation to detention pond's analysis and design, coming from Canada and USA, some important contributions coming from Italy are noticeable. Becciu and Raimondi (2014) [72] derived APM expressions for the overflow spill of stormwater detention ponds. Two management rules regarding the emptying of the pond were considered. Likewise, the probabilities for spilled volumes varied from zero to one, corresponding to no spill and a spill volume equal to the storage capacity of the pond, respectively. Data of rainfall series from Milano-Monviso, Italy, were used. The resulting analytical expressions can act as very valuable tools that can be used to estimate the overflow probability and the probability of a specific spilled volume. Raimondi and Becciu (2015) [73] used rainfall statistics, detention pond outlet operation rules, storage volume and maximum outflow to derive APM expressions for the pre-filling probability of detention ponds. As in their previous paper, the same management rules regarding the pond's emptying were considered. The results can be used to estimate the pond's volume and out flow rate as a function of pre-filling probability. A comparison of the analytical results with continuous simulation, using case study rainfall data from Monviso, Milano (Italy), showed a very good agreement, thus confirming the applicability of the method in the design and performance assessment of stormwater detention basins. Becciu and Raimondi (2015) [74] derived a similar expression for the PDF of a detention pond's spilled volume in order to evaluate its efficiency. Becciu et al. (2015) [75] also derived APMs of retention time in stormwater detention ponds. The analytical formulae developed can be used for the design of pond storage corresponding to a specified retention time that ensures some pollutants are removed from the pond. The APM expressions were validated against results from a continuous simulation using the case study in Monviso, Milano (Italy), and were found to fit very well. Raimondi et al. (2022) [76] derived APM expressions for the probability of runoff volume and residual storage in sustainable urban drainage systems. The models were applied to two catchments in Genova and Milano (Italy) using rainfall data from Monviso station. In both cases, the results were compared with those from continuous simulation, and were found to be accurate.

Due to the shorter rainfall durations compared to the corresponding dry spell between the rainfall events, some researchers have considered rainfall arrival as a marked Poisson's process, and modeled rainfall characteristics stochastically [77–80]. Wang and Guo (2019) [81] used analytical stochastic models (ASM) to describe the runoff capture efficiency of detention ponds as a power function, rather than linear. The ASM results were compared with the results of an SWMM continuous simulation using a case study catchment area located in Jackson, Mississippi. The values of the root mean square error (RMSE), Nash–Sutcliff efficiency (NSE) and correlation coefficient (R) for runoff capture efficiency were 0.021, 0.994 and 0.9983, while these values for average pond fullness level were 0.012, 0.998 and 0.9997, respectively. This indicates the applicability of ASMs.

Stormwater retention basins can also be analyzed by using stochastic water balance to develop analytical models. Parolari et al. (2018) [82] developed a stochastic water balance model of stormwater retention ponds under passive and active outlet conditions. Analytical expressions of the steady-state and joint PDF of water level and valve closure time, which can be used to define the water level and flow duration curves of the basin, were derived. The model's performance was tested by taking observations of water levels from a retention pond located in Ann Arbor, MI, USA. He results show that the model accurately predicts the water level PDF, which can be used to form a basis for evaluating the changes in rainfall–runoff due to climate change and land-use.

Stormwater detention tanks are used to mitigate the impact of sewer overflow. Balistrocchi et al. (2009) [48] applied APMs to develop a CDF of the overflow volume and pollutant load distribution of a sewer tank. Weibull distribution was used to model rainfall characteristics. Analytical expressions of performance indices such as the decrease in the annual runoff volume and ratio of pollutant load captured by the tank were derived. The model was verified with SWMM continuous simulation, using the urban catchment of Brescia, Italy, and the results were found to be satisfactory. Andres-Domenech et al. (2010) [50] derived analytical PDFs of the number of overflows, volume of overflows and overflow reduction efficiency of a stormwater tank. Rainfall data from Valencia and Santander, Spain using different probability distributions were tested. Pareto and Gamma-2 PDFs were found to fit well. The analytical results regarding long-term volumetric flow and overflow reduction efficiencies were compared with those of IW continuous simulation, and were found to be similar. Becciu and Raimondi (2012) [83] developed APM expressions for the pre-filling probability of stormwater tanks. The effects of minimum inter-event time definition on outflow rate and storage volume were investigated using rainfall data from Monviso, Milano, Italy. The results of the APM were compared to the results of continuous simulation, and it was shown that the APM underestimated the pre-filling probability due to some assumptions made in the development of the model. Thus, the model needs to be refined further. Stormwater tanks, designed using APM, have also been found to be capable of improving the quality of sewer discharges from catchments along the Tyrrhenian coast of Italy [84].

Detaining runoff in stormwater detention ponds for a longer period improves the quality of the treated runoff, but this poses the risk of overflow from subsequent rainfall, which may generate runoff. There is an optimal detention time in the facilities such that the trade-off between runoff and pollution control is addressed [18]. There is also a need to minimize the cost of building the facility, while at the same time achieving the objectives. Papa and Adams (1997) [85] used APM expressions to develop a dynamic programming model for the optimization of the cost of building detention ponds in multiple parallel catchments, subject to meeting runoff quality control constraints. Shamsudin et al. (2014) [86] used long-term rainfall data to obtain the rainfall characteristics and develop APM parameters for a catchment in Malaysia. The APM was coded via particle swam optimization (PSO) to develop a methodology that addresses the trade-off between the runoff and pollution control performances of detention ponds. The detention pond's volume and outlet were appropriately sized such that a least-cost design was obtained.

Dan'azumi et al. (2013a) [35] developed APM parameters relevant to the rainfall characteristics of Malaysian cities, and used the parameters to develop an optimization algorithm via PSO that can be used to optimize detention time in wet detention ponds such that they give the best pollution control performance [87]. Behera and Teegavarapu (2015) [88] used the APM expressions of pollution control in extended wet detention ponds to compare the results of three optimization techniques: dynamic programming (DP), non-linear programming (NLP) and genetic algorithm (GA). They sought to obtain the optimal values of pollution control, pond depth, storage volume and release rate of ponds treating urban stormwater from multiple sub-catchments releasing their outflow into a common downstream point, such that the quality control target at the downstream river could be met at minimum cost. The results show that the NLP and GA provided an improved solution compared to the DP.

4.2. *Rainwater Harvesting System/Rainwater Tanks*

Rainwater tanks, consisting of rain-barrels and cisterns, are rainwater harvesting systems (RHS) that store rainwater for household use and reduce the volume of runoff generated from urban surfaces. The use of rainwater tanks reduces water consumption from municipal supply, and thus reduces the water bill. The water stored in the tank can be used for gardening and toilet flushing, thus reducing municipal water consumption. Some rainwater tanks have two compartments: the rainwater tank itself and an infiltration facility, which aids in groundwater recharge [89,90].

Raimondi and Becciu (2014a) [89] developed APM expressions to estimate the probability of meeting the water demand using rainwater tanks as a function of household population and number of storm events occurring, using long-term rainfall data from 35 years at the Milano-Monviso station. The results of the study can be used to determine whether it is efficient to use rainwater harvesting alone, or in combination with municipal water supply. Raimondi and Becciu (2014b) [90] developed APMs for the design of multi-use rainwater tanks. These rainwater tanks were designed to have two basins: a rainwater basin and an infiltration basin. A trade-off between the risk of water shortage in the basin and the risk of overflow was studied. The results of a case study in a catchment in Milan, Italy, show that the probability of complete rainwater use in a household depends on the period of regulation, with weekly regulation yielding a higher probability compared to daily regulation. Additionally, the probability of overflow was high for a small storage volume and low infiltration rate. Becciu et al. (2016) [91] improved on their previous models by considering the effect of re-filling during the regulation period, and developed an analytical expression to estimate the CDF of active storage in the rainwater tank. The results were compared with those of a continuous simulation model using data from Milano, Italy, and there was a good agreement.

Guo and Baetz (2007) [78] derived an analytical expression that could be used to design rainwater storage units in green buildings, focusing on the rate of water use in the building, the climate characteristics of the area and the reliability of the system. The APM was applied to a hypothetical catchment in Chicago and Montana, USA, and it was shown that the APM provided an efficient approach to designing the system. De Paola and De Martino (2013) [92] studied the efficiency of four stormwater tank configurations using SWMM, and applied the semi-probabilistic approach to determine the qualitative and quantitative stormwater capture efficiencies of the most efficient tank configuration. It was concluded that the analytical approach provided similar results to continuous simulation. Kim et al. (2012) [93] used mass balance equations for each component of a rainwater tank to develop APM expressions for the rainfall–runoff reduction in an RHS. The PDF and CDF of runoff from the catchment and the RHS were derived, and the expected value of runoff volume was determined. The model was applied to a dormitory building in Seoul (Korea) to design an RHS and to estimate the runoff reduction achieved as a result of it. Di Chiano et al. (2023) [94] used APM expressions to derive the CDF of active storage in RHS. Active storage was considered as a function of rainfall moments, water demand and mean number of chained events under deficit conditions. The results of the model were compared with those of continuous simulation, using rainfall data from Monviso, Milano (Italy), focusing on a case study of RHS in Milan. An average normalized RMSE of 0.033, under three demand conditions, was obtained between the APM and the continuous simulation, suggesting a very good prediction.

Stochastic mass balance equations of RHS have been used to develop analytical models for RHS systems. Guo and Guo (2018a) [95] derived an ASM that could be used to determine the size of an RHS using a differential mass balance equation. Analytical expressions of a rainwater tank's efficiency in terms of water supply reliability, required storage volume and its runoff reduction benefits were derived. The stochastic models, developed using rainfall data from five different climates (Atlanta, Concord, Detroit, Flagstaff and Billings) in the USA, were validated against the results obtained from SWMM continuous simulation and also those of Guo and Baetz (2007) [78]. The values of mean Nash–Sutcliffe efficiency (NSE), root mean square error (RMSE) and correlation coefficient of 0.98, 0.035 and 0.99, respectively, were obtained, indicating a good result. Pelak and Porporato (2016) [96] modeled rainfall as a marked Poisson's process, and developed an analytical expression that optimizes the volume of a rainwater harvesting system at minimum cost. The volume was expressed as a function of rainfall parameters, roof area, water use rate, and the cost of the cistern and that of the external water source. The cost consists of fixed and distributed costs. The result of the study can be used to size an RHS in any climate. This will help reduce urban stormwater runoff and water consumption from public mains. Sim and Kim (2020) [97] used stochastic mass balance to develop an analytical model for the quantification of the water supply and stormwater interception efficiency of an RHS. In the study, the sensitivity of the RHS to climate change was evaluated, and the model was assessed using rainfall data from Busan (Korea). The results of the analytical model were compared with those derived using multiple regression. The R^2 and RMSE values for water supply and stormwater interception efficiency ranged from 0.91 to 0.96 and 0.026 to 0.033, respectively. Cheng et al. (2021) [98] also used water balance to develop a stochastic model of an RHS. Due to the random occurrence of rainfall, the reliability of the model was expressed in terms of the fraction of time for which the RHS satisfies water demand. The model was applied to three RHSs in Toronto, Canada, and was found to have high accuracy.

4.3. Green Roofs

A green roof is a rooftop garden. These are used to provide shade, reduce the temperatures of the roof surface and surrounding air, and to moderate the heat island effect [99]. Green roofs comprise four layers: a vegetation layer, a substrate layer, a drainage layer and a waterproof layer. Some green roofs have a water storage layer combined with the drainage layer for holding more rainwater. Vegetation is planted on top of the substrate layer, where rainwater is retained. Excess rainwater from the roof is drained through the drainage layer [100].

Researchers also explored the application of APMs to green roof design and analysis. Zhang and Guo (2013a) [79] derived analytical expressions for runoff generation from green roofs. The results obtained from the analytical models were compared with those of continuous simulation using the LID module of SWMM, and also from the field results derived from a real case study in Portland, USA. The results of the APM were found to be in good agreement with both. Additionally, Guo et al. (2014) [101] derived analytical expressions for long-term average runoff reduction rates (defined as the ratio of total runoff captured to that of total runoff generated) and the irrigation water requirement of green roofs. The performance values of the APM in terms of runoff reduction rates and irrigation time fraction at different growing medium depths under semi-arid climate (Atlanta, USA) and humid climate (Billings, USA) conditions were compared with those from continuous simulation using SWMM, and it was concluded that the APM can be used as an alternative to SWMM in the planning, design and management of green roofs.

Guo (2016) [102] further refined the work of [101] by considering rainfall occurrence as a stochastic process to derive a stochastic differential equation of green roofs. The stochastic water balance equation was formulated to determine the mean and PDF of the moisture contents of green roofs. The accuracy of the model, in terms of runoff reduction rates and irrigation time fraction, was evaluated by testing the results against those of SWMM continuous simulation using four sets of rainfall data from Billings, Phoenix, Atlanta and Boston (USA), and using sandy and loamy soils as growing media. The comparison of results between SWMM and the stochastic model implied the good correlation coefficients of 0.993 and 0.995, respectively, for runoff reduction rates and irrigation time fraction. Most studies assume that the RHS is dry at the beginning of the rainfall event. However, some moisture retention is possible in the roof at the beginning of the next rainfall period. Raimondi and Becciu (2020) [100] considered the possibility of pre-filling from previous rainfall events to develop an APM for the design of green roofs. The APM was tested against the results of continuous simulation, using rainfall data from the Milano Monviso (Italy) station. The results show that the model compared well with continuous simulation. Thus, the APM can be used for the optimization of the design of green roofs. Raimondi et al. (2022a) [103] extended their work from 2020 to develop an APM that could be used to determine the thickness of the substrate layer of green roofs as a function of runoff reduction. The results of the study compare well with those obtained from continuous simulation.

Raimondi et al. (2022b) [104] used APM to develop a model for the survival of vegetation on green roofs without the need for irrigation. The thickness of substrate medium and risk of vegetation withering were combined to design the green roof. The APM was tested using two green roofs in Milano and Calabria (Italy). The results from the analytical model compared excellently with those of continuous simulation. Guo et al. (2022) [105] used a stochastic model of rainfall to model the hydrologic and hydraulic process occurring on green roofs. Both the saturation excess runoff and infiltration excess runoff were considered, and analytical equations that can be used for the quantification of the performance of green roofs were derived. The results of the analytical model were compared with those of continuous simulation, and were found to be accurate. Raimondi et al. (2023) [106] used APMs to determine the probability that runoff from a green roof will exceed a certain threshold, given the substrate thickness, climatic variable and moisture content of the roof. The reduced retention capacity of the system due to previous rainfall was also considered. The analytical model was tested using a case study in Milano (Italy), and the results were similar between the model and the continuous simulation.

4.4. Filtration Practices

Filtration practices are surface or underground practices that reduce the volume of runoff by infiltration through the soil. They provide a performance that is independent of local conditions, and their designs are applicable to roadside and congested urban conditions. According to [107], bioretention cells and sand filters are amongst the filtration practices commonly used for small to medium catchment basins, because they usually occupy only 2 to 3% of the drainage area, and hence are suitable in dense urban settings. Sand and gravel filters are also commonly used as filtration practices for the management of urban stormwater [108]. Other filtration practices include pervious pavements, etc.

4.4.1. Bioretention Cells/Biofilters/Rain Gardens/Impervious Area Disconnection

Bioretention systems are shallow landscaped depressions, commonly located in parking lots or within residential land-use areas, that are designed to incorporate many of the pollutant removal mechanisms that are operated in forested ecosystems. They are also known as biofilters or rain gardens [13]. Stormwater treatment in a bioretention cell is achieved through sedimentation, filtration, soil adsorption, micro-biological decay processes and the uptake of pollutants by plants [109]. The components of a bioretention area include a grass buffer strip, planting soil, plant material, a ponding area with surface mulch, an underground sand bed, an organic layer and infiltration chambers [110].

Daly et al. (2011) [13] tracked the water balance of a biofilter by considering its inflow variability, filter media and vegetation type. An analytical model for the long-term PDF of soil moisture content of the filter, and the statistics of outflow, evapotranspiration and overflow, were derived. The total nitrogen removal performance was also estimated from the model. The results of the analytical model were tested with real data collected from a biofilter in Malborne, Australia, and it was shown that the model could be used to assess the performance of biofilters across different climates.

Zhang and Guo (2014) [111] modeled runoff from both pervious and impervious urban surfaces to develop closed-form APM expressions for the stormwater runoff capture efficiency of bioretention cells. The results obtained were compared to those of an SWMM continuous simulation model, and close agreement was observed, thus validating the APM expression. However, some assumptions were made regarding the amount of water present in the cells before any rainfall event, which need to be considered in extreme cases. Accordingly, Guo et al. (2020) [112] refined the work of [111] to consider wider ranges of application. A dynamic water balance was considered to stochastically model the hydrology of bioretention cells. Analytical expressions for the long-term runoff capture efficiency, the fraction of time for which the cell processes runoff, the average water stored inside a cell, and the storage capacity required to achieve capture efficiency were derived. These four performance indices generated by the ASM were compared with the results of continuous simulation, and close agreements were obtained, thus verifying the applicability of the ASM.

The resilience and reliability of using bioretention cells as runoff control systems was studied by [113]. APM expressions were used to evaluate the resilience indices related to the system's robustness, rapidity and serviceability in the context of extreme runoff events. The results of the APM were compared with those generated using the continuous simulation SWMM. Resilience indices of 0.66 to 1.0 and 0.73 to 1.0, respectively, were observed for the APM and SWMM. The reliability index found ranged from 56% to 100% and 60% to 100% for the APM and SWMM, respectively.

Impervious area disconnection is a system that works in a similar way to bioretention cells. Runoff from urban surfaces (roof tops, pavements) is made to pass through pervious surfaces (grassed area), where processes such as infiltration and pollutant removal occur, thus reducing the volume of surface runoff. The time of concentration in the catchment is also reduced, thereby reducing the peak discharge from the catchment. Wang et al. (2019) [81] determined the effect of impervious area disconnection on runoff reduction from two urban catchments in the USA. Two different catchments' soils (sandy and loamy) were used. The runoff reduction due to impervious area disconnection was examined using different imperviousness ratios. The results of the APM and SWMM were compared, and in all cases, impervious area disconnection was found to significantly reduce the volume of runoff to the sewer system, and the APM results compared very well with those of the SWMM. Zhang and Guo (2013b) [80] studied the hydrologic operations of a rain garden to derive analytical expressions for its long-term runoff capture efficiency. The results from the APM model were compared with those of SWMM simulations, and a very good agreement between the APM and continuous simulation results was observed. The APM was applied to rain gardens in Atlanta and Flagstaff in USA to demonstrate the sensitivity of runoff capture efficiency to specific model parameters.

4.4.2. Infiltration Trenches/Basins

Infiltration trenches are rectangular excavations with void-forming materials, such as gravel aggregates, which are designed to receive, filter, store and infiltrate urban stormwater. They aid in reducing urban runoff and improving groundwater recharge. They also assist in sediment and heavy metal removal from stormwater [114]. Guo and Gao (2016) [115] derived analytical expression for the total annual overflow volume and total runoff reduction rate of infiltration basins. The results of the APM were compared with those of SWMM continuous simulation using rainfall data from Atlanta and Billings (USA), and the results were found to be consistent, with a relative difference of less than 10%. Guo and Guo (2018b) [116] derived APM expressions for the overflow frequency and stormwater capture efficiency of non-vegetated infiltration facilities, such as infiltration trenches, infiltration chambers, dry wells, etc. In deriving the expressions, infiltration was assumed to occur at the bottom only. The results from the analytical expressions were compared with those of SWMM simulations in relation to a case study of a catchment in Concord, New Hampshire (USA), using sandy and loamy soils, and the two sets were found to be in good agreement. The average absolute difference and average relative difference between the APM and SWMM were found to be 0.04% and 5%, respectively. Wang and Guo (2020a) [117] analyzed the water balance of infiltration-based BMPs by considering infiltration through their sides and bottom, in an attempt to overcome the shortcomings of [108]. The mean degree of saturation and mean runoff capture efficiency were derived, and the results of the analytical model were compared with those of SWMM. Two soils, sandy and loamy, were used, and the rainfall data from two climate conditions (Billings and Jackson) were used to develop the APM model. The results were found to be reasonably comparable, with the largest absolute relative difference being less than 10%.

Following the design guidelines released by the Atlanta and New Durham authorities, Wang and Guo (2020b) [118] applied the analytical models they had developed earlier in [117] to a practical design analysis of infiltration trenches. Runoff values, generated using rainfall data from hypothetical catchments, in the two locations were assumed, and the performance of the trench was assessed as a function of its soil type, footprint dimensions, drain time and infiltration conditions. The results of the runoff reduction ratio indicate that the conditions of infiltration through the sides, the bottom, and both combined have profound effects on the runoff reduction ratio, by up to 15%. The runoff reduction ratio was found to be most highly affected by changes in soil type and trench dimensions.

4.4.3. Permeable Pavements

Pervious pavements consist of pavements made with porous blocks or porous asphalt that permits water to infiltrate. Pervious pavements may also be made from impervious blocks that are fitted in such a way that water can pass between them. They can be used in road surfaces with light traffic or in car parks. The infiltration rate through the pavement may be as high as 1000 mm/h in new developments, although this value may reduce to 10% of the original value over the lifetime of the pavement [119]. Zhang and Guo (2015) [120] derived analytical expressions for the runoff capture efficiency of a permeable pavement as an LID system to mitigate the impact of urban stormwater. SWMM simulations were run on the modeled pavement in order to validate the APM expressions, and the results showed little discrepancy. It is recommended that the APM results be compared with those of a case study on real-life pavements.

Stochastic differential equations of permeable pavements were used by Guo et al. (2018) [121] to model the dynamic water balance of their system. Rainfall and the corresponding net inflow were represented as a marked Poisson process to develop the PDF of inflow volume, and to derive analytical expressions for the long-term stormwater capture efficiency and moisture content of permeable pavements. The results of the APM were compared with those simulated using SWMM using data from the four climates of Atlanta, New Durham, Charlotte and Flagstaff (USA), and were found to be very similar.

Three runoff control systems—bioretention cell, permeable pavement and green roofs—were compared to determine the most cost-effective. The runoff reduction efficiencies and licecycle costs of implementing each of them were considered. APM expressions were combined with a genetic algorithm for the optimization. The objective function was to maximize runoff reduction capacity and minimize the lifecycle cost. The results show that the bioretention cell had the greatest runoff reduction capability, but given the high land cost in urbanized areas, permeable pavements are the most reasonable option [122].

4.5. Vegetated Open Channel Practices

These are systems designed to treat stormwater runoff in a swale or channel formed by check dams or other processes. Usually, they do not allow for quantity control, and are therefore combined with other stormwater BMPs to meet regulations. These systems directly receive runoff from an impervious surface; they have a temporary ponding time of less than 48 h and feature a 6-inch drop onto a protected shelf to minimize the clogging potential of the inlet [4]. Up to the time of submitting this review, no publications have been found on the application of APMs to vegetated open channel practices. This issue can be explored by future researchers. Two of the common types of vegetated open channel practices include grassed swales (dry/wet) and grassed channels.

According to [123], grassed swales are broad, shallow earthen channels used to treat stormwater runoff using flood-tolerant and erosion-resistant grasses. Filtering via these practices occurs through the vegetation, a subsoil matrix, and infiltration into the underlying soils. Grassed swales feature gentle longitudinal slopes, with check dams perpendicular to the flow so as to slow down the flows and allow the particulates to settle. There are two types of grass swales—dry swales, which have a filter bed of prepared soil laid over an under-drain system, and wet swales, which are designed to sustain moisture conditions that support wetland vegetation [124].

Grassed channels are used in pretreatment practices that provide nominal treatment, because they lack the filter media present in grassed swales. They act by allowing the infiltration of some runoff from small storms into areas with pervious soils, and are therefore most highly applicable to other structural stormwater BMPs [123]. They help in reducing the effect of imperviousness, and provide aesthetic benefits. Grassed channels are designed for use on <4% flat slopes with infiltration rates greater than 0.27 inches per hour. The stormwater runoff takes an average of 5 min to flow from the top to the bottom of the channel. For efficient usage, the channel should be used to treat small drainage areas of less than 5 acres. For the effective removal of particles, the grass of the channel should be maintained at a height of 3 to 4 inches [4].

4.6. Other Stormwater BMPs

Other types of stormwater BMPs that are used to control urban stormwater runoff include: constructed wetlands, dry wells, artificial marshes, oil/greet separators, catch basins, etc. [119,125]. Their effectiveness can be represented via a decrease in the SCS curve number of the basins. Perez-Pedini et al. (2005) [126] determined the optimal number and location of infiltration facilities in a watershed for the purpose of peak flow reduction at the watershed outlet. The watershed was discretized into 4553 hydrologic response units, whereby each unit represents a 120×120 m plot of the watershed. Different types of infiltration-based BMPs were conceptualized as binary integers that decrease the curve number of hydrologic response unit by five. The results of the optimization show that the optimal number and location of infiltration-based BMPs depends on various factors, such as flow travel time, catchment network connectivity, land-use, contributing area, and distance to the channel. APMs of stormwater management can be applied to constructed wetlands, dry wells, artificial marshes, catch basins, etc., in future studies.

5. Recommendations for Future Direction

Urban stormwater raises flood and water pollution problems for many communities across the world, and the cost of the damage cannot be easily quantified. This paper has reviewed the literature on APM applications in urban stormwater management. Once derived, the APM models can easily be co-opted into any optimization frameworks, thereby giving the freedom to maximize benefits and minimize cost. The following recommendations are given:

(a) The APM parameters were obtained from analyses of the long-term data on rainfall depth, duration and inter-event time. To make them more applicable, it is necessary to develop a comprehensive database of APM parameters describing rainfall characteristics in cities across the world, for the purpose of runoff control systems design;

(b) Most rain-gauge stations, particularly in developing countries, record daily rainfall only. Urban catchments have shorter times of concentration, and studies in these parts of the world have to rely on rainfall disaggregation techniques, whereby daily rainfall is broken down to an hourly or even sub-hourly time scale, which may raise some reliability problems. There is a need for a database of finer-resolution rainfall data. The provision of a large network of hourly and sub-hourly rain-gauge data will not only be useful to urban hydrologists, but also to other professionals. It will also help in reducing the uncertainties caused by rainfall disaggregation. Another source of uncertainty is the spatial distribution of rain-gauge stations used to develop rainfall characteristics and APM parameters. Research is required into the effects of the spatial distribution of rain-gauge stations on the reliability of rainfall characteristics;

(c) The APM parameters were derived based on minimum inter-event times of 2 h, 6 h, 12 h and 24 h. In the case of small urban catchments, with faster concentration, it is recommended that a database of APM parameters based on a smaller discretized inter-event time, such as 5 min, 15 min, 30 min or 1 h, be developed. This requires the archiving of rainfall data at a sub-hourly resolution, which could then be used to develop its own database;

(d) There is uncertainty about the inter-event time value to be used in rainfall event aggregation from a continuous time series. This calls for further research on its reliability;

(e) The APMs are mostly based on the exponential distribution of rainfall characteristics. Rainfall characteristics were also found to follow other distributions, such as Gamma, Weibull, and log-normal. A distribution fit test for other PDFs needs to be undertaken in different climates;

(f) A decision support system that incorporates meteorological, catchment and runoff control systems' characteristics altogether needs to be developed, which can then eventually be used in the design and real-time control of these systems;

(g) The design of some systems, such as rainwater tanks, involves the consideration of rainfall variability vis-à-vis water demand and the cost of municipal water consumption. Likewise, designing detention ponds for runoff quantity and pollution control involves conflicting objectives. There is a need for studies that embed APMs into optimization techniques so as to derive optimum benefits from the runoff control systems at the least cost;

(h) Climate change is known to affect the design of stormwater conveyance and storage systems. There is a research gap regarding the effect of climate change on runoff control systems designed using APMs. The impact of climate change on the reliability of the systems needs to be investigated, so as to ensure their design functions are met;

(i) There is a research gap regarding the APMs related to the runoff reduction efficiency and pollution control performance of vegetated open channel technologies, such as swales, grass channel, etc.;

(j) Different runoff control systems have been reviewed in this paper. Some systems may be more suitable to specific climates. There is a research gap in the determination of the best system for each specific geographical area.

6. Conclusions

Urban stormwater runoff is detrimental to downstream drainage systems and to receiving water bodies. The risks range from flooding to water pollution. This paper has reviewed the literature on runoff control systems, such as detention basins, rain gardens, rainwater harvesting system, bioretention cells, pervious pavements, infiltration trenches, etc. The design of runoff control systems can be carried out using the traditional design storm approach, continuous simulation and APMs. The major flaw of the design storm approach is its inability to capture the effects of inter-event time. That is, the design storm assumes that the recurrence interval of runoff is the same as that of the rainfall that causes it. The continuous simulation approach, on the other hand, is laborious and time-consuming, thus making it unsuitable for use at the planning stage of a runoff control project. APMs, however, are more compact, easy to use, and offer a direct way to conduct sensitivity analyses in routine planning projects. Moreover, APMs are flexible and can be co-opted into an optimization framework. Despite their simplicity, the APMs provide results that are as accurate as those of continuous simulation. This paper offers an extensive review of the applications of APMs to urban stormwater management.

Author Contributions: S.D. and A.A. jointly contributed to the production of this review paper. All authors have read and agreed to the published version of the manuscript.

Funding: This research was funded by Prince Sattam bin Abdulaziz University project number PSAU/2022/01/23514.

Acknowledgments: The authors extend their appreciation to Prince Sattam bin Abdulaziz University for funding this research work through the project number PSAU/2022/01/23514. We also acknowledge the inputs of anonymous reviewers.

Conflicts of Interest: The authors declare no conflict of interest. The funders had no role in the design of the study; in the collection, analyses, or interpretation of data; in the writing of the manuscript; or in the decision to publish the results.

Abbreviations

ACRU	Agricultural Catchments Research Unit
APM	Analytical Probabilistic Models
ASM	Analytical Stochastic Models
ASTORM	Analytical STORM
BMPs	Best Management Practices
CDF	Cumulative Distribution Function
COD	Chemical Oxygen Demand
DP	Dynamic Programming
EMC	Event Mean Concentration
EX-SUDS	Extended Sustainable Urban Drainage System
GA	Genetic Algorithm
HEC-HMS	Hydrologic Engineering Center's Hydrologic Modeling System
IDF	Intensity–Duration–Frequency
MIDUSS	Microcomputer Interactive Design of Urban Stormwater Drainage Systems
NLP	Non-Linear Programming
NSE	Nash–Sutcliff Efficiency
OTTHYMO	Ottawa Hydrological Model
PC-SWMM	Personal Computer–Storm Water Management Model
PDF	Probability Density Function
PSO	Particle Swam Optimization
RMSE	Root Mean Square Error
SCN-CN	Soil Conservation Service–Curve Number
STORM	Stormwater Management Software
SUDS	Sustainable Urban Drainage system
SWMM	Storm Water Management Model

References

1. Steg, R. *Summary of the Stormwater Sources in the Flatheadd Lake Basin*; Montana Operations Office, USEPA: Montana, WY, USA, 2010.
2. McCuen, R.H. *Hydrologic Analysis and Design*; Prentice Hall: Upper Saddle River, NJ, USA, 2004.
3. Birkinshaw, S.J.; Kilsby, C.; O'donnell, G.; Quinn, P.; Adams, R.; Wilkinson, M.E. Stormwater Detention Ponds in Urban Catchments—Analysis and Validation of Performance of Ponds in the Ouseburn Catchment, Newcastle upon Tyne, UK. *Water* **2021**, *13*, 2521. [CrossRef]
4. GSMM. General Application Structural Stormwater Controls. Georgia Stormwater Management Manual, Vol. 2. Atlanta Re-gional Commission, GA, 2016. Available online: http://www.georgiastormwater.com/ (accessed on 3 October 2022).
5. Guo, Y. Hydrologic Design of Urban Flood Control Detention Ponds. *J. Hydrol. Eng.* **2001**, *6*, 472–479. [CrossRef]
6. Methods, H.; Durrans, S.R. *Stormwater Conveyance Modeling and Design*; Bently Institute Press: Exton, PA, USA, 2001.
7. Jia, H.; Hu, J.; Huang, T.; Chen, A.S.; Ma, Y. Urban Runoff Control and Sponge City Construction. *Water* **2022**, *14*, 1910. [CrossRef]
8. Jeng, H.A.C.; Englande, A.J.; Bakeer, R.M.; Bradford, H.B. Impact of urban stormwater runoff on estuarine environmental quality. *Estuar. Coast. Shelf Sci.* **2005**, *63*, 513–526. [CrossRef]
9. Todeschini, S. Hydrologic and Environmental Impacts of Imperviousness in an Industrial Catchment of Northern Italy. *J. Hydrol. Eng.* **2016**, *21*, 05016013. [CrossRef]
10. Schütte, S.; Schulze, R. Projected impacts of urbanisation on hydrological resource flows: A case study within the uMngeni Catchment, South Africa. *J. Environ. Manag.* **2017**, *196*, 527–543. [CrossRef] [PubMed]
11. USEPA. *Preliminary Data Summary of Urban Stormwater Best Management Practices*; Report No EPA-821-R-99-012; United States Environmental Protection Agency: Washington, DC, USA, 1999.
12. Sharifi, S.; Massoudieh, A.; Kayhanian, M. A Stochastic Stormwater Quality Volume-Sizing Method with First Flush Emphasis. *Water Environ. Res.* **2011**, *83*, 2025–2035. [CrossRef]
13. Daly, E.; Deletic, A.; Hatt, B.E.; Fletcher, T.D. Modelling of stormwater biofilters under random hydrologic variability: A case study of a car park at Monash University, Victoria (Australia). *Hydrol. Process.* **2011**, *26*, 3416–3424. [CrossRef]
14. Li, Y.; Babcock, R.W. Green roof hydrologic performance and modeling: A review. *Water Sci. Technol.* **2013**, *69*, 727–738. [CrossRef]
15. Krasnogorskaya, N.; Longobardi, A.; Mobilia, M.; Khasanova, L.F.; Shchelchkova, A.I. Hydrological Modeling of Green Roofs Runoff by Nash Cascade Model. *Open Civ. Eng. J.* **2019**, *13*, 163–171. [CrossRef]
16. Berndtsson, J.C. Green roof performance towards management of runoff water quantity and quality: A review. *Ecol. Eng.* **2010**, *36*, 351–360. [CrossRef]
17. Qi, W.; Ma, C.; Xu, H.; Chen, Z.; Zhao, K.; Han, H. A review on applications of urban flood models in flood mitigation strategies. *Nat. Hazards* **2021**, *108*, 31–62. [CrossRef]
18. Ellis, J. Sustainable surface water management and green infrastructure in UK urban catchment planning. *J. Environ. Plan. Manag.* **2013**, *56*, 24–41. [CrossRef]
19. Maes, J. and Jacobs, S. Nature-based solutions for Europe's sustainable development. *Conserv. Lett.* **2017**, *10*, 121–124. [CrossRef]
20. O'Donnell, E.; Thorne, C.; Ahilan, S.; Arthur, S.; Birkinshaw, S.; Butler, D.; Dawson, D.; Everett, G.; Fenner, R.; Glenis, V.; et al. The blue-green path to urban flood resilience. *Blue-Green Syst.* **2019**, *2*, 28–45. [CrossRef]
21. Yin, D.; Xu, C.; Jia, H.; Yang, Y.; Sun, C.; Wang, Q.; Liu, S. Sponge City Practices in China: From Pilot Exploration to Systemic Demonstration. *Water* **2022**, *14*, 1531. [CrossRef]
22. Jia, H.; Wang, Z.; Zhen, X.; Clar, M.; Yu, S.L. China's sponge city construction: A discussion on technical approaches. *Front. Environ. Sci. Eng.* **2017**, *11*, 18. [CrossRef]
23. Debo, T.N.; Reese, A.J. *Municipal Stormwater Management*; Lewis Publishers: Boca Raton, FL, USA, 2003.
24. Chen, J.; Adams, B.J. A derived probability distribution approach to stormwater quality modeling. *Adv. Water Resour.* **2007**, *30*, 80–100. [CrossRef]
25. Adams, B.J.; Papa, F. *Urban Stormwater Management Planning with Analytical Probabilistic Models*; John Wiley & Sons: New York, NY, USA, 2000.
26. Chen, J.; Adams, B.J. Development of analytical models for estimation of urban stormwater runoff. *J. Hydrol.* **2007**, *336*, 458–469. [CrossRef]
27. Todeschini, S.; Papiri, S.; Ciaponi, C. Performance of stormwater detention tanks for urban drainage systems in northern Italy. *J. Environ. Manag.* **2012**, *101*, 33–45. [CrossRef]
28. Mobley, J.T.; Culver, T.B. Design of outlet control structures for ecological detention ponds. *J. Water Res. Plan. Manag. ASCE* **2014**, *140*, 250–257. [CrossRef]
29. Souza, F.P.; Costa, M.E.L.; Koide, S. Hydrological Modelling and Evaluation of Detention Ponds to Improve Urban Drainage System and Water Quality. *Water* **2019**, *11*, 1547. [CrossRef]
30. Mielke, P.W.; Benjamin, J.R.; Cornell, C.A. Probability, Statistics and Decision for Civil Engineers. *J. Am. Stat. Assoc.* **1971**, *66*, 923. [CrossRef]
31. Eagleson, P.S. Dynamics of flood frequency. *Water Resour. Res.* **1972**, *8*, 878–898. [CrossRef]
32. Adams, B.J.; Fraser, H.G.; Howard, C.D.D.; Hanafy, M.S. Meteorological Data Analysis for Drainage System Design. *J. Environ. Eng.* **1986**, *112*, 827–848. [CrossRef]
33. Bacchi, B.; Balistrocchi, M.; Grossi, G. Proposal of a semi-probabilistic approach for storage facility design. *Urban Water J.* **2008**, *5*, 195–208. [CrossRef]

34. Branham, T.L.; Behera, P.K. Development of a Rainfall Statistical Analysis Tool for Analytical Probabilistic Models for Urban Stormwater Management Analysis. In Proceedings of the World Environmental and Water Resources Congress, Providence, RI, USA, 16–20 May 2010. [CrossRef]
35. Dan'azumi, S.; Shamsudin, S.; Aris, A. Development of analytical probabilistic model parameters for urban stormwater management. *Sains Malays.* **2013**, *42*, 325–332.
36. Guo, Y.; Adams, B.J. Hydrologic analysis of urban catchment with event based probabilistic models, Runoff volume. *Water Resour. Res.* **1998**, *34*, 3421–3432. [CrossRef]
37. Guo, Y.; Adams, B.J. Hydrologic analysis of urban catchment with event based probabilistic models, 2. Peak discharge rate. *Water Resour. Res.* **1998**, *34*, 3433–3443. [CrossRef]
38. Chen, J.; Adams, B.J. Urban Storm Water Control Evaluation with Analytical Probabilistic Models. *J. Water Resour. Plan. Manag.* **2005**, *131*, 362–374. [CrossRef]
39. Chen, J.; Adams, B.J. Analysis of storage facilities for urban stormwater quantity control. *Adv. Water Resour.* **2005**, *28*, 377–392. [CrossRef]
40. Chen, J.; Adams, B.J. Analytical urban storm water quality models based on pollutant buildup and wash-off processes. *J. Environ. Eng. ASCE* **2006**, *132*, 1314–1330. [CrossRef]
41. Chen, J.; Adams, B.J. Urban stormwater quality control analysis with detention ponds. *Water Environ. Res.* **2006**, *78*, 744–753. [CrossRef] [PubMed]
42. Guo, Y.; Adams, B.J. An analytical probabilistic approach to sizing flood control detention facilities. *Water Resour. Res.* **1999**, *35*, 2457–2468. [CrossRef]
43. Guo, Y.; Adams, B.J. Analysis of detention ponds for stormwater quality control. *Water Resour. Res.* **1999**, *35*, 2447–2456. [CrossRef]
44. Li, J.Y.; Adams, B.J. Probabilistic Models for Analysis of Urban Runoff Control Systems. *J. Environ. Eng.* **2000**, *126*, 217–224. [CrossRef]
45. Behera, P.K.; Adams, B.J.; Li, J.Y. Runoff Quality Analysis of Urban Catchments with Analytical Probabilistic Models. *J. Water Resour. Plan. Manag.* **2006**, *132*, 4–14. [CrossRef]
46. Quader, A.; Guo, Y. Peak Discharge Estimation Using Analytical Probabilistic and Design Storm Approaches. *J. Hydrol. Eng.* **2006**, *11*, 46–54. [CrossRef]
47. Hassini, S.; Guo, Y. Exponentiality Test Procedures for Large Samples of Rainfall Event Characteristics. *J. Hydrol. Eng.* **2016**, *21*, 04016003. [CrossRef]
48. Balistrocchi, M.; Grossi, G.; Bacchi, B. An analytical probabilistic model of the quality efficiency of a sewer tank. *Water Resour. Res.* **2009**, *45*, W12420. [CrossRef]
49. Palynchuk, B.; Guo, Y. Threshold analysis of rainstorm depth and duration statistics at Toronto, Canada. *J. Hydrol.* **2008**, *348*, 535–545. [CrossRef]
50. Andrés-Doménech, I.; Montanari, A.; Marco, J.B. Stochastic rainfall analysis for storm tank performance evaluation. *Hydrol. Earth Syst. Sci.* **2010**, *14*, 1221–1232. [CrossRef]
51. Szeląg, B.; Suligowski, R.; Studziński, J.; De Paola, F. Application of logistic regression to simulate the influence of rainfall genesis on storm overflow operations: A probabilistic approach. *Hydrol. Earth Syst. Sci.* **2020**, *24*, 595–614. [CrossRef]
52. Lee, M.; An, H.; Jeon, S.; Kim, S.; Jung, K.; Park, D. Development of an analytical probabilistic model to estimate runoff event volumes in South Korea. *J. Hydrol.* **2022**, *612*, 128129. [CrossRef]
53. Kim, S.; Han, S. Urban Stormwater Capture Curve Using Three-Parameter Mixed Exponential Probability Density Function and NRCS Runoff Curve Number Method. *Water Environ. Res.* **2010**, *82*, 43–50. [CrossRef]
54. Guo, Y.; Liu, S.; Baetz, B.W. Probabilistic rainfall-runoff transformation considering both infiltration and saturation excess runoff generation processes. *Water Resour. Res.* **2012**, *48*, W06513. [CrossRef]
55. Hassini, S.; Guo, Y. Derived flood frequency distributions considering individual event hydrograph shapes. *J. Hydrol.* **2017**, *547*, 296–308. [CrossRef]
56. Hassini, S.; Guo, Y. Analytical Hydrological Model for the Planning and Design of Low-Impact Development Practices. In *Recent Advances in Environmental Science from the Euro-Mediterranean and Surrounding Regions*, 2nd ed.; EMCEI 2019. Environmental Science and Engineering; Springer: Cham, Switzerland, 2021. [CrossRef]
57. Hassini, S.; Guo, Y. Analytical Derivation of Urban Flood Frequency Models Accounting Saturation-Excess Runoff Generation. *J. Hydrol.* **2020**, *584*, 124713. [CrossRef]
58. Hassini, S.; Guo, Y. Analytical Derivation of Urban Runoff-Volume Frequency Models. *J. Sustain. Water Built Environ.* **2022**, *8*, 04021022. [CrossRef]
59. Guo, Y.; Zhuge, Z. Analytical probabilistic flood routing for urban stormwater management purposes. *Can. J. Civ. Eng.* **2008**, *35*, 487–499. [CrossRef]
60. Guo, Y.; Hansen, D.; Li, C. Probabilistic approach to estimating the effects of channel reaches on flood frequencies. *Water Resour. Res.* **2009**, *45*, W08404. [CrossRef]
61. Guo, Y.; Markus, M. Analytical Probabilistic Approach for Estimating Design Flood Peaks of Small Watersheds. *J. Hydrol. Eng.* **2011**, *16*, 847–857. [CrossRef]
62. Guo, Y.; Dai, J. Expanded analytical probabilistic stormwater models for use in watershed and master drainage planning. *Can. J. Civ. Eng.* **2009**, *36*, 933–943. [CrossRef]

63. Shishegar, S.; Duchesne, S.; Pelletier, G. Optimization methods applied to stormwater management problems: A review. *Urban Water J.* **2018**, *15*, 276–286. [CrossRef]
64. Gayer, J. Managing Urban Runoff in the Light of Integrated Water Resources Management. Ph.D. Thesis, Faculty of Water Management and Melioration, Corvinus University of Budapest, Budapest, Hungary, 2004.
65. Drake, J.; Young, D.; McIntosh, N. Performance of an Underground Stormwater Detention Chamber and Comparison with Stormwater Management Ponds. *Water* **2016**, *8*, 211. [CrossRef]
66. Chin, D. *Water Resources Engineering*; Prentice Hall: Hoboken, NJ, USA, 2006.
67. Shishegar, S.; Duchesne, S.; Pelletier, G. An integrated optimization and rule-based approach for predictive real time control of urban stormwater management systems. *J. Hydrol.* **2019**, *577*, 124000. [CrossRef]
68. Hogan, D.M.; Walbridge, M.R. Best Management Practices for Nutrient and Sediment Retention in Urban Stormwater Runoff. *J. Environ. Qual.* **2007**, *36*, 386–395. [CrossRef]
69. Papa, F.; Adams, B.J.; Guo, Y. Detention time selection for stormwater quality control ponds. *Can. J. Civ. Eng.* **1999**, *26*, 72–82. [CrossRef]
70. Wang, G.-T.; Chen, S.; Barber, M.E.; Yonge, D.R. Modeling Flow and Pollutant Removal of Wet Detention Pond Treating Stormwater Runoff. *J. Environ. Eng.* **2004**, *130*, 1315–1321. [CrossRef]
71. Papa, F.; Adams, B.J.; Bryant, G.J. Models for Wwater Qquality Ccontrol by Sstonnwater Pponds. *J. Water Manag. Model.* **1997**, *5*, R195-01. [CrossRef]
72. Becciu, G.; Raimondi, A. Probabilistic analysis of spills from stormwater detention facilities. *Urban Water II* **2014**, *139*, 159–170. [CrossRef]
73. Raimondi, A.; Becciu, G. On pre-filling probability of flood control detention facilities. *Urban Water J.* **2013**, *12*, 344–351. [CrossRef]
74. Becciu, G.; Raimondi, A. Probabilistic Analysis of the Retention Time in Stormwater Detention Facilities. *Procedia Eng.* **2015**, *119*, 1299–1307. [CrossRef]
75. Becciu, G.; Raimondi, A.; Brebbia, C. Probabilistic modeling of the efficiency of a stormwater detention facility. *Int. J. Sustain. Dev. Plan.* **2015**, *10*, 1–11. [CrossRef]
76. Raimondi, A.; Di Chiano, M.G.; Marchioni, M.; Sanfilippo, U.; Becciu, G. Probabilistic modeling of sustainable urban drainage systems. *Urban Ecosyst.* **2022**, 1–10. [CrossRef]
77. Ross, S.M. *Introduction to Probability Models*; Academic Press: New York, NY, USA, 2007.
78. Guo, Y.; Baetz, B.W. Sizing of Rainwater Storage Units for Green Building Applications. *J. Hydrol. Eng.* **2007**, *12*, 197–205. [CrossRef]
79. Zhang, S.; Guo, Y. Analytical Probabilistic Model for Evaluating the Hydrologic Performance of Green Roofs. *J. Hydrol. Eng.* **2013**, *18*, 19–28. [CrossRef]
80. Zhang, S.; Guo, Y. Explicit Equation for Estimating Storm-Water Capture Efficiency of Rain Gardens. *J. Hydrol. Eng.* **2013**, *18*, 1739–1748. [CrossRef]
81. Wang, J.; Guo, Y. Stochastic analysis of storm water quality control detention ponds. *J. Hydrol.* **2019**, *571*, 573–584. [CrossRef]
82. Parolari, A.J.; Pelrine, S.; Bartlett, M.S. Stochastic water balance dynamics of passive and controlled stormwater basins. *Adv. Water Resour.* **2018**, *122*, 328–339. [CrossRef]
83. Becciu, G.; Raimondi, A. Factors affecting the pre-filling probability of water storage tanks. *WIT Trans. Ecol. Environ.* **2012**, *164*, 473–484. [CrossRef]
84. De Paola, F.; Ranucci, A. Analysis of Spatial Variability for Stormwater Capture Tank Assessment. *Irrig. Drain.* **2012**, *61*, 682–690. [CrossRef]
85. Papa, F.; Adams, B.J. Application of derived probability and dynamic programming techniques to planning regional storm-water management systems. *Water Sci. Technol.* **1997**, *36*, 227–234. [CrossRef]
86. Shamsudin, S.; Dan'azumi, S.; Aris, A.; Yusop, Z. Optimum combination of pond volume and outlet capacity of a storm-water detention pond using particle swarm optimization. *Urban Water J.* **2014**, *11*, 127–136. [CrossRef]
87. Dan'Azumi, S.; Shamsudin, S.; Aris, A. Optimization of pollution control performance of wet detention ponds in tropical urban catchments using particle swarm optimization. *J. Hydroinform.* **2012**, *15*, 529–539. [CrossRef]
88. Behera, P.K.; Teegavarapu, R.S.V. Optimization of a Stormwater Quality Management Pond System. *Water Resour. Manag.* **2014**, *29*, 1083–1095. [CrossRef]
89. Raimondi, A.; Becciu, G. Probabilistic Modeling of Rainwater Tanks. *Procedia Eng.* **2014**, *89*, 1493–1499. [CrossRef]
90. Raimondi, A.; Becciu, G. Probabilistic Design of Multi-use Rainwater Tanks. *Procedia Eng.* **2014**, *70*, 1391–1400. [CrossRef]
91. Becciu, G.; Raimondi, A.; Dresti, C. Semi-probabilistic design of rainwater tanks: A case study in Northern Italy. *Urban Water J.* **2016**, *15*, 192–199. [CrossRef]
92. De Paola, F.; De Martino, F. Stormwater Tank Performance: Design and Management Criteria for Capture Tanks Using a Continuous Simulation and a Semi-Probabilistic Analytical Approach. *Water* **2013**, *5*, 1699–1711. [CrossRef]
93. Kim, H.; Han, M.; Lee, J.Y. The application of an analytical probabilistic model for estimating the rainfall–runoff reductions achieved using a rainwater harvesting system. *Sci. Total. Environ.* **2012**, *424*, 213–218. [CrossRef] [PubMed]
94. Di Chiano, M.G.; Marchioni, M.; Raimondi, A.; Sanfilippo, U.; Becciu, G. Probabilistic Approach to Tank Design in Rainwater Harvesting Systems. *Hydrology* **2023**, *10*, 59. [CrossRef]

95. Guo, R.; Guo, Y. Stochastic modelling of the hydrologic operation of rainwater harvesting systems. *J. Hydrol.* **2018**, *562*, 30–39. [CrossRef]
96. Pelak, N.; Porporato, A. Sizing a rainwater harvesting cistern by minimizing costs. *J. Hydrol.* **2016**, *541*, 1340–1347. [CrossRef]
97. Sim, I.; Kim, S. Probabilistic solution to rainwater harvesting system and its impact on climate change. *Desalination Water Treat.* **2020**, *200*, 407–421. [CrossRef]
98. Cheng, G.; Huang, G.; Guo, Y.; Baetz, B.W.; Dong, C. Stochastic Rainwater Harvesting System Modeling Under Random Rainfall Features and Variable Water Demands. *Water Resour. Res.* **2021**, *57*, e2021WR029731. [CrossRef]
99. USEPA. Using Green Roofs to Reduce Heat Islands. 2023. Available online: https://www.epa.gov/heatislands/using-green-roofs-reduce-heat-islands (accessed on 14 March 2023).
100. Raimondi, A.; Becciu, G. Performance of Green Roofs for Rainwater Control. *Water Resour. Manag.* **2020**, *35*, 99–111. [CrossRef]
101. Guo, Y.; Zhang, S.; Liu, S. Runoff Reduction Capabilities and Irrigation Requirements of Green Roofs. *Water Resour. Manag.* **2014**, *28*, 1363–1378. [CrossRef]
102. Guo, Y. Stochastic Analysis of Hydrologic Operation of Green Roofs. *J. Hydrol. Eng.* **2016**, *21*, 04016016. [CrossRef]
103. Raimondi, A.; Marchioni, M.; Sanfilippo, U.; Stroppa, F.F.; Becciu, G. Probabilistic estimation of runoff from green roofs. *Int. J. Comput. Methods Exp. Meas.* **2022**, *10*, 13–25. [CrossRef]
104. Raimondi, A.; Marchioni, M.; Sanfilippo, U.; Becciu, G. Vegetation Survival in Green Roofs without Irrigation. *Water* **2021**, *13*, 136. [CrossRef]
105. Guo, R.; Guo, Y.; Liu, S. Analytical Equations for Direct Quantification of Green Roofs' Hydrologic Performance Statistics. *J. Hydrol. Eng.* **2022**, *27*, 04022003. [CrossRef]
106. Raimondi, A.; Sanfilippo, U.; Marchioni, M.; Di Chiano, M.; Becciu, G. Influence of climatic parameters on the probabilistic design of green roofs. *Sci. Total Environ.* **2023**, *865*, 161291. [CrossRef]
107. FHWA. Stormwater Best Management Practices in Ultra-Urban Setting: Selection and Monitoring. Washington DC, Federal Highway Administration, U.S. Department of Transportation, 2000. Available online: https://rosap.ntl.bts.gov/view/dot/25727. (accessed on 12 November 2022).
108. Błażejewski, R.; Murat-Błażejewska, S. Analytical solutions of a routing problem for storm water in a detention basin. *Hydro-Log. Sci. J.* **2003**, *48*, 665–671. [CrossRef]
109. Weiss, P.T.; Gulliver, J.; Erickson, A.J. Cost and Pollutant Removal of Storm-Water Treatment Practices. *J. Water Resour. Plan. Manag.* **2007**, *133*, 218–229. [CrossRef]
110. VASM. *Virginia Stormwater Handbook*; Virginia Department of Conservation and Recreation, Division of Soil and Water Conservation: Richmond, VA, USA, 1999; Volume II. Available online: http://observatoriaigua.uib.es/repositori/suds_virginiab.pdf (accessed on 5 October 2022).
111. Zhang, S.; Guo, Y. Stormwater Capture Efficiency of Bioretention Systems. *Water Resour. Manag.* **2013**, *28*, 149–168. [CrossRef]
112. Guo, R.; Guo, Y.; Zhang, S.; Zhu, D.Z. A Tool for Water Balance Analysis of Bioretention Cells. *J. Sustain. Water Built Environ.* **2020**, *6*, 04020013. [CrossRef]
113. Islam, A.; Hassini, S.; El-Dakhakhni, W. Resilience Quantification of Low-Impact Development Systems Using SWMM and a Probabilistic Approach. *J. Sustain. Water Built Environ.* **2022**, *8*, 04022013. [CrossRef]
114. Fach, S.; Dierkes, C. On-site infiltration of road runoff using pervious pavements with subjacent infiltration trenches as source control strategy. *Water Sci. Technol.* **2011**, *64*, 1388–1397. [CrossRef] [PubMed]
115. Guo, Y.; Gao, T. Analytical Equations for Estimating the Total Runoff Reduction Efficiency of Infiltration Trenches. *J. Sustain. Water Built Environ.* **2016**, *2*, 06016001. [CrossRef]
116. Guo, R.; Guo, Y. Analytical Equations for Use in the Planning of Infiltration Facilities. *J. Sustain. Water Built Environ.* **2018**, *4*, 06018001. [CrossRef]
117. Fadhelab, S.; Rico, M.A.; Hana, D. Sensitivity of peak flow to the change of rainfall temporal pattern due to warmer climate. *J. Hydrol.* **2018**, *560*, 546–559. [CrossRef]
118. Wang, J.; Guo, Y. Proper Sizing of Infiltration Trenches Using Closed-Form Analytical Equations. *Water Resour. Manag.* **2020**, *34*, 1–13. [CrossRef]
119. Butler, D.; Davies, J.W. *Urban Drainage*, 2nd ed.; Spon Press, Tailor & Francis Group: New York, NY, USA, 2004.
120. Zhang, S.; Guo, Y. An analytical equation for evaluating the stormwater capture efficiency of permeable pavement systems. *J. Irrig. Drain. Eng. ASCE* **2015**, *131*, 06014004-1–06014004-9.
121. Guo, R.; Guo, Y.; Wang, J. Stormwater capture and antecedent moisture characteristics of permeable pavements. *Hydrol. Process.* **2018**, *32*, 2708–2720. [CrossRef]
122. Zeng, J.; Huang, G.; Mai, Y.; Chen, W. Optimizing the cost-effectiveness of low impact development (LID) practices using an analytical probabilistic approach. *Urban Water J.* **2020**, *17*, 136–143. [CrossRef]
123. SMRC. *Assorted Fact Sheets: Stormwater Management Practices*; Stormwater Manager's Resource Center, Center for Watershed Protection: Ellicot City, MD, USA, 2020; Available online: http://www.stormwatercenter.net (accessed on 5 July 2020).
124. USEPA. *Free Water Surface Wetlands for Wastewater Treatments: A Technology Assessment*; EPA 832 S 99 032; Office of Water, USEPA: Washington, DC, USA, 1999. Available online: http://www.epa.gov/owow/wetlands/pdf/FW_Surface_Wetlands.pdf. (accessed on 9 August 2020).

125. AWWA. *National Water Quality Management Strategy: Australian Guidelines for Urban Stormwater Management*; Australia and New-Zealand; Australian Water Works Association: St. Leonards, NSW, Australia, 2000.
126. Perez-Pedini, C.; Limbrunner, J.F.; Vogel, R.M. Optimal Location of Infiltration-Based Best Management Practices for Storm Water Management. *J. Water Resour. Plan. Manag.* **2005**, *131*, 441–448. [CrossRef]

water

MDPI

Review

Integrating Non-Targeted Ecosystem Services into Assessment of Natural Stormwater Treatment Systems

Jennifer T. Le [1,*,†], Jennifer P. Gonzalez [1], Richard T. Carson [2], Richard F. Ambrose [3] and Lisa A. Levin [1]

1 Center for Marine Biodiversity and Conservation, Scripps Institution of Oceanography, University of California San Diego, San Diego, CA 92093, USA
2 Department of Economics, University of California San Diego, San Diego, CA 92093, USA
3 Department of Environmental Health Sciences, University of California Los Angeles, Los Angeles, CA 90095, USA
* Correspondence: jtl025@ucsd.edu
† Current Address: The Bureau of Ocean Energy Management, Sterling, VA 20166, USA.

Abstract: Natural stormwater treatment systems (NTS) are built ecosystems designed to capture and treat stormwater runoff via natural processes. Although NTS design typically targets water services, the biological communities associated with NTS (i.e., plants, animals, and microbes) can provide non-targeted functions that can result in ecosystem services, such as biodiversity, pollination, and climate regulation, or in some cases disservices. Additional co-benefits of NTS include recreation, education and outreach opportunities, and aesthetic value. A review of NTS ecosystem services and co-benefits is provided with specific examples from Los Angeles County, highlighting the need for ecosystem services indicators, standard measurements, and monitoring. As NTS become globally widespread, best practices must include the ability to holistically assess NTS performance in ways that extend beyond water treatment services. Three models are presented that can be used to evaluate NTS performance. Such information can be important in advancing NTS design, choosing spatial placement, and making choices between NTS and more traditional stormwater treatment options.

Keywords: urban runoff; urban ecology; nature-based solutions; natural treatment systems; biofilters; ecosystem services; monitoring and evaluation; planning and management

Citation: Le, J.T.; Gonzalez, J.P.; Carson, R.T.; Ambrose, R.F.; Levin, L.A. Integrating Non-Targeted Ecosystem Services into Assessment of Natural Stormwater Treatment Systems. *Water* **2023**, *15*, 1460. https://doi.org/10.3390/w15081460

Academic Editors: Dafang Fu, Haifeng Jia, Jiangyong Hu and Wei-Shan Chen

Received: 9 February 2023
Revised: 31 March 2023
Accepted: 3 April 2023
Published: 8 April 2023

1. Introduction

Development and urbanization transform landscapes by replacing vegetation with impermeable surfaces. Subsequent precipitation events can lead to altered levels of water infiltration, modified water flows, and introduction of contaminants into stormwater runoff [1,2]. As a result, flooding and property damage, increased safety and health risks, and environmental damage can occur [3]. Urban planners and developers have traditionally addressed stormwater runoff issues by building drainage systems that connect directly to large bodies of water (e.g., streams, oceans) or treatment plants.

Natural stormwater treatment systems (NTS) are an alternative urban runoff management strategy [4]. NTS (also referred to as low-impact development or sustainable drainage systems) are human-made systems that use natural processes (e.g., gravity-driven hydrology, plant absorption of water and nutrients) to capture and treat urban runoff. We use the term NTS to encompass the different types of installations following current usage by collaborators [5,6]. They come in different forms: bioretention systems (e.g., biofilters, bioswales, rain gardens), infiltration basins and trenches, permeable pavements, dry wells and ponds, treatment wetlands, and combinations of these. They are part of a larger framework that now is often referred to as the sponge city [7]. The sponge city concept was developed in China to address urban water issues through "natural storage, natural infiltration, and natural purification" [8]. A sponge city seeks to retain rainwater, prevent flooding, and increase water quality through natural and semi-natural solutions [7,9]. The focus of this paper is on bioretention systems (i.e., biofilters) because they are explicitly intended to host biological communities as part of their design and less oriented toward hardscape infrastructure; however, they can offer numerous, often unrecognized benefits beyond the water services targeted in their initial design.

NTS can support diverse ecosystem services, defined as the direct and indirect benefits humans obtain from ecosystems [10,11] (Figure 1). Non-water (i.e., non-targeted) ecosystem services associated with NTS have long been acknowledged [5,12] and are only beginning to be meaningfully considered by spatial planning [13,14], engineering-based design, monitoring programs [15], and economic valuation efforts [16], which have been generally limited to targeted water functions. Ecosystem services described for other green spaces (e.g., parks, green roofs, urban forests) include offsetting carbon emissions [17,18], cooling local temperatures [19], cultural services (e.g., recreation, education and outreach, aesthetic value; [20]), and benefits to human health and well-being [21]. The incorporation of ecosystem services and other co-benefits into NTS design can present urban planners and developers with additional benefits, costs, and tradeoffs to consider [22,23].

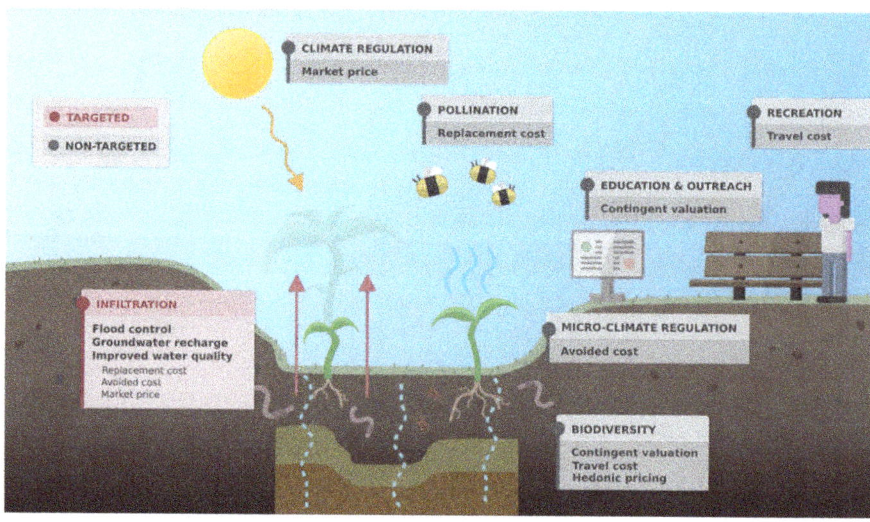

Figure 1. Diagram of targeted and non-targeted ecosystem services associated with natural stormwater treatment systems (a bioretention system pictured here) and methods used to assign a value to them.

We use Los Angeles County to focus this first comprehensive overview of NTS services because it hosts a dense human population near the coast where urban runoff has a large influence on local water quality [24], is in a Mediterranean climatic zone where periodic droughts are common, and expects to experience hydrological changes due to climate change [25]. Los Angeles County also has policy in place to encourage NTS infrastructure. California has passed several propositions to protect water supply and quality (e.g., AB-1471 2014). In 2004, Los Angeles passed Proposition O, which allowed the city to issue up to 500 million USD to fund projects that improve local water quality. In 2012, Los Angeles adopted its Low Impact Development Ordinance (RA-2012-0175) requiring development and redevelopment projects that alter impervious areas to mitigate runoff by capturing precipitation and utilizing natural resources where possible. A subsequent low impact development standards manual was published in 2014 [26]. As a result of these environmental conditions and political momentum, green infrastructure and NTS have been broadly distributed throughout Los Angeles County (Figure 2 [27]; https://www.arcgis.com/apps/dashboards/c41e288cd1cf4bf18b28404423c58735 (accessed on 31 January

2023)) and continue to be implemented. This study uses a small subset of Los Angeles NTS to illustrate potential ecosystem services.

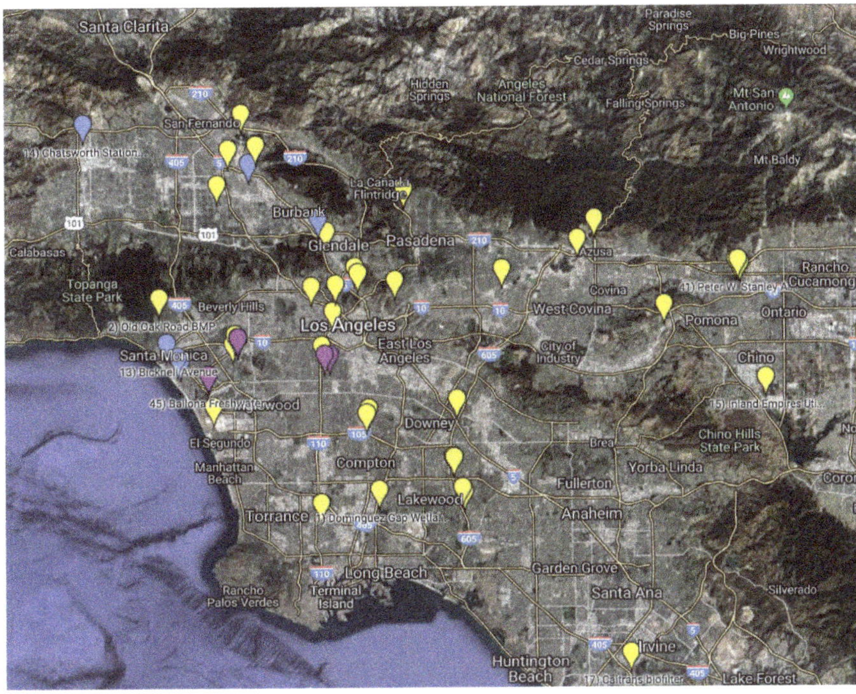

Figure 2. Location of natural stormwater treatment systems in Los Angeles County [27]. Blue indicates sites visited in 2015. Purple indicates sites visited in 2016. View the interactive Google map here: https://goo.gl/76sswN (accessed on 31 January 2023).

This paper is a narrowly focused literature review on targeted and non-targeted ecosystem services associated with NTS using Los Angeles County to provide specific geographic context. As there is an existing wealth of literature on targeted water services [28], they are reviewed here in summary for presentation alongside non-targeted ecosystem services. No attempt is made to systematically review the larger literature that is growing rapidly and, at some point, will be able to support informative meta-analyses concerning specific structures and functions underlying NTS. However, there are elements of a semi-systematic review in its broad coverage of multidisciplinary research [29]. The present lack of quantitative hydrologic and ecological NTS information in the context we examine highlights the need for robust data collection. Measurement of ecosystem services and indicators is seen as the starting point. However, U.S. regulatory frameworks, specifically a series of Presidential executive orders [30], require cost–benefit assessment of different options so the translation from biophysical impacts to the monetary valuation comparisons policymakers need is emphasized. An ecosystem services framework can offer alternative design options that maximize overall NTS value and introduce areas of improvement for urban stormwater management [31].

2. Targeted Water Services

2.1. Stormwater Infiltration

NTS are designed to capture stormwater runoff for infiltration or reuse and have been shown to be effective [32], although methods to evaluate infiltration vary [33]. Most bioretention systems are oriented vertically, using gravity to direct water flow through several layers that generally consist of a ponding area with vegetation, porous filter media, and a drainage zone (Figure 3). Infiltration rates can vary widely depending on variables such as size [34], age [35], filter media [36,37], and other design factors [38]. For example, NTS with internal water storage or submerged zones can have elevated levels of infiltration, evapotranspiration, and nutrient removal (specifically denitrification) but may not be suitable for areas with impermeable soils [39]. Another example is an underdrain that can help ensure desired water retention times but can be prone to clogging [40]. Vegetation also plays a role by intercepting precipitation and water flows [41] and preventing clogging of filter media to maintain infiltration capacity [42,43]. Additionally, with proper design, evapotranspiration by plants can contribute significantly to reduced volumes of stormwater runoff [44,45].

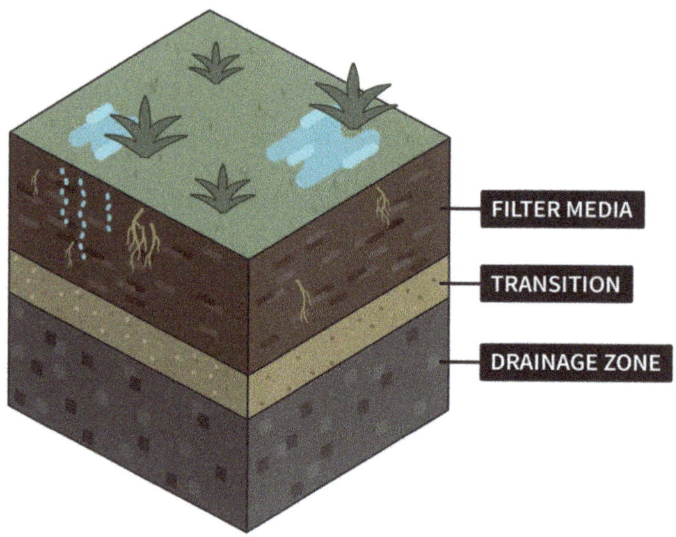

Figure 3. Diagram of a bioretention system with vertical layers.

Stormwater infiltration is important in urban areas with low permeable surface area that are prone to flooding and changes in groundwater recharge [3,46]. Los Angeles lacks the infrastructure to handle large volumes of stormwater [47], and flooding can occur with heavy rainfall resulting from atmospheric rivers. This issue is occurring in a region that experiences episodic drought [48], highlighting the need for proper stormwater management and use, which can include NTS as a strategy that provides additional benefits.

2.1.1. Flood Control

Flood risk is measured as a combination of hazard (e.g., runoff volume, base and peak flows, flood plain) and vulnerability (e.g., infrastructure, population density). Fluvial flooding is caused by overflow of river or lake systems, whereas pluvial flooding is caused by extreme or heavy rainfall. By altering landscapes and hydrology, urbanization can lead to an increased flood risk for both types [49]. NTS have been shown to significantly reduce runoff volume and the magnitude of high-flow events by capturing and storing stormwater runoff. Three bioretention cells in humid continental Ohio reduced both runoff by 45.6% on average and peaks flows ranging from 24 to 96%, performing best when peak flows occurred before the bulk of rainfall volume [50]. During the 13 months of observation, 140 storm events were recorded with antecedent dry periods ranging from 0.3 to 20 days and maximum peak flows of 105 L s^{-1} [50]. In humid subtropical North Carolina, 125 storm events were observed over 22 months with antecedent dry periods ranging from 0.3 to 31.3 days and maximum peak flows of approximately 38 L s^{-1} [51]. The study found that a Filterra™ system in an impervious parking lot captured 72% of runoff volume and reduced peak flows by 57% [51]. Hatt et al. (2009) found an average of 33% reduction in runoff volume, as well as an 80% reduction in peak flow reductions in three bioretention systems in Australia. Data were collected during 32 natural and simulated storm events with maximum peak flows of approximately 40 L s^{-1} and an average return interval of 3 months for simulated events [42]. In a more recent field study, Bonneau et al. (2020) reported a 35% reduction in stormwater volume and an 80% reduction in peak flow in a bioretention basin in Australia during rainfall events over 3 years, with average peak flows of approximately 33 L s^{-1} [52]. Relative to Australia, southern California generally has larger NTS built exclusively for groundwater recharge and longer dry periods that can decrease contaminant removal rates [53].

Most Los Angeles NTS exhibited visual indicators of flood control services: permeable surfaces that allow for water infiltration, graded landscaping to help direct runoff, and ponding areas designed to temporarily store water. For example, Elmer Avenue Green Street was constructed specifically to address street flooding during precipitation events by incorporating bioswales, biofilters, permeable pavements, and rain barrels [54]. However, additional data are needed to confirm whether they are effective. Spatially explicit measurements for flood hazard may be relevant (e.g., soil porosity, basin size, changes to magnitude and timing of flood peaks; [55]), as well as those for flood resilience (e.g., population and area exposed to changes in flood frequency; [56]). There may be a return period at which NTS infiltration rates and storage capacity can no longer provide flood control services, which likely depends on NTS characteristics such as size, filter media, and maintenance frequency [57]. Additionally, Los Angeles is expected to experience a sea level rise in some areas, which may make some types of NTS (e.g., treatment wetlands) that can serve as living shorelines more suitable than others [58].

The value of flood control has been studied in the context of wetlands [59], generally by assessing differences in property damage along a spectrum of wetland area (e.g., reduction in damages as proxied by replacement cost). Brander et al. (2006) conducted a meta-analysis to estimate the value of flood control by wetlands that resulted in a median of 20-30 USD (1995) per hectare annually [60]. Watson et al. (2016) estimated a net present value of less than 100 USD per hectare of wetland annually, which may be lower than comparable studies due to the small beneficiary population [61]. NTS generally operate on a more localized scale than major wetland restoration projects, so one unknown is how different NTS designs vary in their spatial extent of flood control.

2.1.2. Groundwater Recharge

Groundwater is used by more than half the U.S. population, and its recharge is an essential component of the water cycle [62]. Major sources of recharge in urban environments include runoff infiltration and leakages from the water supply and sewage systems [46]. Urbanization can decrease groundwater recharge due to the installation of impermeable surfaces [1]. However, the importation of large volumes of water to meet demand in highly populated urban areas can also lead to significant leakages and water recharge.

NTS can contribute to groundwater recharge by providing permeable surfaces and pore space in their filter media that allow stormwater runoff to pass into the soil subsurface. The Avalon Green Alley project in south Los Angeles is designed to allow water to percolate into the soil and recharge the water table [63]. Beneath the permeable pavement of the green alley, catch basins store stormwater temporarily, where it can be bioremediated before flowing into surrounding soil. Quantifying groundwater recharge can be challenging, but several methods exist to do so, such as physical measurements [64] and numerical modeling techniques [65].

The value of groundwater recharge can be calculated using various methods; the most appropriate is dependent on the end-use of the water. For example, in 2010, California withdrew 12,700 million gallons of groundwater per day, of which approximately 22.2% was used for domestic purposes [62]. The value of groundwater recharge can therefore be associated with the price of water to consumers. Replacement costs can also be used to assign a value to groundwater recharge. Artificial groundwater recharge is the spread of water on land to increase infiltration or the injection of water directly into the aquifer; its cost can provide an estimate of the value of groundwater recharge. These actions have associated costs and can be used to estimate a value for the same service performed by NTS.

2.2. Improved Water Quality

As stormwater runoff flows over urban surfaces, it can acquire contaminants such as polycyclic aromatic hydrocarbons (PAH), pesticides, pharmaceuticals, and other organic pollutants [66]. The "first flush" phenomenon is defined as elevated levels of stormwater pollution at the beginning of storm events [67,68]. First flush can be difficult to model because it is dependent on a variety of factors, such as climate, precipitation, catchment size, and land use type [68]. Southern California receives little precipitation and has long periods without rainfall, even during the "wet" season, which allows for the build-up of contaminants and results in first flush. These contaminants can then enter larger bodies of water (e.g., rivers, oceans) and degrade the local environment [69]. Many highly urban areas, such as Los Angeles, are coastal; this can exacerbate the issue of water quality because stormwater runoff can drain directly into the ocean with little opportunity (i.e., distance) for treatment. Treatment can depend on NTS water retention times, which are influenced by storage basin depth and type and amount of filter media [70]. Short retention times do not allow for adequate biofiltration, but long retention times can present hazards in the form of flood risk and mosquito reproduction [71], so there are tradeoffs to consider.

There is a range of common urban pollutants in stormwater runoff; these include heavy metals (e.g., PAH, Cd, Cu), total suspended solids and nutrients (mainly nitrogen and phosphorus), and pathogens. Müller et al. (2020) provides a comprehensive review of sources [72]. Heavy metals can persist in the environment and accumulate in sediment, plants, and animals, leading to the degradation of environmental and human health [73]. The input of excess nutrients and organics into streams, lakes, and the ocean (i.e., eutrophication) can be damaging to local ecosystems by generating harmful algal blooms, hypoxia, and anoxia [74]. Pathogens also pose a risk to human health through exposure (e.g., from recreation activities and stormwater reuse). Lim et al. (2015) found that captured stormwater can be used for toilet flushing with acceptable risk, but it does not meet required standards for showering and food-crop irrigation [75]. Beach closures and advisories are often the result of bacteria levels exceeding water quality standards [76]. In addition to the threat to human health, the resulting closures have associated economic costs, e.g., less use of parking lots, restaurants, and shopping [77].

NTS remove contaminants through several pathways. Physical filtration removes debris, particle-bound contaminants, and suspended solids, which can lead to clogging and subsequent deterioration of bioremediation functions. Informed NTS design, such as plant selection, can help maintain infiltration capacity [43]. Contaminant molecules can also be removed from runoff by adsorption (sticking of contaminant molecules onto porous surfaces with uneven or leftover attractive forces [78]) and assimilation (uptake of contaminants by plants; [79]). Adsorption can be enhanced with soil amendments, such as biochar; the type of amendment may depend on the targeted contaminant [78]. A significant portion of bioremediation is performed by soil microbial communities that can be stimulated by moisture [80]. In drought-prone areas such as southern California, extended dry conditions can decrease microbial abundance and activity, which can negatively impact enzyme activity and nutrient cycling [81]. Drought-resistant microbes can be introduced to NTS soil to help promote plant growth and continued phytoremediation under stressful conditions [82]. NTS can prevent contaminants from traveling further to pollute local bodies of water, but they can also concentrate pollutants in plants and filter media [83]. These concentrated contaminants can leach into surrounding soils, e.g., due to lower oxygen levels that increase metal solubility. As a result, some maintenance is likely required to prevent build up and transport of contaminants into the environment [28]. Maintenance costs, such as for plant or substrate renewal, can be uncertain [84] but potentially minimized through good design and regular inspection [85].

Contaminant removal may be the most well-studied service provided by NTS. Figure 4 shows examples of systems installed to remove debris and contaminants from runoff. Water infiltration systems can be installed into sidewalk tree wells to capture urban runoff before it can reach larger bodies of water. The Filterra™ system (Contech Engineered Solutions, West Chester, PA, USA) has been shown to effectively filter suspended solids, phosphorous, nitrogen, and zinc, with mean removal efficiency of 92%, 54%, 33%, and 66%, respectively [51]. The removal rates of heavy metals (copper, zinc, lead, and cadmium) by bioretention systems can be quite high (above 90%) [75], but not in all cases. Smolek et al. (2017) found that the Filterra™ system did not remove copper [51]. The type of filter media can influence the removal of metals, which accumulate in the top 5 cm of soil [75]. Heavy metal concentrations are unlikely to exceed soil quality standards for human health over the NTS lifetime [83]. However, ecological thresholds may be reached to varying extents. For example, Waara and Johansson (2022) measured concentrations of cadmium, copper, zinc, silver, and antimony that put aquatic organisms at risk in Swedish stormwater ponds [86]. In contrast, heavy metal concentrations in stormwater ponds in Denmark did not pose a risk to habitat function but were reflected in biomass so ecological risk could increase over time [87]. Sediment and nutrient removal by NTS have also been found to be relatively high (more than 50% in most cases) in several laboratory experiments [88,89]. However, at a watershed-level, NTS may be better suited for reducing contaminant load (the rate at which contaminants are discharged) into local waters rather than meeting concentration-based water quality standards [90]. Although these laboratory experiments provide a deep understanding of how NTS may function under specified conditions (e.g., dry vs. wet conditions, high vs. low temperatures), in situ studies are also needed to put laboratory experiments into context. How NTS operate over both long (e.g., months to seasons) and short (e.g., period of a storm) time scales are an open question, as is how the timing of water quality measurements (e.g., beginning or end of the storm, post-saturation) can influence perceived effectiveness.

Figure 4. Examples of Los Angeles systems designed to remove debris and contaminants from runoff: (**a**) biofilter along Elmer Avenue Paseo, (**b**) biofilter in the Avalon Green Alley, and (**c**) Filterra™ bioretention system on Grand Boulevard.

There are many studies that estimate the value of improved water quality using a variety of methods, with the two most common being a production function approach and contingent valuation. A production function approach can value improved water quality by comparing the cost of alternative methods for contaminant removal, such as a stormwater treatment plant. This is the approach generally used by Stanford's Natural Capital project [91] coupled with a measure of replacement cost. Several of their water-related modules of the InVEST workbench (i.e., urban flood risk mitigation, urban stormwater retention, water purification) can be useful starting points for examining the impacts of different NTS and can often accept location-specific parameter values [92–95]. Contingent valuation, including discrete choice experiments, asks a sample of respondents about their willingness to pay (WTP) for a spectrum of water quality [96,97]. The specific variant of a method is dependent on the fate of the stormwater and the final ecosystem service it provides, e.g., improved intake water quality for drinking water or improved water quality in lakes, rivers, and streams that can host higher levels of biodiversity.

3. Non-Targeted Ecosystem Services

NTS can provide a range of non-targeted ecosystem services, linked to their utilization of natural structures and functions, and other co-benefits [98,99]. Filter media and plant communities, which increase infiltration and remove contaminants, host biodiversity that contributes to ecological processes that can result in beneficial ecosystem services [6]. These ecosystem services (or costs, in the case of disservices) are not generally considered during the design and assessment of NTS. One unique feature of NTS is that they are human made so can be designed to provide specific benefits, which are discussed here to further expand design options that can enhance NTS value.

3.1. Biodiversity

Vegetated NTS act as man-made ecosystems that contain a diversity of plants, animals, and microbes from which ecosystem services can be generated [5,100] (Figure 5). They provide patches of habitat within an urban landscape and potentially act as corridors through which organisms can move. This can be important for population connectivity and resilience in a changing environment due to removal of natural habitat, habitat fragmentation, and climate change [101].

Figure 5. Examples of biodiversity that can support ecosystem services in Los Angeles natural stormwater treatment systems: (**a**) plant diversity at the South Los Angeles Wetland Park, (**b**) Bicknell manzanita tree as a bee attractant, and (**c**) signage illustrating the biodiversity of the Ballona Freshwater Marsh.

Whereas NTS biodiversity (here loosely defined to encompass plant, animal, and microbial species richness, abundance, and distribution) can be measured using a range of accepted methods (e.g., visual surveys, fauna collections, eDNA surveys), how this biodiversity translates into ecosystem services is more complex and will need targeted studies. There can be synergies between NTS design and landscape architecture, such as planning larger project areas [102], adaptive and flexible systems [103], and the redevelopment of un- or underutilized land [104,105]. Plants act as ecosystem engineers in bioretention systems, influencing both hydrological and ecological features and, therefore, plant selection can affect many of the services discussed in this paper [105]. Vegetation captures precipitation [41], undergoes evapotranspiration [44,45], maintains media porosity with roots, and assimilates pollutants [43]. Plants determine photosynthesis and respiration rates, organic matter in soil, and, ultimately, carbon sequestration and storage in NTS [106]. Additionally, plant communities influence microbial and infaunal communities, which subsequently impact function. Despite this, plant selection in NTS design typically emphases stress tolerance and competitiveness rather than specific factors influencing ecosystem services [107]. Microbial communities in NTS are most often assessed in the literature with respect to the harmful taxa present (e.g., fecal-indicator bacteria; [108]) or as functional groups (i.e., denitrification, greenhouse gas emissions). Fauna, except in wetland settings, are usually not considered in NTS design, despite their role in facilitating targeted functions and services. Mehring et al. (2016) identify common biofilter taxa, such as Megadrilacea (earthworms), Enchytraeidae (potworms), and Collembola (springtails) [109]. Earthworms are known to increase water infiltration via burrows [110], whereas springtails can impact plant growth and nutrient cycling [6]. As a result, these soil invertebrates can be considered ecosystem engineers that move and aerate soil, shaping the microbial, floral, and faunal communities from which more ecosystem services can stem. Because urban NTS can receive more water than native ecosystems, they can host elevated biodiversity of local soil invertebrates relative to natural habitats [111]. Higher biodiversity can be beneficial if they enhance service provision or detrimental if they involve invasive species that disrupt function. Wetland bird species contribute to local diversity and provide recreational services in the form of birdwatching. NTS also have the potential to provide bird habitat patches within an urban landscape [112].

Studies exist on the economic value of biodiversity in urban and engineered settings (e.g., constructed wetlands, agriculture) [113]. These employ the standard suite of non-market valuation techniques [114]: the production function approach, contingent valuation, travel cost analysis, and/or the hedonic property pricing methods to estimate society's willingness to pay (WTP) for biodiversity and its conservation. The production function approach tends to be used when the NTS alters an input to some process such as sewage treatment so that it is possible to estimate the reduction in cost of that activity relative to the status quo without the NTS. Contingent valuation is a stated preference approach that is widely employed because it can largely emulate the other techniques by constructing the relevant market in a survey context. Travel cost analysis is heavily used when NTS influence the quantity or quality of outdoor recreation. This method is based on the notion that it costs time and money to travel to the recreational site. Variation in those costs acts like a market price influencing the amount of the recreational activity undertaken. Hedonic pricing tends to be used when an ecosystem service is incorporated in housing or land prices in a manner similar to school quality. The value of NTS biodiversity has not been assessed, but studies in urban ecosystems may be a useful starting point. For example, Dupras et al. (2014) estimated that urban forests in the Greater Montreal area in Canada host biodiversity that creates a value of CAD (2013) 2623 per hectare annually [115]. Biodiversity associated with other human-made ecosystems, such as agricultural land [116], may also be relevant.

3.2. *Climate Regulation Related to Carbon*

Carbon dioxide emissions are the largest contributor to anthropogenic climate change [117] and, as a result, climate-regulating services related to carbon (i.e., carbon sequestration and storage) have become increasingly important, especially in urban areas that contribute disproportionately to global emissions [118]. NTS plants have the potential to contribute to this effort by converting atmospheric carbon dioxide into biomass through photosynthesis. How long this carbon is subsequently stored is dependent on several factors. Although some carbon is quickly rereleased during respiration, some is stored as plant biomass and soil detrital compounds. Turnover rates vary with type of biomass, soil moisture, soil oxygenation, soil organic matter, and microbial communities [119,120]. Some bioretention systems contain saturated or submerged zones designed to create anaerobic conditions for denitrification [88,121] but may also help prevent microbial breakdown of organic matter.

Quantifying carbon sequestration and storage would necessitate measurements that include net carbon fluxes, soil and plant carbon density, and biomass turnover rates. This has been performed in urban green spaces and could also be performed in NTS. Nowak et al.

(2013) estimated annual carbon sequestration in U.S. urban forests to be 25.6 million tons [122], which at 36 USD (2015) per ton of carbon [123], has a value of over 900 million USD (2015) annually. On smaller scales, green roofs have also been shown to sequester carbon and decrease carbon emissions due to lowered electricity usage for cooling [124,125]. NTS likely operate on scales more similar to green roofs than urban forests. The Stanford Natural Capital InVEST Workbench [91] has developed modules that allow for assessing carbon sequestration that may also be useful in specific contexts [126–128]. The monetary valuation of any carbon sequestered from the U.S. regulatory vantage point is driven by the standardized social cost per ton of carbon promulgated by the U.S. Environmental Protection Agency [129].

3.3. Micro-Climate Regulation

Micro-climate regulation is complex and highly dependent on geographic context. In addition to physically storing carbon, increased green space can reduce air and surface temperatures [19,130], reducing electricity use and emissions of greenhouse gases. The urban heat island effect occurs due to increased air temperature in urban settings relative to undeveloped areas as a result of replacing vegetation with pavement [131,132]. Pavements, such as asphalt and cement, have lower surface albedo than natural vegetation and therefore absorb more radiative energy. Non-vegetated surfaces also convert absorbed radiative energy into convective heat, which contributes to the urban heat island effect at higher rates than vegetation; these rates are dependent on the thermal properties of the material [133]. Evapotranspiration also contributes to plant regulation of micro-climates by increasing the amount of water in the air, i.e., humidity. In tropical contexts, increased humidity can counteract air cooling effects [134]. However, in arid southern California, the decrease in air temperature can be more important than the increased humidity in preventing heat impacts and mortality [135]. Additionally, vegetation can provide shade. Even small green spaces (e.g., green roofs) can have a significant impact on microclimate [19]. Factors that affect microclimate regulation include UV intensity, wind, and size of the green space.

The value of microclimate regulation by urban green space can, in many instances, be calculated using avoided costs methods, which can be seen here as a variant of the production function method. For example, if a green space makes an area cooler, people are likely not to run their air conditioning for as long or as intensively, in the same way as would be achieved by better insulation. The energy savings produced could be used to assign an economic value to the microclimate regulating services provided by the green space.

3.4. Pollination

Animal-mediated pollination is an important ecological process that supports many benefits. For example, bee pollination can increase agricultural crop quantity, food quality, and market value [136]. However, bee populations have been declining due to changing land use and management, pesticides, pollution, higher pathogen prevalence, and climate change [137]. Additionally, development and urbanization cause habitat fragmentation, which can lead to changes in species and functional diversity [138]. NTS can provide patches of habitat and refuge for animals within urban areas. Pollinators, such as bees and birds, can connect these habitat patches as they move among plants to feed and collect pollen (Figure 6). Increased suitable habitat and connectivity may facilitate the recovery of pollinator populations and create more resilient communities that can recover from disturbances, e.g., disease or long periods of intense drought [139]. Habitat patches have been shown to maintain distinct bee communities that, in aggregate, retain a significant amount of local species diversity [140].

Pollination services are typically seen as an input to the agricultural process, with improvements to pollination associated with increases in agricultural productivity [141]. They have also been evaluated in urban settings. For instance, Breeze et al. (2015) estimated the WTP for non-market pollination services in the United Kingdom using a contingent valuation survey deploying a discrete choice experiment [142]. Their estimates suggest taxpayers are willing to pay GBP 13.4 annually per person to maintain these benefits. Visual indicators for potential pollination services include the presence of animal-pollinators (e.g., bees, birds) as well as flowering plants. Quantification of this service requires data on the frequency of animal-to-flower visits as well as connectivity among NTS and other areas.

Figure 6. Flowering plants that could provide pollination services observed in bioretention systems at: (**a**) Elmer Avenue, (**b**) Bicknell Avenue, (**c**) Grand Boulevard, and (**d**) Los Angeles Zoo.

3.5. Other Co-Benefits

The following co-benefits have been separated from other ecosystem services because their provision does not directly employ ecological processes. However, biodiversity and ecological processes associated with NTS can enhance these co-benefits.

3.5.1. Recreation

NTS can have built-in public spaces that human communities can utilize. Walking trails, bike lanes, benches, and wildlife viewing sites can provide recreational services (Figure 7). The recreational use of urban green spaces can depend on amenities such as dog areas, playgrounds, barbeque areas, and landscaping [143]. Urban green spaces have been linked to improvements in physical and mental health [21,144,145]. They have also been linked to more active and healthy lifestyles overall [146]. In Los Angeles County, which is the most populous county in California [147], these types of spaces can be important in the midst of a densely populated, heavily urbanized area. The travel cost method is most often used to calculate the value of recreational services [148] but may not be appropriate for hyper-local neighborhood amenities if there is little to no cost associated with access. Contingent valuation and hedonic pricing have also been employed [149]. Sociocultural and socioeconomic characteristics may play a role: people of color and lower-income communities utilize urban green spaces at lower rates than other groups [150,151]. In Porto, Portugal, Graça et al. (2018) found that lower socioeconomic areas have the most green space, but they are unlikely to be developed in ways that provide services to the community [152]. This suggests the possibility that poorer communities can be served by NTS that enhance existing green spaces in a way that make them more accessible.

Figure 7. Visual indicators of recreation at Los Angeles natural stormwater treatment systems: (**a**) walking trail on Elmer Avenue Paseo, and (**b**,**c**) signage and drinking fountains at South Los Angeles Wetlands Park.

3.5.2. Education and Outreach

NTS present an opportunity for education and public outreach regarding stormwater issues, pollution, watershed and urban ecology, urban planning and management, and climate change. For example, Ocean View Growing Grounds (OVGG, San Diego, CA, USA) is a community garden within a food desert (defined as an area that lacks access to fresh produce and whole foods (often in low socioeconomic areas)) that utilizes bioswales to prevent flooding. UC San Diego researchers have partnered with community leaders to host outreach events about hydrology, soil, and urban ecology. Greater understanding and awareness of NTS and the services they provide may lead to safer gardening practices and more efficient water use. In Los Angeles, many of the sites have educational information posted about the project, their goals, and their motivations (Figure 8). As part of the Elmer Avenue Green Street project, local residents participated in the planning and design processes, actively engaging in stormwater issues and how to address them.

Figure 8. Examples of education and outreach services provided by natural stormwater treatment systems: (**a**) South Los Angeles Wetlands Park, (**b**) Ballona Freshwater Marsh, and (**c**) students conducting fieldwork as part of the National Science Foundation Partnerships for International Research and Education.

3.5.3. Aesthetic Value and Other Non-Use Values

Aesthetic value has been attributed to urban green spaces [153], which can increase surrounding property value [154]. NTS may provide similar services to enhance visual, auditory, and olfactory environmental features that have been linked to human well-being [155,156]. Green roofs and walls can dampen road traffic noise by up to 7.5 dBA [157]. Urban green spaces provide habitat for "natural" sounds (e.g., birds), which are preferred over traffic noises [158]. Exposure to green spaces has also been associated with better cognitive development [159] and reduced physiological stress [158].

3.6. Disservices and Unanticipated Costs

NTS can potentially introduce undesired services, such as installation of unattractive elements, accumulation of pollutants, and proliferation of disease vectors. The aesthetics of NTS may not be favorable to all people, especially when they are not regularly maintained. Regular maintenance can not only help plant survival but also prevent build-up of debris and pollutants, which NTS are designed to intercept. Heavy metals can leach into surrounding soil and groundwater [42,160]. It is also possible that NTS, despite having vegetation, can be sites of net greenhouse gas emissions (carbon dioxide and methane) rather than carbon sinks [161]. Ponding areas, which allow microbial communities to remove contaminants, can also provide habitat for mosquitoes that pose health hazards in the form of infectious diseases and allergies. For example, two Los Angeles NTS sites had advisories about West Nile virus and its carriers (August 2016). Vegetation and soil media can also provide habitat for urban pests, such as rodents and ticks [162].

Increases in urban green space have been linked to decreases in violent crime [163]; however, the potential of urban parks to help foster various types of crime is well known. An account of assault was cited in the South Los Angeles Wetland Park (http://abc7.com/news/arrest-made-in-connection-to-sex-assault-attempts-in-south-la-park/1297020/ (accessed on 31 January 2023)). These issues can generally be avoided at some cost, whether it is more frequent monitoring, improved lighting, or increased maintenance. The main warning here is that whereas these costs are not usually prohibitively expensive, they do need to be considered during planning and budgeting processes.

4. Potential Quantitative Models to Assess Ecosystem Services Performance in Natural Stormwater Treatment Systems

One approach to better internalize ecosystem services associated with NTS into decision making is to (1) determine the quantity or rate at which the ecosystem service is provided (i.e., biophysical value) and (2) determine the value (whether economic or not) of that quantity or rate. Although non-targeted ecosystem services are acknowledged [12], field studies that quantify these services are still lacking [164]. Here, we highlight three potential models for ecosystem services performance assessment in NTS that can be used as a starting point and improved upon as more data become available.

Benefit transfer approaches use regression models, often built on meta-analyses, to provide an estimate of the dependent variable (i.e., the rate or value of an ecosystem service). Examples of potential explanatory variables include study site size, geographic location, scale of the study (e.g., local, regional, national, global), gross domestic product per capita, valuation method, and sample size. These preliminary estimates can then be used further, such as in cost–benefit analysis, because primary studies are not always feasible.

Stochastic frontier analysis (SFA) and data envelopment analysis (DEA) are production function modelling techniques that create a "best practice frontier" as a benchmark of efficiency [165,166]. This benchmark is generated using input and output data from multiple sources (e.g., firms, projects, NTS), creating one scenario (normalized to be the best practice frontier) against which the efficiency for other scenarios can be compared and evaluated. Thus, these approaches provide a measure of how efficient a particular NTS is at providing an ecosystem service relative to a benchmark and by extension to competing NTS configurations. SFA includes a stochastic error term that may make it suitable for urban systems that are subject to environmental variation, e.g., drought and precipitation events. However, SFA requires an a priori assumption regarding the production function form, i.e., the mathematical relationship between inputs and outputs (e.g., Cobb–Douglas, Leontief), which can have a substantive influence on results if it does not provide a reasonably good approximation to the actual production function.

DEA is a nonparametric approach that does not require prior knowledge of the production function form but has the weakness that it does not formally accommodate idiosyncratic random shocks. DEA handles multiple outputs better than SFA, which may make it more appropriate for NTS-associated ecosystem services. A recent comparison of DEA and SFA for rice production is contained in [167].

The benefit transfer approaches, SFA, and DEA results can be used to evaluate how well an NTS site is performing and may help point to how to improve its functioning.

Table 1 provides a characterization of benefit transfer, DEA, and SFA from the perspective of NTS ecosystem services.

Table 1. Advantages and disadvantages of benefit transfer approaches, stochastic frontier analysis, and data envelopment analysis for assessing the value of ecosystem services in natural stormwater treatment systems. Adapted from [168].

	Benefit Transfer Approaches	**Stochastic Frontier Analysis**	**Data Envelopment Analysis**
Assumptions	Policy site is equitable to study site	A priori production function	Deterministic approach
Error	due to differences in site characteristics	Incorporated as stochastic variable	cannot be separated from inefficiency
Multiple outputs	Single ecosystem service	Weighted basket of ecosystem services	Allows for multiple ecosystem services
Relationship to independent variables	Relationship built into regression model	Relationship built into regression model	Need additional regression model
Interpretation of results	Associated value and drivers	Benchmark for efficiency	Benchmark for efficiency

We would like to have been able to directly compare results from an SFA and DEA analysis for our Los Angeles sites. This cannot be easily performed due to the lack of readily available data. There are no standard data collection or monitoring programs for Los Angeles NTS. This makes it difficult to compare across time and across sites, and to date the authors know of no monitoring of ecosystem services associated with NTS. More intensive monitoring and a standardized set of protocols for reporting such data could help identify transparent and effective management strategies [169], e.g., timing of maintenance, and improve the above quantitative models for more accurate estimates of value and efficiency. In situ, mesocosm studies are also needed to evaluate NTS performance under actual environmental conditions [79]. Other questions that still need to be addressed include how networks of NTS compare with single systems regarding both targeted and non-targeted ecosystem services and how NTS operate over time.

Nonmonetary valuation approaches may be required for some ecosystem services, such as cultural ones that are difficult or inappropriate to assign a monetary value. For example, a deliberative mapping survey was used to assess the local perception of and assign a social value to landscape services provided by the Mekong Delta in Vietnam [170]. Other non-monetary valuation approaches include interviews, oral histories, scaling (e.g., "very good" through "very bad", numerical 1–10), and ranking options into an order [171]. Another alternative to monetary valuation is the use of benefit-relevant indicators that link ecological responses to management decisions and human well-being [172]. These indicators can support environmental decision making both on their own and as a step towards valuation [172].

5. Conclusions

Because NTS, such as bioretention systems and treatment wetlands, rely on natural structures and functions, they provide built ecosystems that can support a host of targeted and non-targeted benefits. A lack of standardized monitoring data and programs makes it difficult to assess whether these systems are generating the services they are built to provide as well as any potential co-benefits. The proposed ecosystem service measurements and indicators presented above are a first step towards quantification, verification of their utility, and identification of design improvements. Although there are few data on the quantification of these NTS benefits specifically, examples from relevant systems can be used to get an idea of how these processes may work and on what scale. There also exist many valuation techniques that can be employed to assign a value to these ecosystem services for incorporation into urban design and management given the widespread use of cost–benefit analyses in decision making. Most NTS are implemented to meet water quality regulations and, although expensive, some of these costs may be offset by the value of non-targeted ecosystem services provided. However, long-term monitoring is necessary to assess whether the benefits of non-targeted services actually accrue, and research is needed to determine how to optimize their collective performance. Better understanding the co-benefits provided could also garner support for the use of NTS from local governing bodies and the public. In places such as southern California, where water issues are in sharp focus in policy debates, effective management of stormwater runoff can reduce flooding and enhance the water supply while improving microclimate, wildlife habitat, biodiversity, pollination, recreation, education, or other valuable services. Consideration of non-targeted ecosystem services by NTS can provide one tool to help urban planners, community groups, and developers make better decisions.

Author Contributions: Conceptualization, L.A.L., R.F.A., R.T.C. and J.T.L.; field investigation, J.P.G., J.T.L. and L.A.L.; writing—original draft preparation, J.T.L. and J.P.G.; writing—review and editing, J.T.L., L.A.L., R.T.C. and R.F.A.; funding acquisition, L.A.L., R.T.C. and R.F.A. All authors have read and agreed to the published version of the manuscript.

Funding: This work was supported by the University of Southern California Sea Grant [grant numbers USC NOAA 61207781, USC NOAA 75199714] and the University of California Office of the President [grant number UCOP MRP-17-455083].

Data Availability Statement: No new data were created or analyzed in this study. Data sharing is not applicable to this article.

Acknowledgments: The authors would like to thank Dale Squires and Andrew Mehring for insightful comments and suggestions, Jessica Aceret for illustrations, Brandon Winfrey for field site access, and Katie Galloway and Laura Walsh for field assistance. We thank the editors and two anonymous reviewers for their thoughts and insight and alerting us to new references.

Conflicts of Interest: The authors declare no conflict of interest.

References

1. Walsh, C.J.; Roy, A.H.; Feminella, J.W.; Cottingham, P.D.; Groffman, P.M.; Morgan, R.P., II. The Urban Stream Syndrome: Current Knowledge and the Search for a Cure. *J. N. Am. Benthol. Soc.* **2005**, *24*, 706–723. [CrossRef]
2. McGrane, S.J. Impacts of Urbanisation on Hydrological and Water Quality Dynamics, and Urban Water Management: A Review. *Hydrol. Sci. J.* **2016**, *61*, 2295–2311. [CrossRef]
3. Hossain, M.K.; Meng, Q. A Fine-Scale Spatial Analytics of the Assessment and Mapping of Buildings and Population at Different Risk Levels of Urban Flood. *Land Use Policy* **2020**, *99*, 104829. [CrossRef]
4. Wendling, L.A.; Holt, E.E. Integrating Engineered and Nature-Based Solutions for Urban Stormwater Management. In *Women in Water Quality. Women in Engineering and Science*; O'Bannon, D., Ed.; Springer: Berlin/Heidelberg, Germany, 2020.
5. Rippy, M.A.; Pierce, G.; Feldman, D.; Winfrey, B.; Mehring, A.S.; Holden, P.A.; Ambrose, R.; Levin, L.A. Perceived Services and Disservices of Natural Treatment Systems for Urban Stormwater: Insight from the next Generation of Designers. *People Nat.* **2022**, *4*, 481–504. [CrossRef]
6. Mehring, A.S.; Levin, L.A. Potential Roles of Soil Fauna in Improving the Efficiency of Rain Gardens Used as Natural Stormwater Treatment Systems. *J. Appl. Ecol.* **2015**, *52*, 1445–1454. [CrossRef]
7. Hamidi, A.; Ramayandi, B.; Sorial, G. Sponge City-An Emerging Concept in Sustainable Water Resource Management: A Scientometric Analysis. *Resour. Environ. Sustain.* **2021**, *1*, 100028. [CrossRef]
8. Yin, D.; Xu, C.; Jia, H.; Yang, Y.; Sun, C.; Wang, Q.; Liu, S. From Pilot Exploration to Systemic Demonstration. *Water* **2022**, *14*, 1531. [CrossRef]
9. Li, H.; Ding, L.; Ren, M.; Li, C.; Wang, H. Sponge City Construction in China: A Survey of the Challenges and Opportunities. *Water* **2017**, *9*, 594. [CrossRef]

10. Millennium Ecosystem Assessment. In *Ecosystems and Human Well-Being: Synthesis*; Island Press: Washington, DC, USA, 2005; ISBN 1597260401.
11. Haines-Young, R.; Potschin, M. *CICES V5. 1. Guidance on the Application of the Revised Structure*; Fabis Consulting Ltd.: Nottingham, UK, 2018.
12. U.S. Environmental Protection Agency. *Benefits of Low Impact Development: How LID Can Protect Your Community's Resources*; U.S. Environmental Protection Agency: Washington, DC, USA, 2012.
13. Liu, Z.; Yang, Y.; Hou, J.; Jia, H. Decision-Making Framework for GI Layout Considering Site Suitability and Weighted Multi-Function Effectiveness: A Case Study in Beijing Sub-Center. *Water* **2022**, *14*, 1765. [CrossRef]
14. Jia, H.; Liu, Z.; Xu, C.; Chen, Z.; Zhang, X.; Xia, J.; Yu, S.L. Adaptive Pressure-Driven Multi-Criteria Spatial Decision-Making for a Targeted Placement of Green and Grey Runoff Control Infrastructures. *Water Res.* **2022**, *212*, 118126. [CrossRef]
15. Chen, Y.; Wang, Y.; Liew, J.H.; Wang, P.L. Development of a Methodological Framework for Evaluating Biodiversity of Built Urban Green Infrastructures by Practitioners. *J. Clean. Prod.* **2021**, *303*, 127009. [CrossRef]
16. Jia, H.; Yao, H.; Tang, Y.; Yu, S.L.; Zhen, J.X.; Lu, Y. Development of a Multi-Criteria Index Ranking System for Urban Runoff Best Management Practices (BMPs) Selection. *Environ. Monit. Assess.* **2013**, *185*, 7915–7933. [CrossRef] [PubMed]
17. Kavehei, E.; Jenkins, G.A.; Lemckert, C.; Adame, M.F. Carbon Stocks and Sequestration of Stormwater Bioretention/Biofiltration Basins. *Ecol. Eng.* **2019**, *138*, 227–236. [CrossRef]
18. Shafique, M.; Xue, X.; Luo, X. An Overview of Carbon Sequestration of Green Roofs in Urban Areas. *Urban For. Urban Green.* **2020**, *47*, 126515. [CrossRef]
19. Erlwein, S.; Zölch, T.; Pauleit, S. Regulating the Microclimate with Urban Green in Densifiying Cities: Joint Assessment on Two Scales. *Build. Environ.* **2021**, *205*, 108233. [CrossRef]
20. Dickinson, D.C.; Hobbs, R.J. Cultural Ecosystem Services: Characteristics, Challenges and Lessons for Urban Green Space Research. *Ecosyst. Serv.* **2017**, *25*, 179–194. [CrossRef]
21. Houlden, V.; Weich, S.; Jarvis, S.; Rees, K. The Relationship between Greenspace and the Mental Wellbeing of Adults: A Systematic Review. *PloS ONE* **2018**, *13*, e0203000. [CrossRef]
22. BenDor, T.K.; Shandas, V.; Miles, B.; Belt, K.; Olander, L. Ecosystem Services and U.S. Stormwater Planning: An Approach for Improving Urban Stormwater Decisions. *Environ. Sci. Policy* **2018**, *88*, 92–103. [CrossRef]
23. Elliot, R.; Motzny, A.; Majd, S.; Chavez, F.; Laimer, D.; Orlove, B.; Culligan, P. Identifying Linkages between Urban Green Infrastructure and Ecosystem Services Using an Expert Opinion Methodology. *Ambio* **2020**, *49*, 569–583. [CrossRef]
24. Kessouri, F.; McWilliams, J.C.; Bianchi, D.; Sutula, M.; Renault, L.; Deutsch, C.; Feely, R.A.; McLaughlin, K.; Ho, M.; Howard, E.M.; et al. Coastal Eutrophication Drives Acidification, Oxygen Loss, and Ecosystem Change in a Major Oceanic Upwelling System. *Proc. Natl. Acad. Sci. USA* **2021**, *118*, 1–8. [CrossRef]
25. Huang, X.; Swain, D.L. Climate Change Is Increasing the Risk of a California Megaflood. *Sci. Adv.* **2022**, *8*, 1–14. [CrossRef] [PubMed]
26. County of Los Angeles Department of Public Works. Low Impact Development Standards Manual. 2014. Available online: https://dpw.lacounty.gov/ldd/lddservices/docs/Low_Impact_Development_Standards_Manual.pdf (accessed on 31 January 2023).
27. Levin, L.; Le, J.; Gonzalez, J.; Ambrose, R. Biofilter Catalog and Database for the Los Angeles Region. 2017. Available online: https://dornsife.usc.edu/assets/sites/291/docs/Publications/Levin_et_al._2017_Biofilter_Report.pdf (accessed on 31 January 2023).
28. Yang, F.; Fu, D.; Zevenbergen, C.; Rene, E.R. A Comprehensive Review on the Long-Term Performance of Stormwater Biofiltration Systems (SBS): Operational Challenges and Future Directions. *J. Environ. Manag.* **2022**, *302*, 113956. [CrossRef]
29. Snyder, H. Literature Review as a Research Methodology: An Overview and Guidelines. *J. Bus. Res.* **2019**, *104*, 333–339. [CrossRef]
30. Dudley, S.; Mannix, B. Improving Regulatory Benefit-Cost Analysis. *J. Law Polit.* **2018**, *34*, 1.
31. Caro, C.; Marques, J.C.; Cunha, P.P.; Teixeira, Z. Ecosystem Services as a Resilience Descriptor in Habitat Risk Assessment Using the InVEST Model. *Ecol. Indic.* **2020**, *115*, 106426. [CrossRef]
32. Tirpak, R.A.; Afrooz, A.N.; Winston, R.J.; Valenca, R.; Schiff, K.; Mohanty, S.K. Conventional and Amended Bioretention Soil Media for Targeted Pollutant Treatment: A Critical Review to Guide the State of the Practice. *Water Res.* **2021**, *189*, 116648. [CrossRef]
33. Rammal, M.; Berthier, E. Runoff Losses on Urban Surfaces during Frequent Rainfall Events: A Review of Observations and Modeling Attempts. *Water* **2020**, *12*, 2777. [CrossRef]
34. Berretta, C.; Aiello, A.; Jensen, H.S.; Al, E. Influenc of Design and Media Amendments on the Performance of Stormwater Biofilters. *Proc. ICE-Water Manag.* **2018**, *171*, 87–98.
35. Le Coustumer, S.; Fletcher, T.D.; Deletic, A.; Barraud, S.; Poelsma, P. The Influence of Design Parameters on Clogging of Stormwater Biofilters: A Large-Scale Column Study. *Water Res.* **2012**, *46*, 6743–6752. [CrossRef]
36. Sileshi, R.; Pitt, R.E.; Clark, S.E. Statistical Analyses of Flow Rates of Stormwater Treatment Bioretention Media. *Water Environ. Res.* **2019**, *91*, 877–887. [CrossRef]
37. Skorobogatov, A.; He, J.; Chu, A.; Valeo, C.; van Duin, B. The Impact of Media, Plants and Their Interactions on Bioretention Performance: A Review. *Sci. Total Environ.* **2020**, *715*, 136918. [CrossRef] [PubMed]

38. Askarizadeh, A.; Rippy, M.A.; Fletcher, T.D.; Feldman, D.L.; Peng, J.; Bowler, P.; Mehring, A.S.; Winfrey, B.K.; Vrugt, J.A.; Aghakouchak, A.; et al. From Rain Tanks to Catchments: Use of Low-Impact Development To Address Hydrologic Symptoms of the Urban Stream Syndrome. *Environ. Sci. Technol.* **2015**, *49*, 11264–11280. [CrossRef] [PubMed]
39. Minnesota Pollution Control Agency. MPCA Bioretention Terminology. *Minnesota Stormwater Man.* **2015**, 1–9. Available online: https://stormwater.pca.state.mn.us/index.php?title=Main_Page (accessed on 31 January 2023).
40. Bouwer, H. Artificial Recharge of Groundwater: Hydrogeology and Engineering. *Hydrogeol. J.* **2002**, *10*, 121–142. [CrossRef]
41. Berland, A.; Shiflett, S.A.; Shuster, W.D.; Garmestani, A.S.; Goddard, H.C.; Herrmann, D.L.; Hopton, M.E. The Role of Trees in Urban Stormwater Management. *Landsc. Urban Plan.* **2017**, *162*, 167–177. [CrossRef]
42. Hatt, B.E.; Fletcher, T.D.; Deletic, A. Hydrologic and Pollutant Removal Performance of Stormwater Biofiltration Systems at the Field Scale. *J. Hydrol.* **2009**, *365*, 310–321. [CrossRef]
43. Payne, E.G.I.; Pham, T.; Deletic, A.; Hatt, B.E.; Cook, P.L.M.; Fletcher, T.D. Which Species? A Decision-Support Tool to Guide Plant Selection in Stormwater Bio Fi Lters. *Adv. Water Resour.* **2018**, *113*, 86–99. [CrossRef]
44. Sharkey, L.J. *The Performance of Bioretention Areas in North Caroline: A Study of Water Quality, Water Quantity, and Soil Media*; North Caroline State University: Raleigh, NC, USA, 2006.
45. Thom, J.K.; Szota, C.; Coutts, A.M.; Fletcher, T.D.; Livesley, S.J. Transpiration by Established Trees Could Increase the Efficiency of Stormwater Control Measures. *Water Res.* **2020**, *173*, 115597. [CrossRef]
46. Minnig, M.; Moeck, C.; Radny, D.; Schirmer, M. Impact of Urbanization on Groundwater Recharge Rates in Dübendorf, Switzerland. *J. Hydrol.* **2018**, *563*, 1135–1146. [CrossRef]
47. Sheng, J.; Wilson, J.P. Watershed Urbanization and Changing Flood Behavior across the Los Angeles Metropolitan Region. *Nat. Hazards* **2009**, *48*, 41–57. [CrossRef]
48. Seager, R.; Ting, M.; Li, C.; Naik, N.; Cook, B.; Nakamura, J.; Liu, H. Projections of Declining Surface-Water Availability for the Southwestern United States. *Nat. Clim. Chang.* **2013**, *3*, 482–486. [CrossRef]
49. Park, K.; Lee, M.H. The Development and Application of the Urban Flood Risk Assessment Model for Reflecting upon Urban Planning Elements. *Water* **2019**, *11*, 920. [CrossRef]
50. Winston, R.J.; Dorsey, J.D.; Hunt, W.F. Quantifying Volume Reduction and Peak Flow Mitigation for Three Bioretention Cells in Clay Soils in Northeast Ohio. *Sci. Total Environ.* **2016**, *553*, 83–95. [CrossRef] [PubMed]
51. Smolek, A.P.; Anderson, A.R.; Hunt, W.F. Hydrologic and Water-Quality Evaluation of a Rapid-Flow Biofiltration Device. *J. Environ. Eng.* **2018**, *144*, 1–13. [CrossRef]
52. Bonneau, J.; Fletcher, T.D.; Costelloe, J.F.; Poelsma, P.J.; James, R.B.; Burns, M.J. The Hydrologic, Water Quality and Flow Regime Performance of a Bioretention Basin in Melbourne, Australia. *Urban Water J.* **2020**, *17*, 303–314. [CrossRef]
53. Ambrose, R.F.; Winfrey, B.K. Comparison of Stormwater Biofiltration Systems in Southeast Australia and Southern California. *WIREs Water* **2015**, *2*, 131–146. [CrossRef]
54. Belden, E.; Antos, M.; Kristy, M.; Steele, N.L.C. Sustainable Infrastructure: The Elmer Avenue Neighborhood Retrofit. 2012. Available online: https://urbancoast.org/wp-content/uploads/2014/10/V3_13_EdwardBelden.pdf (accessed on 31 January 2023).
55. Quinn, N.; Bates, P.D.; Neal, J.; Smith, A.; Wing, O.; Sampson, C.; Smith, J.; Heffernan, J. The Spatial Dependence of Flood Hazard and Risk in the United States. *Water Resour. Res.* **2019**, *55*, 1890–1911. [CrossRef]
56. Nofal, O.M.; van de Lindt, J.W. Understanding Flood Risk in the Context of Community Resilience Modeling for the Built Environment: Research Needs and Trends. *Sustain. Resilient Infrastruct.* **2022**, *7*, 171–187. [CrossRef]
57. Qin, Y. Urban Flooding Mitigation Techniques: A Systematic Review and Future Studies. *Water* **2020**, *12*, 3579. [CrossRef]
58. Aerts, J.C.J.H.; Barnard, P.L.; Botzen, W.; Grifman, P.; Hart, J.F.; De Moel, H.; Mann, A.N.; de Ruig, L.T.; Sadrpour, N. Pathways to Resilience: Adapting to Sea Level Rise in Los Angeles. *Ann. N. Y. Acad. Sci.* **2018**, *1427*, 1–90. [CrossRef]
59. Barbier, E.B. *The Value of Coastal Wetland Ecosystem Services*; Elsevier: Amsterdam, The Netherlands, 2018; ISBN 9780444638939.
60. Brander, L.M.; Van Beukering, P.; Cesar, H.S.J. The Recreational Value of Coral Reefs: A Meta-Analysis. *Ecol. Econ.* **2007**, *63*, 209–218. [CrossRef]
61. Watson, K.B.; Ricketts, T.; Galford, G.; Polasky, S.; O'Niel-Dunne, J. Quantifying Flood Mitigation Services: The Economic Value of Otter Creek Wetlands and Floodplains to Middlebury, VT. *Ecol. Econ.* **2016**, *130*, 16–24. [CrossRef]
62. Maupin, M.A.; Kenny, J.F.; Hutson, S.S.; Lovelace, J.K.; Barber, N.L.; Linsey, K.S. Estimated Use of Water in the United States in 2010. 2014; U.S. Geological Survey Circular 1405. Available online: https://pubs.usgs.gov/circ/1405/pdf/circ1405.pdf (accessed on 31 January 2023).
63. Lindt, R.; Callahan, C.; DeShazo, J.R.; Bieber, E. Lessons Learned from Previous Projects for Green Alley Development in Los Angeles & Beyond. LA, USA. 2014. Available online: https://www.tpl.org/wp-content/uploads/2015/05/ca-green-alley-avalon-green-alleys-demo-project.pdf (accessed on 31 January 2023).
64. Kazmierczak, J.; Muller, S.; Nilsson, B.; Postma, D.; Czejak, J.; Sebok, E.; Jessen, S.; Karan, S.; Stenvig Jensen, C.; Edelvang, K.; et al. Groundwater Flow and Heterogeneous Discharge into a Seepage Lake: Combined Use of Physical Methods and Hydrochemical Tracers. *Water Resour. Res.* **2016**, *52*, 9109–9130. [CrossRef]
65. Reitz, M.; Sanford, W.E.; Senay, G.B.; Cazenas, J. Annual Estimates of Recharge, Quick-Flow Runoff, and Evapotranspiration for the Contiguous U.S. Using Empirical Regression Equations. *J. Am. Water Resour. Assoc.* **2017**, *53*, 961–983. [CrossRef]

66. Masoner, J.R.; Kolpin, D.W.; Cozzarelli, I.M.; Barber, L.B.; Burden, D.S.; Foreman, W.T.; Forshay, K.J.; Furlong, E.T.; Groves, J.F.; Hladik, M.L.; et al. Urban Stormwater: An Overlooked Pathway of Extensive Mixed Contaminants to Surface and Groundwaters in the United States. *Environ. Sci. Technol.* **2019**, *53*, 10070–10081. [CrossRef] [PubMed]
67. Deletic, A. The First Flush Load of Urban Surface Runoff. *Water Res.* **1998**, *32*, 2462–2470. [CrossRef]
68. Maniquiz-Redillas, M.; Robles, M.E.; Cruz, G.; Reyes, N.J.; Kim, L.H. First Flush Stormwater Runoff in Urban Catchments: A Bibliometric and Comprehensive Review. *Hydrology* **2022**, *9*, 63. [CrossRef]
69. Fanelli, R.M.; Prestegaard, K.L.; Palmer, M.A. Urban Legacies: Aquatic Stressors and Low Aquatic Biodiversity Persist despite Implementation of Regenerative Stormwater Conveyance Systems. *Freshw. Sci.* **2019**, *38*, 818–833. [CrossRef]
70. Guo, J.C.Y.; Luu, T.M. Operation of Cap Orifice in a Rain Garden. *J. Hydrol. Eng.* **2015**, *20*, 1–6. [CrossRef]
71. Hunt, W.F.; Greenway, M.; Moore, T.C.; Brown, R.A.; Kennedy, S.G.; Line, D.E.; Lord, W.G. Constructed Storm-Water Wetland Installation and Maintenance: Are We Getting It Right? *J. Irrig. Drain. Eng.* **2011**, *137*, 469–474. [CrossRef]
72. Müller, A.; Österlund, H.; Marsalek, J.; Viklander, M. The Pollution Conveyed by Urban Runoff: A Review of Sources. *Sci. Total Environ.* **2020**, *709*, 136125. [CrossRef] [PubMed]
73. Kumar, M.; Gogoi, A.; Kumari, D.; Borah, R. Review of Perspective, Problems, Challenges, and Future Scenario of Metal Contamination in the Urban Environment. *J. Hazard. Toxic Radioact. Waste* **2017**, *21*, 290–307. [CrossRef]
74. Hallett, C.S.; Valesini, F.J.; Clarke, K.R.; Hoeksema, S.D. Effects of a Harmful Algal Bloom on the Community Ecology, Movements and Spatial Distributions of Fishes in a Microtidal Estuary. *Hydrobiologia* **2016**, *763*, 267–284. [CrossRef]
75. Lim, H.S.; Lim, W.; Hu, J.Y.; Ziegler, A.; Ong, S.L. Comparison of Filter Media Materials for Heavy Metal Removal from Urban Stormwater Runoff Using Biofiltration Systems. *J. Environ. Manag.* **2015**, *147*, 24–33. [CrossRef] [PubMed]
76. Searcy, R.T.; Boehm, A.B. A Day at the Beach: Enabling Coastal Water Quality Prediction with High-Frequency Sampling and Data-Driven Models. *Environ. Sci. Technol.* **2021**, *55*, 1908–1918. [CrossRef]
77. Pendleton, L.; Kildow, J. The Non-Market Value of California's Beaches. *Shore Beach* **2006**, *74*, 34–37.
78. Rathi, B.S.; Kumar, P.S. Application of Adsorption Process for Effective Removal of Emerging Contaminants from Water and Wastewater. *Environ. Pollut.* **2021**, *280*, 116995. [CrossRef]
79. Payne, E.G.I.; Fletcher, T.D.; Russell, D.G.; Grace, M.R.; Cavagnaro, T.R.; Evrard, V.; Deletic, A.; Hatt, B.E.; Cook, P.L.M. Temporary Storage or Permanent Removal? The Division of Nitrogen between Biotic Assimilation and Denitrification in Stormwater Biofiltration Systems. *PLoS ONE* **2014**, *9*, e90890. [CrossRef]
80. Badin, A.; Monier, A.; Volatier, L.; Geremia, R.; Delolme, C.; Bedell, J. Structural Stability, Microbial Biomass and Community Composition of Sediments Affected by the Hydric Dynamics of an Urban Stormwater Infiltration Basin. *Environ. Microbiol.* **2011**, *61*, 885–897. [CrossRef]
81. Bogati, K.; Walczak, M. The Impact of Drought Stress on Soil Microbial Community, Enzyme Activities and Plants. *Agronomy* **2022**, *12*, 189. [CrossRef]
82. Ma, Y.; Rajkumar, M.; Zhang, C.; Freitas, H. Inoculation of Brassica Oxyrrhina with Plant Growth Promoting Bacteria for the Improvement of Heavy Metal Phytoremediation under Drought Conditions. *J. Hazard. Mater.* **2016**, *320*, 36–44. [CrossRef] [PubMed]
83. Al-Ameri, M.; Hatt, B.; Le Coustumer, S.; Fletcher, T.; Payne, E.; Deletic, A. Accumulation of Heavy Metals in Stormwater Bioretention Media: A Field Study of Temporal and Spatial Variation. *J. Hydrol.* **2018**, *567*, 721–731. [CrossRef]
84. Irvine, K.N.; Chua, L.H.C.; Hua'an, Z.; Qi, L.E.; Xuan, L.Y. Nature-Based Solutions to Manage Particle-Bound Metals in Urban Stormwater Runoff: Current Design Practices and Knowledge Gaps. *J. Soils Sediments* **2022**, 1–18. [CrossRef]
85. Al-Rubaei, A.M.; Engström, M.; Viklander, M.; Blecken, G.T. Long-Term Hydraulic and Treatment Performance of a 19-Year Old Constructed Stormwater Wetland—Finally Maturated or in Need of Maintenance? *Ecol. Eng.* **2016**, *95*, 73–82. [CrossRef]
86. Waara, S.; Johansson, F. Ecological Risk Assessment of Trace Elements Accumulated in Stormwater Ponds within Industrial Areas. *Environ. Sci. Pollut. Res.* **2022**, *29*, 27026–27041. [CrossRef]
87. Søberg, L.C.; Vollertsen, J.; Blecken, G.T.; Nielsen, A.H.; Viklander, M. Bioaccumulation of Heavy Metals in Two Wet Retention Ponds. *Urban Water J.* **2016**, *13*, 697–709. [CrossRef]
88. Zinger, Y.; Blecken, G.T.; Fletcher, T.D.; Viklander, M.; Deletić, A. Optimising Nitrogen Removal in Existing Stormwater Biofilters: Benefits and Tradeoffs of a Retrofitted Saturated Zone. *Ecol. Eng.* **2013**, *51*, 75–82. [CrossRef]
89. Kuoppamäki, K.; Pflugmacher Lima, S.; Scopetani, C.; Setälä, H. The Ability of Selected Filter Materials in Removing Nutrients, Metals, and Microplastics from Stormwater in Biofilter Structures. *J. Environ. Qual.* **2021**, *50*, 465–475. [CrossRef]
90. Boehm, A.B.; Bell, C.D.; Fitzgerald, N.J.M.; Gallo, E.; Higgins, C.P.; Hogue, T.S.; Luthy, R.G.; Portmann, A.C.; Ulrich, B.A.; Wolfand, J.M. Biochar-Augmented Biofilters to Improve Pollutant Removal from Stormwater-Can They Improve Receiving Water Quality? *Environ. Sci. Water Res. Technol.* **2020**, *6*, 1520–1537. [CrossRef]
91. Stanford Natural Capital Project. Available online: https://naturalcapitalproject.stanford.edu/software/invest (accessed on 4 March 2023).
92. Ouyang, Z.; Zheng, H.; Xiao, Y.; Polasky, S.; Liu, J.; Xu, W.; Wang, Q.; Zhang, L.; Xio, Y.; Rao, E.; et al. Improvements in Ecosystem Services from Investments in Natural Capital. *Science* **2016**, *352*, 1455–1459. [CrossRef]
93. Redhead, J.W.; Stratford, C.; Sharps, K.; Jones, L.; Ziv, G.; Clarke, D.; Oliver, T.H.; Bullock, J.M. Empirical Validation of the InVEST Water Yield Ecosystem Service Model at a National Scale. *Sci. Total Environ.* **2016**, *569–570*, 1418–1426. [CrossRef]

94. Benra, F.; De Frutos, A.; Gaglio, M.; Álvarez-Garretón, C.; Felipe-Lucia, M.; Bonn, A. Mapping Water Ecosystem Services: Evaluating InVEST Model Predictions in Data Scarce Regions. *Environ. Model. Softw.* **2021**, *138*, 104982. [CrossRef]
95. Salata, S.; Garnero, G.; Barbieri, C.A.; Giaimo, C. The Integration of Ecosystem Services in Planning: An Evaluation of the Nutrient Retention Model Using InVEST Software. *Land* **2017**, *6*, 48. [CrossRef]
96. Carson, R.T.; Mitchell, R.C. The Value of Clean Water: The Public's Willingness to Pay for Boatable, Fishable, and Swimmable Quality Water. *Water Resour. Res.* **1993**, *29*, 2445–2454. [CrossRef]
97. Chatterjee, C.; Triplett, R.; Johnson, C.K.; Ahmed, P. Willingness to Pay for Safe Drinking Water: A Contingent Valuation Study in Jacksonville, FL. *J. Environ. Manag.* **2017**, *203*, 413–421. [CrossRef] [PubMed]
98. Levin, L.A.; Mehring, A.S. Optimization of Bioretention Systems through Application of Ecological Theory. *Wiley Interdiscip. Rev. Water* **2015**, *2*, 259–270. [CrossRef]
99. Hobbie, S.E.; Grimm, N.B. Nature-Based Approaches to Managing Climate Change Impacts in Cities. *Philos. Trans. R. Soc. B Biol. Sci.* **2020**, *375*, 20190124. [CrossRef]
100. Science for Environment Policy. In *Ecosystem Services and Biodiversity*; 2015.
101. Aavik, T.; Helm, A. Restoration of Plant Species and Genetic Diversity Depends on Landscape-Scale Dispersal. *Restor. Ecol.* **2018**, *26*, S92–S102. [CrossRef]
102. Kiers, A.H.; Krimmel, B.; Larsen-Bircher, C.; Hayes, K.; Zemenick, A.; Michaels, J. Different Jargon, Same Goals: Collaborations between Landscape Architects and Ecologists to Maximize Biodiversity in Urban Lawn Conversions. *Land* **2022**, *11*, 1665. [CrossRef]
103. Dreiseitl, H. *Water and Sustainable Design*; 2020.
104. Tan, K.W. A Greenway Network for Singapore. *Landsc. Urban Plan.* **2006**, *76*, 45–66. [CrossRef]
105. Song, Y.; Kirkwood, N.; Maksimović, Č.; Zhen, X.; O'Connor, D.; Jin, Y.; Hou, D. Nature Based Solutions for Contaminated Land Remediation and Brownfield Redevelopment in Cities: A Review. *Sci. Total Environ.* **2019**, *663*, 568–579. [CrossRef] [PubMed]
106. Heimann, M.; Reichstein, M. Terrestrial Ecosystem Carbon Dynamics and Climate Feedbacks. *Nature* **2008**, *451*, 289–292. [CrossRef] [PubMed]
107. Krauss, L.; Rippy, M. Adaptive Strategy Biases in Engineered Ecosystems: Implications for Plant Community Dynamics and the Provisioning of Ecosystem Services to People. *People Nat.* **2022**, *4*, 1644–1663. [CrossRef]
108. Kranner, B.; Afrouz, A.; Fitzgerald, N.; Boehm, A. Fecal Indicator Bacteria and Virus Removal in Stormwater Biofilters: Effects of Biochar, Media Saturation, and Field Conditioning. *PLoS ONE* **2019**, *14*, e0222719. [CrossRef]
109. Mehring, A.S.; Hatt, B.E.; Kraikittikun, D.; Orelo, B.D.; Rippy, M.A.; Grant, S.B.; Gonzalez, J.P.; Jiang, S.C.; Ambrose, R.F.; Levin, L.A. Soil Invertebrates in Australian Rain Gardens and Their Potential Roles in Storage and Processing of Nitrogen. *Ecol. Eng.* **2016**, *97*, 138–143. [CrossRef]
110. Capowiez, Y.; Gilbert, F.; Vallat, A.; Poggiale, J.C.; Bonzom, J.M. Depth Distribution of Soil Organic Matter and Burrowing Activity of Earthworms—Mesocosm Study Using X-Ray Tomography and Luminophores. *Biol. Fertil. Soils* **2021**, *57*, 337–346. [CrossRef]
111. Ge, B.; Mehring, A.S.; Levin, L.A. Urbanization Alters Belowground Invertebrate Community Structure in Semi-Arid Regions: A Comparison of Lawns, Biofilters and Sage Scrub. *Landsc. Urban Plan.* **2019**, *192*, 103664. [CrossRef]
112. Evans, B.A.; Gawlik, D.E. Urban Food Subsidies Reduce Natural Food Limitations and Reproductive Costs for a Wetland Bird. *Sci. Rep.* **2020**, *10*, 1–12. [CrossRef]
113. Hanley, N.; Perrings, C. The Economic Value of Biodiversity. *Annu. Rev. Resour. Econ.* **2019**, *11*, 355–375. [CrossRef]
114. Champ, P.; Boyle, K.; Brown, T. (Eds.) *A Primer on Nonmarket Valuation*, 2nd ed.; Springer International Publishing: Dordrecht, The Netherlands, 2017.
115. Dupras, J.; Alam, M.; Reveret, J.P. Economic Value of Greater Montreal's Non-Market Ecosystem Services in a Land Use Management and Planning Perspective. *Can. Geogr.* **2014**, *59*, 93–106. [CrossRef]
116. Gallai, N.; Salles, J.M.; Settele, J.; Vaissière, B.E. Economic Valuation of the Vulnerability of World Agriculture Confronted with Pollinator Decline. *Ecol. Econ.* **2009**, *68*, 810–821. [CrossRef]
117. IPCC. Climate Change 2021: The Physical Science Basis. 2021. Available online: https://www.ipcc.ch/report/ar6/wg1/ (accessed on 31 January 2023).
118. Global Platform for Sustainable Cities São Paolo Statement on Urban Sustainability: A Call to Integrate Our Responses to Climate Change, Biodiversity Loss, and Social Inequality. Available online: https://www.thegpsc.org (accessed on 4 October 2019).
119. Galbraith, D.; Malhi, Y.; Affum-Baffoe, K.; Castanho, A.; CE, D. Residence Times of Woody Biomass in Tropical Forests. *Plant Ecol. Divers.* **2013**, *6*, 139–157. [CrossRef]
120. Pronk, G.J.; Mellage, A.; Milojevic, T.; Smeaton, C.M.; Engel, K.; Neufeld, J.D.; Rezanezhad, F.; Van Cappellen, P. Carbon Turnover and Microbial Activity in an Artificial Soil under Imposed Cyclic Drainage and Imbibition. *Vadose Zone J.* **2020**, *19*, 1–16. [CrossRef]
121. Li, K.; Fang, F.; Wang, H.; Wang, C.; Chen, Y.; Guo, J.; Wang, X.; Jiang, F. Pathways of N Removal and N2O Emission from a One-Stage Autotrophic N Removal Process under Anaerobic Conditions. *Sci. Rep.* **2017**, *7*, 1–10. [CrossRef]
122. Nowak, D.J.; Greenfield, E.J.; Hoehn, R.E.; Lapoint, E. Carbon Storage and Sequestration by Trees in Urban and Community Areas of the United States. *Environ. Pollut.* **2013**, *178*, 229–236. [CrossRef] [PubMed]
123. U.S. Environmental Protection Agency. *Social Cost of Carbon*; U.S. Environmental Protection Agency: Washington, DC, USA, 2016.
124. Getter, K.; Rowe, D.; Robertson, G.; Cregg, B.; Andresen, J. Carbon Sequestration Potential of Extensive Green Roofs. *Environ. Sci. Technol.* **2009**, *43*, 7564–7570. [CrossRef]

125. Seyedabadi, M.R.; Eicker, U.; Karimi, S. Plant Selection for Green Roofs and Their Impact on Carbon Sequestration and the Building Carbon Footprint. *Environ. Chall.* **2021**, *4*, 100119. [CrossRef]
126. Lyu, R.; Mi, L.; Zhang, J.; Xu, M.; Li, J. Modeling the Effects of Urban Expansion on Regional Carbon Storage by Coupling SLEUTH-3r Model and InVEST Model. *Ecol. Res.* **2019**, *34*, 380–393. [CrossRef]
127. Babbar, D.; Areendran, G.; Sahana, M.; Sarma, K.; Raj, K.; Sivadas, A. Assessment and Prediction of Carbon Sequestration Using Markov Chain and InVEST Model in Sariska Tiger Reserve, India. *J. Clean. Prod.* **2021**, *278*, 123333. [CrossRef]
128. González-García, A.; Arias, M.; García-Tiscar, S.; Alcorlo, P.; Santos-Martín, F. National Blue Carbon Assessment in Spain Using InVEST: Current State and Future Perspectives. *Ecosyst. Serv.* **2022**, *53*, 101397. [CrossRef]
129. U.S. Environmental Protection Agency Social Cost of Greenhouse Gases. Available online: https://www.epa.gov/environmental-economics/scghg (accessed on 4 March 2023).
130. Li, P.; Wang, Z.H. Environmental Co-Benefits of Urban Greening for Mitigating Heat and Carbon Emissions. *J. Environ. Manag.* **2021**, *293*, 112963. [CrossRef]
131. Oke, T. The Energetic Basis of the Urban Heat Island. *Q. J. R. Meteorol. Soc.* **1982**, *108*, 1–24. [CrossRef]
132. Stewart, I.D.; Oke, T.R. Local Climate Zones for Urban Temperature Studies. *Bull. Am. Meteorol. Soc.* **2012**, *93*, 1879–1900. [CrossRef]
133. Stache, E.; Schilperoort, B.; Ottelé, M.; Jonkers, H.M. Comparative Analysis in Thermal Behaviour of Common Urban Building Materials and Vegetation and Consequences for Urban Heat Island Effect. *Build. Environ.* **2022**, *213*, 108489. [CrossRef]
134. Meili, N.; Acero, J.A.; Peleg, N.; Manoli, G.; Burlando, P.; Fatichi, S. Vegetation Cover and Plant-Trait Effects on Outdoor Thermal Comfort in a Tropical City. *Build. Environ.* **2021**, *195*, 107733. [CrossRef]
135. Kalkstein, L.S.; Eisenman, D.P.; de Guzman, E.B.; Sailor, D.J. Increasing Trees and High-Albedo Surfaces Decreases Heat Impacts and Mortality in Los Angeles, CA. *Int. J. Biometeorol.* **2022**, *66*, 911–925. [CrossRef]
136. Porto, R.G.; de Almeida, R.F.; Cruz-Neto, O.; Tabarelli, M.; Viana, B.F.; Peres, C.A.; Lopes, A.V. Pollination Ecosystem Services: A Comprehensive Review of Economic Values, Research Funding and Policy Actions. *Food Secur.* **2020**, *12*, 1425–1442. [CrossRef]
137. Dicks, L.V.; Breeze, T.D.; Ngo, H.T.; Senapathi, D.; An, J.; Aizen, M.A.; Basu, P.; Buchori, D.; Galetto, L.; Garibaldi, L.A.; et al. A Global-Scale Expert Assessment of Drivers and Risks Associated with Pollinator Decline. *Nat. Ecol. Evol.* **2021**, *5*, 1453–1461. [CrossRef]
138. Haddad, N.M.; Brudvig, L.A.; Clobert, J.; Davies, K.F.; Gonzalez, A.; Holt, R.D.; Lovejoy, T.E.; Sexton, J.O.; Austin, M.P.; Collins, C.D.; et al. Habitat Fragmentation and Its Lasting Impact on Earth's Ecosystems. *Sci. Adv.* **2015**, *1*, e1500052. [CrossRef]
139. Kremen, C.; Williams, N.M.; Aizen, M.A.; Gemmill-Herren, B.; LeBuhn, G.; Minckley, R.; Packer, L.; Potts, S.G.; Roulston, T.; Steffan-Dewenter, I.; et al. Pollination and Other Ecosystem Services Produced by Mobile Organisms: A Conceptual Framework for the Effects of Land-Use Change. *Ecol. Lett.* **2007**, *10*, 299–314. [CrossRef]
140. Hung, K.-L.J.; Kingston, J.M.; Albrecht, M.; Holway, D.A.; Kohn, J.R. The Worldwide Importance of Honey Bees as Pollinators in Natural Habitats. *Proc. R. Soc. B Biol. Sci.* **2018**, *285*, 20172140. [CrossRef]
141. IPBES Conceptual Framework for the Intergovernmental Science-Policy Platform on Biodiversity and Ecosystem Services. *Curr. Opin. Environ. Sustain.* **2014**, *IPBES-2/4*, 231–269.
142. Breeze, T.D.; Bailey, A.P.; Potts, S.G.; Balcombe, K.G. A Stated Preference Valuation of the Non-Market Benefits of Pollination Services in the UK. *Ecol. Econ.* **2015**, *111*, 76–85. [CrossRef]
143. Veitch, J.; Ball, K.; Crawford, D.; Abbott, G.; Salmon, J. Park Improvements and Park Activity: A Natural Experiment. *Am. J. Prev. Med.* **2012**, *42*, 616–619. [CrossRef]
144. Van Den Berg, A.E.; Maas, J.; Verheij, R.A.; Groenewegen, P.P. Social Science & Medicine Green Space as a Buffer between Stressful Life Events and Health Q. *Soc. Sci. Med.* **2010**, *70*, 1203–1210. [CrossRef] [PubMed]
145. Hartig, T.; Mitchell, R.; De Vries, S.; Frumkin, H. Nature and Health. *Annu. Rev. Public Health* **2014**, *35*, 207–228. [CrossRef]
146. Coombes, E.; Jones, A.P.; Hillsdon, M. The Relationship of Physical Activity and Overweight to Objectively Measured Green Space Accessibility and Use. *Soc. Sci. Med.* **2010**, *70*, 816–822. [CrossRef]
147. U.S. Census Bureau Population Density Data. Available online: https://www.census.gov/quickfacts/fact/table/losangelescountycalifornia,CA/PST045222 (accessed on 13 January 2023).
148. More, T.A.; Stevens, T.; Allen, P.G. Valuation of Urban Parks. *Landsc. Urban Plan.* **1988**, *15*, 139–152. [CrossRef]
149. Łaszkiewicz, E.; Czembrowski, P.; Kronenberg, J. Can Proximity to Urban Green Spaces Be Considered a Luxury? Classifying a Non-Tradable Good with the Use of Hedonic Pricing Method. *Ecol. Econ.* **2019**, *161*, 237–247. [CrossRef]
150. Ward, C.D.; Parker, C.M.; Shackleton, C.M. The Use and Appreciation of Botanical Gardens as Urban Green Spaces in South Africa. *Urban For. Urban Green.* **2010**, *9*, 49–55. [CrossRef]
151. Byrne, J.; Wolch, J.; Zhang, J. Planning for Environmental Justice in an Urban National Park. *J. Environ. Plan. Manag.* **2009**, *52*, 365–392. [CrossRef]
152. Graça, M.; Alves, P.; Gonçalves, J.; Nowak, D.J.; Hoehn, R.; Farinha-marques, P.; Cunha, M. Landscape and Urban Planning Assessing How Green Space Types a Ff Ect Ecosystem Services Delivery in Porto, Portugal. *Landsc. Urban Plan.* **2018**, *170*, 195–208. [CrossRef]
153. Southon, G.E.; Jorgensen, A.; Dunnett, N.; Hoyle, H.; Evans, K.L. Biodiverse Perennial Meadows Have Aesthetic Value and Increase Residents' Perceptions of Site Quality in Urban Green-Space. *Landsc. Urban Plan.* **2017**, *158*, 105–118. [CrossRef]

154. Wolch, J.R.; Byrne, J.; Newell, J.P. Urban Green Space, Public Health, and Environmental Justice: The Challenge of Making Cities "Just Green Enough". *Landsc. Urban Plan.* **2014**, *125*, 234–244. [CrossRef]
155. Deng, L.; Luo, H.; Ma, J.; Huang, Z.; Sun, L.X.; Jiang, M.Y.; Zhu, C.Y.; Li, X. Effects of Integration between Visual Stimuli and Auditory Stimuli on Restorative Potential and Aesthetic Preference in Urban Green Spaces. *Urban For. Urban Green.* **2020**, *53*, 126702. [CrossRef]
156. Reyes-Riveros, R.; Altamirano, A.; De La Barrera, F.; Rozas-Vásquez, D.; Vieli, L.; Meli, P. Linking Public Urban Green Spaces and Human Well-Being: A Systematic Review. *Urban For. Urban Green.* **2021**, *61*, 127105. [CrossRef]
157. Van Renterghem, T.; Hornikx, M.; Forssen, J.; Botteldooren, D. The Potential of Building Envelope Greening to Achieve Quietness. *Build. Environ.* **2013**, *61*, 34–44. [CrossRef]
158. Hedblom, M.; Gunnarsson, B.; Iravani, B.; Knez, I.; Schaefer, M.; Thorsson, P.; Lundström, J.N. Reduction of Physiological Stress by Urban Green Space in a Multisensory Virtual Experiment. *Sci. Rep.* **2019**, *9*, 1–11. [CrossRef]
159. Dadvand, P.; Nieuwenhuijsen, M.J.; Esnaola, M.; Forns, J.; Basagaña, X.; Alvarez-Pedrerol, M.; Rivas, I.; López-Vicente, M.; De Castro Pascual, M.; Su, J.; et al. Green Spaces and Cognitive Development in Primary Schoolchildren. *Proc. Natl. Acad. Sci. USA* **2015**, *112*, 7937–7942. [CrossRef]
160. Hermawan, A.A.; Jung, D.Y.; Talei, A. Removal Process of Nutrients and Heavy Metals in Tropical Biofilters. *E3S Web Conf.* **2018**, *65*, 1–11. [CrossRef]
161. Le, J. *Characterization of Ecosystem Services Associated with Deep-Sea Habitats and Natural Stormwater Treatment Systems and Their Incorporation into Environmental Management*; University of California: San Diego, CA, USA, 2020.
162. Hansford, K.M.; Fonville, M.; Gillingham, E.L.; Coipan, E.C.; Pietzsch, M.E.; Krawczyk, A.I.; Vaux, A.G.C.; Cull, B.; Sprong, H.; Medlock, J.M. Ticks and Borrelia in Urban and Peri-Urban Green Space Habitats in a City in Southern England. *Ticks Tick-Borne Dis.* **2017**, *8*, 353–361. [CrossRef] [PubMed]
163. Bogar, S.; Beyer, K.M. Green Space, Violence, and Crime: A Systematic Review. *Trauma Violence Abus.* **2016**, *17*, 160–171. [CrossRef] [PubMed]
164. Prudencio, L.; Null, S.E. Stormwater Management and Ecosystem Services: A Review. *Environ. Res. Lett.* **2018**, *13*, 033002. [CrossRef]
165. Charnes, A.; Cooper, W.; Rhodes, E. Measuring the Efficiency of Decision Making Units. *Eur. J. Oper. Res.* **1978**, *2*, 429–444. [CrossRef]
166. Aigner, D.; Lovell, C.A.K.; Schmidt, P. Formulation and Estimation of Stochastic Frontier Production Function Models. *J. Econ.* **1977**, *6*, 21–37. [CrossRef]
167. Hossain, M.K.; Kamil, A.A.; Baten, M.A.; Mustafa, A. Stochastic Frontier Approach and Data Envelopment Analysis to To-Tal Factor Productivity and Efficiency Measurement of Bangladeshi Rice. *PLoS ONE* **2012**, *7*, e46081. [CrossRef]
168. Chen, C.; Delmas, M.; Lieberman, M. Production Frontier Methodologies and Efficiency as a Performance Measure in Strategic Management Research. *Strateg. Manag. J.* **2015**, *36*, 19–36. [CrossRef]
169. Brauman, K.A.; Bremer, L.L.; Hamel, P.; Ochoa-Tocachi, B.F.; Roman-Dañobeytia, F.; Bonnesoeur, V.; Arapa, E.; Gammie, G. Producing Valuable Information from Hydrologic Models of Nature-Based Solutions for Water. *Integr. Environ. Assess. Manag.* **2022**, *18*, 135–147. [CrossRef] [PubMed]
170. Loc, H.H.; Diep, N.T.H.; Tuan, V.T.; Shimizu, Y. An Analytical Approach in Accounting for Social Values of Ecosystem Services in a Ramsar Site: A Case Study in the Mekong Delta, Vietnam. *Ecol. Indic.* **2018**, *89*, 118–129. [CrossRef]
171. Hirons, M.; Comberti, C.; Dunford, R. Valuing Cultural Ecosystem Services. *Annu. Rev. Environ. Resour.* **2016**, *41*, 545–574. [CrossRef]
172. Olander, L.P.; Johnston, R.J.; Tallis, H.; Kagan, J.; Maguire, L.A.; Polasky, S.; Urban, D.; Boyd, J.; Wainger, L.; Palmer, M. Benefit Relevant Indicators: Ecosystem Services Measures That Link Ecological and Social Outcomes. *Ecol. Indic.* **2018**, *85*, 1262–1272. [CrossRef]

MDPI
St. Alban-Anlage 66
4052 Basel
Switzerland
www.mdpi.com

Water Editorial Office
E-mail: water@mdpi.com
www.mdpi.com/journal/water

www.ingramcontent.com/pod-product-compliance
Lightning Source LLC
Chambersburg PA
CBHW040744250926
51850CB00007BB/1406
* 9 7 8 3 7 2 5 8 0 6 2 5 6 *